Bioinformatics

FOR

DUMMIES®

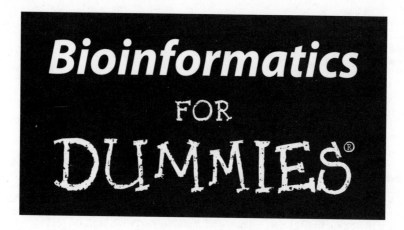

by Jean-Michel Claverie and Cedric Notredame

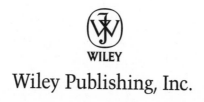

WILEY

Wiley Publishing, Inc.

Bioinformatics For Dummies®

Published by
Wiley Publishing, Inc.
111 River Street
Hoboken, NJ 07030
www.wiley.com

Copyright © 2003 by Wiley Publishing, Inc., Indianapolis, Indiana

Published by Wiley Publishing, Inc., Indianapolis, Indiana

Published simultaneously in Canada

No part of this publication may be reproduced, stored in a retrieval system or transmitted in any form or by any means, electronic, mechanical, photocopying, recording, scanning or otherwise, except as permitted under Sections 107 or 108 of the 1976 United States Copyright Act, without either the prior written permission of the Publisher, or authorization through payment of the appropriate per-copy fee to the Copyright Clearance Center, 222 Rosewood Drive, Danvers, MA 01923, (978) 750-8400, fax (978) 750-4744. Requests to the Publisher for permission should be addressed to the Legal Department, Wiley Publishing, Inc., 10475 Crosspoint Blvd., Indianapolis, IN 46256, (317) 572-3447, fax (317) 572-4447, e-mail: permcoordinator@wiley.com.

Trademarks: Wiley, the Wiley Publishing logo, For Dummies, the Dummies Man logo, A Reference for the Rest of Us!, The Dummies Way, Dummies Daily, The Fun and Easy Way, Dummies.com and related trade dress are trademarks or registered trademarks of Wiley Publishing, Inc., in the United States and other countries, and may not be used without written permission. All other trademarks are the property of their respective owners. Wiley Publishing, Inc., is not associated with any product or vendor mentioned in this book.

For general information on our other products and services or to obtain technical support, please contact our Customer Care Department within the U.S. at 800-762-2974, outside the U.S. at 317-572-3993, or fax 317-572-4002.

Wiley also publishes its books in a variety of electronic formats. Some content that appears in print may not be available in electronic books.

Library of Congress Control Number: 2002114813

ISBN: 0-7645-1696-5

Manufactured in the United States of America

10 9 8 7 6

1B/QV/QY/QU/IN

ⓦWiley Publishing, Inc. is a trademark of Wiley Publishing, Inc.

About the Authors

Jean-Michel Claverie is Research Director at the French National Center for Scientific Research and a consultant in genomics and bioinformatics. He is the founder and current head of the Structural & Genomic Information Laboratory, located in Marseilles, a sunny city on the Mediterranean coast of France. Using science as a pretext to travel, Jean-Michel has held positions in Paris (France); Sherbrooke (PQ, Canada) — where he learned the meaning of snow; the Salk Institute (La Jolla, CA) — where he learned the pleasures of hang gliding; the Pasteur Institute (Paris); Incyte pharmaceutical (Palo Alto, CA); and the National Center for Biotechnology Information (Bethesda, MD). He has been involved with the development of bioinformatics since the early days — his PhD work involved programming an 8K Honeywell computer in binary right from the console switches!

Jean-Michel's current research interests are in comparative and structural genomics, and in the development of bioinformatics methods for the prediction of gene function. He is the author or coauthor of more than 120 scientific publications, and a member of numerous international review panels and scientific councils. He is a reviewing editor for the magazine *Science,* published by the American Association of Advancement of Science. In his spare time, he enjoys the relaxed pace of life in Marseilles, with his wife Chantal and their two sons, Nicholas and Raphael.

Cedric Notredame has a joint appointment as a researcher for the French National Center for Scientific Research and as a Professor of Bioinformatics at Lausanne University in Switzerland. Cedric is also a group leader at the Swiss Bioinformatics Institute and a consultant in bioinformatics. He has been teaching and developing bioinformatics tools for the last ten years. He graduated from Toulouse University, France, and earned his PhD in bioinformatics from the European Bioinformatics Institute in Cambridge (UK), under the supervision of Des Higgins. Cedric specializes in the development of multiple sequence algorithms and has been involved in designing Tcoffee, one of the most accurate multiple-alignment packages available today. He spends as much time as he can under water and holds his breath for affordable waterproof laptop computers.

Dedication

This is for my parents Monique and Jack, for keeping me in school, and for Chantal, for keeping me happy — in and out of the lab. It's also for my daughter Vanessa, and my sons Nicholas and Raphael, for reminding me that not *everything* in life is scientific.

— J-MC

This is for my mother Marie, my daughter Lina, and in memory of my grandparents, Simone and Louis.

— CN

Authors' Acknowledgments

The entire Wiley staff did a great job pulling together to publish this book on tight deadlines. We'd especially like to thank our tireless project editor, Paul Levesque, and Nicole Sholly, who did a great job copyediting a text full of obscure biochemical words.

We'd also like to thank Jim Kent, our technical editor. Jim nailed down major and minor inaccuracies alike. His many suggestions did much to improve the book.

We also have to thank the bioinformatics community for creating the many great Web resources that we describe in this book and for making them available for free over the Internet. We personally know a number of the folks who keep these sites up and running and salute all of them for their hard work, enthusiasm, and dedication. Topping this list are the staff members of the Swiss Bioinformatics Institute, who run the ExPASy and the Swiss EMBnet Web server. They always went out of their way to answer any query regarding their site. The NCBI folks have also been very helpful, and we thank them for that.

We also want to pat each other on the back for making the writing of this book great fun!

Finally, we'd like to thank our families and friends, who put up with missed dinners, extra child care, changing deadlines, late nights, and the many other demands of a project like this. We really appreciate their patience and promise that we won't do another one . . . at least not anytime soon!

Publisher's Acknowledgments

We're proud of this book; please send us your comments through our online registration form located at www.dummies.com/register/.

Some of the people who helped bring this book to market include the following:

Acquisitions, Editorial, and Media Development

Project Editor: Paul Levesque

Acquisitions Editor: Melody Layne

Copy Editor: Nicole Sholly

Technical Editor: Jim Kent

Editorial Manager: Leah Cameron

Permissions Editor: Carmen Krikorian

Media Development Manager: Laura VanWinkle

Media Development Supervisor: Richard Graves

Editorial Assistant: Amanda Foxworth

Cartoons: Rich Tennant (www.the5thwave.com)

Production

Project Coordinator: Maridee Ennis

Layout and Graphics: Beth Brooks, Carrie Foster, Tiffany Muth, Jackie Nicholas, Jeremey Unger

Proofreaders: Tyler Connoley, Robert Springer, TECHBOOKS Production Service

Indexer: TECHBOOKS Production Services

Publishing and Editorial for Technology Dummies

Richard Swadley, Vice President and Executive Group Publisher

Andy Cummings, Vice President and Publisher

Mary C. Corder, Editorial Director

Publishing for Consumer Dummies

Diane Graves Steele, Vice President and Publisher

Joyce Pepple, Acquisitions Director

Composition Services

Gerry Fahey, Vice President of Production Services

Debbie Stailey, Director of Composition Services

Contents at a Glance

Table of Contents

Introduction

*B*ioinformatics is fast becoming an oft-uttered buzzword these days, so this book is for you if you're tired of pretending you know what this buzz is all about. If you expect something complicated, you're in for a (good or bad) surprise: Bioinformatics is nothing but good, sound, regular biology appropriately dressed so that it can fit into a computer.

Bioinformatics is about searching biological databases, comparing sequences, looking at protein structures, and more generally, asking biological questions with a computer. The bioinformatics we show you in this book can save you months of work in the lab at the minute cost of a few hours working on your computer.

Don't look here for long equations and computer geek gibberish. The purpose of this book is to show you quickly and plainly how to use the bioinformatics programs that you need to get your work done. On every page, we give you tricks and treats to get the most out of existing tools. If you didn't know that you can use the most sophisticated programs for free over the Internet and that you can do this (sometimes) without installing anything on your own computer, then stay tuned: You're in for many more good surprises.

What This Book Does for You

This book is here to help you get things done. For every standard bioinformatics task you may want to undertake, you'll find a detailed steps list that you can use to quickly produce the result you need.

To use most of the tools we describe in this book, you do not need to install any program on your computer. Everything we show you here runs over the Internet via your Internet browser.

If you know what you want to do — or at least know the task by name — going through the Table of Contents is the best strategy for finding exactly what you need. If you have an idea of what you want to do but you're not sure how to express it with words, Chapter 2 is here to help you decide which part of the book will suit your needs.

As you're about to find out, bioinformatics is not a standardized world. Depending on the authors and on the Internet sites, the meaning of jargon words tends to change a little. In this book, we give you a glossary with the most standard definitions; yet don't be alarmed if you find that things aren't always as consistent as they should be. When this happens, trust your common sense.

At the end of most chapters you'll find a convenient "Doing It for Free over the Internet" section, where we list a few carefully chosen Web sites that are similar to those we describe in the rest of the chapter. Treat this information as a spare wheel! If the main site is down, this section probably lists a convenient replacement.

Foolish Assumptions

While writing this book, we have assumed that

- ✔ You have a PC running Microsoft Windows.

- ✔ You have an Internet connection (a fast one if possible, but not necessarily).

- ✔ You likely have a background in molecular biology. If you don't — or if you need to brush up on your molecular biology — Chapter 1 gives you a brief overview of the basics.

- ✔ You know how to use an Internet browser but not much more about computers.

- ✔ You don't want to become a bioinformatics guru; you simply want to use the right tools for your problem and not spend days finding out about things you don't need!

- ✔ Most private biotech companies consider it unsafe to send data over the Internet. We assume here that the data you want to analyze over the Internet is not very confidential. Also, some of the "public" databases and services listed in this book require commercial users to enter a license agreement.

How This Book is Organized

Bioinformatics is a broad field, with many nooks and crannies, hills and dales, and other charming features. Rather than present the whole vast discipline in one fell swoop, we've divided our discussion into five (more manageable) parts.

Part 1: Getting Started in Bioinformatics

If you have less than an hour to find out what bioinformatics can do for you, Part I is the right place for you! It tells you everything you need to know in order to actually *do* something with bioinformatics. In Part I, we also remind you of just those bits of molecular biology that you'll need to know when you do sequence analysis. We show you here how to run the main bioinformatics tools so that you know what's in store for you.

Part 11: A Survival Guide to Bioinformatics

If you want to find out everything that's ever been published on your sequence, this part is for you. It shows you how you can deal with the bioinformaticist's bread and butter: *DNA or protein sequences and their databases*. Here we tell you where you can find all the available sequences, and how to find the one you really need among zillions of irrelevant others. We also show you how to gather everything that's known in the universe about this special sequence that interests you so much.

Part 111: Becoming a Pro in Sequence Analysis

If you want to compare sequences, this is the part for you. Here we show you how to search databases for sequences that are similar to yours, as well as show you how to compare two or more sequences. This part also tells you how to gather hints about the function of a gene, through sequence comparisons. Finally we give you pointers on how to produce, edit, and beautify your multiple sequence alignments so that you can show them in presentations and publications.

Part 1V: Becoming a Specialist: Advanced Bioinformatics Techniques

To take full advantage of this part, you should have a pretty good idea of what you're looking for. Heavy stuff is going on here: how to predict a protein structure, how to predict an RNA structure, and how to do phylogenetic analysis. These are complicated subjects; it's simply amazing what you can do with a simple PC, thanks to the Internet resources we describe in this part.

Part V: The Part of Tens

Welcome to our bazaar! If you haven't found what you were looking for in the other parts, you're now in the right place. The wealth of online resources that exist in bioinformatics is extraordinary and almost overwhelming. With every student and their cousins putting their semester reports online, finding exactly what you need with a simple keyword search can be a daunting task. In the Part of Tens, we give you a list of central resources that you can use as a starting point. Chances are that the program or server you're looking for is only one or two clicks away. In this part, we also give you ten important pieces of advice to make sure that your lab work can safely depend on your Internet work.

Icons Used in This Book

Always eager to please, we've decided to use a series of icons in the margins of this book as a way to help you key in on important information. We came up with four, which seemed like a nice, round number.

Some particularly technoid information is coming up. You can skip it and nothing terrible will happen. Yet if you want to be in full control of what you are doing, reading this may help! Your call. . . .

This icon shows you something simple, or smart, or a cute shortcut. In any case, it's something that can save you time and troubles.

There are many booby traps around when you use Internet servers. This icon warns you when some ambiguity surrounds what the server you're using is up to or when disaster is only one (wrong) mouse click away. Treat the Warning icon with respect in a steps list!

This icon indicates something you should remember. It can be one of the few important principles that you need to know or it can be a very special tip — the kind that can save you three days of work or drive you nuts if you forget it. You may assume that the head of your institute/company is where he is because he discovered one of these very special tips for himself!

Where to Go from Here

If you know nothing about bioinformatics, this book is here to reassure you. Bioinformatics is a much simpler subject than you ever thought possible. For most people new to this field, the main difficulty is finding out the kind of questions they can ask with these new tools. If you're a biologist, don't let the computer scare you; bioinformatics is nothing more than good, sound, regular biology hidden inside a computer.

The magic thing about bioinformatics is that, with a simple Internet connection, you can browse databases that contain the sum of our entire human biological knowledge and you can do this with the most sophisticated tools ever developed by mankind. And how much is this going to cost you? Nothing!

If you do molecular biology, this is the equivalent of having an entire lab with expensive, state-of-the-art equipment and staffed by an army of post-docs that go and fetch anything you need any time you need it. The only difference is that you cannot set this lab on fire (even if you try very hard).

If you think of it, it is quite incredible to realize that all this is right here, at your fingertips, one or two mouse clicks away! The Web is borderless; it is color blind and unimpressed by wealth! Whether you come from a rich or a poor country, whether you're a first year student, a scientist, or a Nobel Prize winner, you have access for free to the same high-quality information. No other scientific discipline has ever been so democratically widespread.

This book isn't a textbook but a cookbook! And we take pride in this! It contains many recipes that colleagues showed us over the years or that we discovered ourselves. Accommodating and serving biological data is something very personal, and we're sure that you'll gradually find your own way to do it. In the meantime, if you need a quick fix, you can always use some of the off-the-shelf solutions that we provide here.

No discipline in science has benefited as much as biology from the global village phenomenon of the Internet. Whatever your question, whatever you want to do, starting on the Internet is the proper thing to do. Nonetheless, remember that the best and the worse appear online these days. Do as you do in real life and trust only those sites or institutions that you know well.

This book is as up-to-date as we can make it, but the world doesn't stand still right after we finish correcting the last galley proofs and send *Bioinformatics For Dummies* into the bookstores. For those of you who want up-to-date info on the growing field of bioinformatics (including lists of our favorite bioinformatics links) and don't want to wait until the next edition, check out the Web site associated with this title at www.dummies.com/extras.

Sometimes browsing the Internet gives one the depressing feeling that everything has been done by others and that it's all over. This may be true. Now that the whole world talks together, it is clear that there is a finite number of interesting questions to ask. This is the bad news. The good news is that there are many more answers than there are questions! Never exclude the hypothesis that your answer may be the best in the universe (at least for a few days. . . .)!

Part I
Getting Started in Bioinformatics

The 5th Wave By Rich Tennant

"OK, it's your turn to ask for a tissue sample for DNA analysis."

In this part . . .

Bioinformatics is a new discipline, which means that nobody should feel ashamed if he or she doesn't have a clue what the excitement's all about. Don't worry; after finishing this book, you'll be speaking bioinformatics-speak with the best of them.

We start you off in Part I with a quick reminder of what you need to know about DNA and proteins to make sense of this book. We also give you an overview of the main bioinformatics tools available on the Internet.

We don't give too many details here, but if all you need to know is which Internet page to open and which button to press, come on in, 'cuz we've got just what you need!

Chapter 1

Finding Out What Bioinformatics Can Do for You

In This Chapter

▶ What is bioinformatics?

▶ Understanding the links between modern biology, genomics, and bioinformatics

▶ The kind of biological questions bioinformatics can help you answer quickly

> *Organic chemistry is the chemistry of carbon compounds. Biochemistry is the study of carbon compounds that crawl.*
>
> — Mike Adam

*I*t looks like *bio*logists are colonizing the dictionary with all these *bio*-words: we have bio-chemistry, bio-metrics, bio-physics, bio-technology, bio-hazards, and even bio-terrorism. Now what's up with the new entry in the bio-sweepstakes, bio-informatics?

What Is Bioinformatics?

In today's world, computers are as likely to be used by biologists as by any other highly trained professionals — bankers or flight controllers, for example. Many of the tasks performed by such professionals are common to most of us: We all tend to write lots of memos and send lots of e-mails, many of us use spreadsheets, and we all store immense amounts of never-to-be-seen-again data in complicated file systems.

However, besides these general tasks, biologists also use computers to address problems that are very specific to biologists, which are of no interest to bankers or flight controllers: These specialized tasks, taken together, make

up the field of *bioinformatics*. More specifically, we can define bioinformatics as the computational branch of molecular biology.

Time for a little bit of history. Before the era of bioinformatics, only two ways of performing biological experiments were available: within a living organism (so-called *in vivo*) or in an artificial environment (so-called *in vitro*, from the Latin *in glass*). Taking the analogy further, we can say that bioinformatics is in fact *in silico* biology, from the silicon chips on which microprocessors are built.

This new way of doing biology has certainly become very trendy, but don't think that "trendy" translates into "lightweight" or "flash-in-the-pan." Bioinformatics goes beyond trendy because it's at the center of the most recent developments in biology, such as the deciphering of the human genome (another buzz word), the biotechnologies, new legal and forensic techniques, as well as the medicine of the future.

Because of the centrality of bioinformatics to cutting-edge developments in molecular biology, people from many different fields have been stumbling across the term in a variety of different contexts. If you're a biology, medical, or computer science student, a professional in the pharmaceutical industry, a lawyer or a policeman worrying about DNA testing, a consumer concerned about GMOs (Genetically Modified Organisms), or even a NASDAQ investor interested in start-up companies, you'll already have come across the word *bioinformatics*. If you're good at what you do, you'll want to know what all the fuss is about. This chapter, then, is for you.

Instead of a formal definition that would take hours to cover all the ins and outs of the topic, the best way to get a quick feel for what bioinformatics — or swimming, for that matter — is all about is to jump right into the water: That's what we do next. Go ahead and get your feet wet with some basic molecular biology concepts — and the relevant questions intimately connected with such concepts — that all together define bioinformatics.

Analyzing Protein Sequences

If you eat steak, you're intimately acquainted with proteins. (Your taste buds know them intimately anyway, even if your rational mind hasn't quite mastered the concept.) For you non-steak lovers out there, you'll be pleased to know that proteins abound in fish and vegetables, too. Moreover, all these proteins are made up of the same basic building blocks, called *amino acids*. Amino acids are already quite complex organic molecules, made of carbon, hydrogen, oxygen, nitrogen, and sulfur atoms. So the overall recipe for a protein (the one your rational mind will appreciate, even if your taste buds won't) is something like $C_{1200}H_{2400}O_{600}N_{300}S_{100}$.

The early days of biochemistry were devoted to finding out a better way to represent proteins, preferably in terms of a formula that would explain their biological (or even nutritional) properties. Biochemists realized over time that proteins were huge molecules (macromolecules) made up of large numbers of amino acids (typically from 100 to 500), picked out from a selection of 20 "flavors" with names such as alanine, glycine, tyrosine, glutamine, and so on. Table 1-1 gives you the list of these twenty building blocks, with their full names, three-letter codes, and one-letter codes (the *IUPAC code*, after the *International Union of Pure and Applied Chemistry* committee that designed it).

Table 1-1	The 20 Amino Acids and Their Official Codes		
#	**1-Letter Code**	**3-Letter Code**	**Name**
1	A	Ala	Alanine
2	R	Arg	Arginine
3	N	Asn	Asparagine
4	D	Asp	Aspartic acid
5	C	Cys	Cysteine
6	Q	Gln	Glutamine
7	E	Glu	Glutamic acid
8	G	Gly	Glycine
9	H	His	Histidine
10	I	Ile	Isoleucine
11	L	Leu	Leucine
12	K	Lys	Lysine
13	M	Met	Methionine
14	F	Phe	Phenylalanine
15	P	Pro	Proline
16	S	Ser	Serine
17	T	Thr	Threonine
18	W	Trp	Tryptophan
19	Y	Tyr	Tyrosine
20	V	Val	Valine

Four additional amino acid codes

When you work with databases or analysis programs, you're likely to have some unusual letters popping up now and then in your protein sequences. These letters don't designate actual amino acids, but are used to denote various level of ambiguity — or a total lack of information — about certain positions in the sequence. We've listed these particular letters in the following table.

The B and Z codes, now becoming obsolete, speak to the difficulty of distinguishing between Asp and Asn in the early days of protein sequence determination. The *X* code is still very much used as a placeholder letter, where you don't know the amino acid at a given position in the sequence. Alignment programs use "-" to denote positions apparently missing from the sequence.

Four Codes That Aren't Amino Acids

1-Letter Code	3-Letter Code	Meaning
B	Gln or Glu	Glutamine or glutamic acid
Z	Asn or Asp	Asparagine or aspartic acid
X	Xaa	Any residue
-	— — —	No corresponding residue (gap)

Biochemists then recognized that a given type of protein (such as insulin or myoglobin) always contains precisely the same number of total amino acids (generically called *residues*) in the same proportion. Thus, a better formula for a protein looks like:

insulin = (30 glycine + 44 alanine + 5 tyrosine + 14 glutamine + . . .)

Finally, biochemists discovered that these amino acids are linked together as a chain, and that the true identity of a protein isn't only derived from its composition but also from the precise order of its constituent amino acids. The first amino-acid sequence of a protein — insulin — was determined in 1951. The actual recipe for human insulin, from which all its biological properties derive, is the following chain of 110 residues:

insulin = MALWMRLLPLLALLALWGPDPAAAFVNQHLCGSH-
LVEALYLVCGERGFFYTPKTRREAEDLQVGQVELGGGPGAGSLQPLALEGSLQKR-
GIVEQCCTSICSLYQLENYCN

More than 50 years later, analyzing protein sequences like these remains a central topic of bioinformatics in all laboratories throughout the world. Check

out Chapters 2, 4, and 6 through 11 to quickly figure out how to analyze your protein sequence and become a member of the club!

A brief history of sequence analysis

Besides earning Alfred Sanger his first Nobel Prize, the sequencing of insulin inaugurated the modern era of molecular and structural biology. Traditionally a *soft science* (that is, more tolerant of fuzzy reasoning and hand waving than chemistry or physics), biology got a taste of its first fundamental dataset: molecular sequences. In the early sixties, protein sequences accumulated slowly — perhaps a blessing in disguise, given that computers capable of analyzing them were not yet invented! In this precomputer era, sequences were assembled, analyzed, and compared by (manually) writing them on pieces of papers, taping them side by side on laboratory walls, and/or moving them around for optimal alignment or *pattern matching*.

As soon as computers became available (as big as a locomotive and just as fast, and with 8K RAM!), the first computational biologists started to enter these manual algorithms into the available machines. This was brand new because nobody before them had to manipulate and analyze these kinds of molecular *texts*. Most methods had to be invented from scratch and, in the process, a new area of research — the analysis of protein sequences using computers — was generated. This was the genesis of bioinformatics.

Reading protein sequences from N to C

The twenty amino acids found in proteins have different bodies but all have the same pair of hooks — NH2 and COOH — that are used to form the so-called *peptidic bond* between successive residues in the sequence. Figure 1-1 shows free individual amino acids floating about, displaying their hooks for all to see.

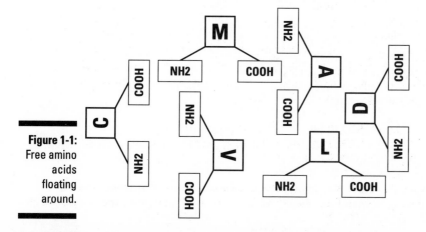

Figure 1-1: Free amino acids floating around.

The protein molecule itself is then made by chemically linking a free NH2 group with a COOH group, forming the peptide bond CO-NH. Figure 1-2 shows a schematic picture of the resulting chain.

Figure 1-2:
Amino acids
chained
together to
constitute a
protein
molecule.

As a result of this chaining process, your protein molecule is going to be left with an unused NH2 at one end and an unused COOH at the other end. These extremities are respectively called the *N-terminus* and *C-terminus* of the protein chain. This is important to know because (by convention) the sequence of a protein (or of a fragment) is always defined (in books, databases, and so on) as the succession of its constituent amino acids listed from the N-terminus to the C-terminus. The sequence of our (short!) demo protein is then:

```
MAVLD= Met-Ala-Val-Leu-Asp= Methionine-Alanine-Valine-
                 Leucine-Aspartic
```

Working with protein 3-D structures

The precise succession of a protein's constituent amino acids is what defines a given protein molecule. However, the biological properties of the protein (for instance, its ability to digest sugar or to become part of a muscle fiber) do not result directly from this chain as a linear object, but rather from the three-dimensional (3-D) shape this ribbon of amino acids adopts in its environment. A protein molecule, once made, is not a chain-like, highly flexible object (think a section of chain-link fence), but rather a compact, well-bundled ball of string. The final 3-D shape of the protein molecule is uniquely dictated by its sequence because some amino-acid types (for instance, *hydrophobic* residues L, V, I) have no desire whatsoever to be at the surface interacting with the surrounding

water, while others (for instance, *hydrophilic* residues D, S, K) are actively look-ing for such an opportunity. The protein chain is also affected by other influ-ences, such as the electric charges carried by some of the amino acids, or their capacity to fit with their immediate neighbors.

The first 3-D structure of a protein was determined in 1958 by Drs. Kendrew and Perutz, using the complicated technique of X-ray crystallography. (Not for the faint of heart. Don't try to understand how it works unless you want to turn professional!) Besides winning one more Nobel Prize for the nascent field of molecular biology, this feat made them realize that proteins have precise and specific shapes, encoded in their sequence of amino acids. Hence, they predicted that proteins with similar sequences would fold into similar shapes and, conversely, that proteins with similar structures would be encoded by similar sequences of amino acids. The function of a protein being a direct con-sequence of its 3-D structure (shape), the logical link was established.

SEQUENCE⇨STRUCTURE⇨FUNCTION

It is now a central concept of molecular biology and bioinformatics.

Playing with protein structure models on a computer screen is, of course, much easier than carrying around a thousand-piece 3-D plastic puzzle. As a consequence, an increasing proportion of the bioinformatics pie is now devoted to the development of tools to navigate between sequences and 3-D structures. (This specialized area is called *structural bioinformatics*.) Thanks to many free resources on the Internet, it is not difficult to display some beautiful protein pictures on your own computer, and start playing with them like in video games. We show you how to do it in Chapter 11.

Before you get a chance to read that chapter, Figure 1-3 gives you an idea of what a 400-amino-acid typical protein 3-D (schematic) structure looks like — when you don't have a color monitor and can't make it move and turn!

Don't forget: Protein molecules, even in their wonderful complexity, are still pretty small. The one in Figure 1-3 would fit in a box whose sides measure 70/1,000,000 millimeters. There are thousands of different proteins in a bac-terium, each of them in thousands of copies — enough to provide evidence to support the notion that Life Is Not Simple!

Figure 1-3:
Example of
protein 3-D
structure
(schematic).

Protein bioinformatics covered in this book

The study of protein sequences can get pretty complicated; so complicated, in fact, that it would take a pretty thick book to cover all aspects of the field. We'd like to take a more selective approach by focusing on those aspects of protein sequences where bioinformatic analyses can be most useful. The following list lets you in on those topics where such an analysis is particularly relevant to protein sequences — and also lets you know in which chapters of this book we cover the topic in greater detail:

- Retrieving protein sequences from databases (Chapters 2, 3, and 4)
- Computing amino-acid composition, molecular weight, isoelectric point, and other parameters (Chapter 6)
- Computing hydrophobicity/hydrophilicity profiles, predicting antigenic sites, locating membrane-spanning segments (Chapter 6)
- Predicting elements of secondary structure (Chapters 6 and 11)
- Predicting the domain organization of proteins (Chapters 6, 7, 9, and 11)
- Visualizing protein structures in 3-D (Chapter 11)
- Predicting a protein's 3-D structure from its sequence (Chapter 11)

✔ Finding all proteins sharing sequence similarity (Chapter 7)

✔ Classifying proteins into families (Chapters 7, 8, and 9)

✔ Finding the best alignment between two or more proteins (Chapters 8 and 9)

✔ Finding evolutionary relationships between proteins, drawing trees (Chapters 7, 9, 11, and 13)

Analyzing DNA Sequences

During the 1950s, when scientists such as Kendrew and Perutz were still struggling to determine the first 3-D structures of proteins, other biologists had already acquired a lot of indirect evidence (by using extremely clever genetics experiments) that *deoxyribonucleic acid* (DNA), the stuff that makes our genes, was also a large macromolecule consisting of a chain of four constituents that they called *nucleotide*. A nucleotide is made up of one phosphate group linked to a pentose sugar which is itself linked to one of 4 types of nitrogenous organic bases symbolized by the four letters A, C, G, and T.

The IUPAC code for DNA sequences

The following table lists the one-letter codes (IUPAC codes) used to work with DNA sequences. Official IUPAC codes, from the *I*nternational *U*nion of *P*ure and *A*pplied *C*hemistry, are defined for all possible 2- and 3-way ambiguities. The table only indicates the ones most frequently used.

Most Common Letters Used for DNA Nucleotide Sequences

1-Letter Code	Nucleotide Name	Category (Comment)
A	Adenine	Purine
C	Cytosine	Pyrimidine
G	Guanine	Purine
T	Thymine	Pyrimidine
N	Any nucleotide (any base)	(n/a)
R	A or G	Purine
Y	C or T	Pyrimidine
-	— — —	None (gap)

However, molecular biologists had to wait until much later — the 1970s, to be more precise — before they could determine the sequence of DNA molecules and get direct access to the gene nucleotide sequences.

This was a revolution (earning A. Sanger his second Nobel Prize!) because the small DNA sequence alphabet — 4 nucleotides compared to 20 amino acids — allowed a much simpler and faster reading, and quickly lent itself to complete automation. Currently, the worldwide rate of DNA sequence determination is orders of magnitude larger than the one of protein sequencing.

Reading DNA sequences the right way

As was the case for the 20 amino acids found in proteins, the 4 nucleotides making DNA have different bodies but all have the same pair of hooks: 5' phosphoryl and 3' hydroxyl (pronounced 5 *prime* and 3 *prime*) by reference to their positions in the deoxyribose sugar molecule, part of the nucleotide chaining device. Figure 1-4 shows what free individual nucleotides look like.

Figure 1-4:
The four
nucleotides
making
DNA.

Forming a bond between the 5' and 3' positions of the constituent nucleotides then makes the DNA molecule. Figure 1-5 shows a schematic representation of the resulting DNA strand.

Figure 1-5:
Chained
nucleotides
constituting
a DNA
strand.

After the nucleotides are linked, the resulting DNA strand exhibits an unused phosphoryl group (PO_4) at the 5' end, and an unused hydroxyl group (OH) at

the 3' end. These extremities are respectively called the *5'-terminus* and the *3'-terminus* of the DNA strand. It is important to remember that a DNA sequence is always defined (in books, databases, articles, and programs) as the succession of its constituent nucleotides *listed from the 5'- to 3'- end*. The sequence of the (short!) DNA strand shown in Figure 1-5 is then:

```
TGACT = Thymine-Guanine-Adenine-Cytosine-Thymine
```

The two sides of a DNA sequence

In the same laboratory where Kendrew and Perutz were trying to figure out the first 3-D structure of a protein, Watson and Crick elucidated — in 1953 — the famous double-helical structure of the DNA molecule. Everybody has in mind this famous spiral staircase picture and the elegance of the DNA *double helix* probably contributed a lot to make it the most popular notion of molecular biology. However, what made this discovery so important — earning one more Nobel Prize for molecular biology — was not the helical shape, but the discovery that the DNA molecule consists of two complementary strands, shown in Figure 1-6.

Figure 1-6: The two complementary strands of a complete DNA molecule.

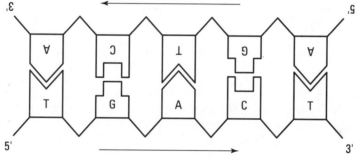

By complementarity, we mean that a thymine (T) on one strand is always facing an adenine (A) (and vice versa), and guanine (G) is always facing a cytosine (C). These couples, A-T and G-C, although not linked by a chemical bond, have a strict one-to-one reciprocal relationship. When you know the sequence of nucleotides along one strand, you can automatically deduce the sequence on the other one. This amazing property — and not the stylish helical structure — is the Rosetta stone that explains everything about DNA sequences. For instance, when living organisms reproduce, each of their genes must be duplicated. In order to do this, nature doesn't go about it like a photocopier would — by making an exact copy. Rather, nature separates the

DNA strands and makes *two complementary* ones, thanks to the magical two-sided structure of DNA molecules.

This double strand structure of DNA makes the definition of a DNA sequence ambiguous: Even with our convention of reading the nucleotides from the 5' end towards the 3' end, you may decide to write down the bottom or the top sequence. Convince yourself that they're both equally valid sequences by turning this book upside down! Thus, at each location, a DNA molecule corresponds to two — totally different — sequences, related by this reverse-and-complement operation. This isn't complicated; simply keep it in mind every time you work with DNA sequences.

Fortunately, most database mining programs, such as BLAST, know about this and take both strands into account when reporting their results. But some programs don't bother and only analyze the sequence you gave them. In cases where both strands matter, always make sure that a complete analysis has been performed. We discuss these details further in Chapters 3, 5, and 7.

Palindromes in DNA sequences

Newcomers to DNA sequence analysis are usually confused by the notion of reverse complementary sequences. However, in due time you'll be able to quickly recognize that the two sequences

```
ATGCTGATCTTGGCCATCAATG and CATTGATGGCCAAGATCAGCAT
```

correspond to facing strands of the same DNA molecule.

One fascinating property of DNA complementarity is the fact that regions of DNA may correspond to sequences that are identical when read from the two complementary strands. Figure 1-7 helps illustrate this magic trick.

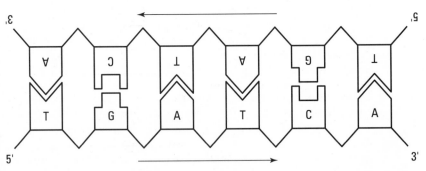

Figure 1-7:
How two comple-mentary strands can be read the same way.

Such sequences are called *palindromes*, borrowed from the term for a word or sentence that reads the same in both directions, such as "Madam I'm Adam" or "A man, a plan, a canal: Panama." Palindromic sequences aren't merely a curiosity; they play important biological roles. For instance, most DNA cutting enzymes (so-called restriction enzymes) have palindromic target sequences. Other palindromic sequences serve as *binding sites*, where regulatory proteins work to turn genes on and off. Palindromic sequences also have a strong influence on the 3-D structure of DNA molecules. (And not just DNA. See the next section for more on palindromic sequences in RNA.) Looking for exact or approximate palindromes in DNA sequences is a classic bioinformatic exercise.

Analyzing RNA Sequences

DNA (deoxyribonucleic acid) is the most dignified member of the nucleic acid family of macromolecules. Its sole and only task is to ensure — forever — the conservation of the genetic information for its organism. It is thus very stable and resistant, and lies well protected in the nucleus of each cell. *Ribonucleic acid* (RNA) is a much more active member of the nucleic acid family, being synthesized and degraded constantly as it makes copies of genes available to the cell factory.

In the context of bioinformatics, the only important differences between RNA and DNA are that 1) RNA differs from DNA by one nucleotide, and 2) RNA comes as a single strand. The one-letter IUPAC codes for RNA sequences are shown in Table 1-2.

Table 1-2	Most Common Letters Used for RNA Nucleotide Sequences	
1-Letter Code	*Nucleotide Base Name*	*Category (comment)*
A	Adenine	Purine
C	Cytosine	Pyrimidine
G	Guanine	Purine
U	Uracil	Pyrimidine
N	Any nucleotide	Purine or Pyrimidine
R	A or G	Purine
Y	C or U	Pyrimidine
-	— — —	None (gap)

Some programs automatically handle the U-instead-of-T conversion, and many do not even distinguish between the two classes of nucleic acids. Don't be surprised if a database entry displays RNA sequences (such as messenger RNA) with a T instead of a U. In fact, like proteins, RNA sequences are encoded in the DNA. For this reason, people have adopted the habit of working with the RNA *gene* (DNA) sequences rather than the RNA sequences.

RNA structures: Playing with sticky strands

Even though RNA molecules consist of a single strand of nucleotides, their natural urge for pairing with complementary sequences is still there. Think of this single strand as free-floating Scotch tape: You know that it won't take long for that tape to become a messy ball, until no sticky part remains exposed. This is exactly what happens to the single-stranded RNA molecule — give or take a bit for the sake of poetic license — although more specifically, as shown in Figure 1-8.

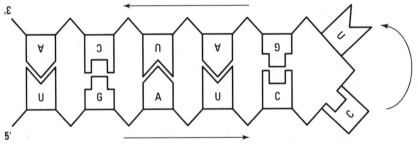

Figure 1-8: How RNA turns itself into a double-stranded structure.

Now you understand why we insisted on the notion of strand complementarity (refer to Figure 1-6). Single-stranded RNA molecules are pairing different regions of their sequence to form stable double-helical structures — admittedly, less regular, but nonetheless quite similar to the double-helical structure of DNA. Once synthesized, each RNA molecule quickly adopts a compact fold — attempting to pair as many nucleotides as possible — while respecting the flexibility and geometrical constraints of the chain. Hairpin shapes, as shown in Figure 1-8, are the basic elements of RNA secondary structure, comprised of *loops* (the unpaired C-U in Figure 1-8) and *stems* (the paired regions).

Just for fun, verify for yourself that a palindromic RNA sequence results in a perfect hairpin, with no loop. While attempting to pair as many nucleotides as possible, the RNA chain folds in space, resulting into a specific 3-D structure,

dictated by its sequences. Like with proteins, the linear sequence of the building blocks dictates the final 3-D shape. The biological function of RNA molecules also derives from their 3-D shapes.

Computing (predicting) the final fold of an RNA molecule from its sequence is a challenging problem that drove many new developments in bioinformatics. Go directly to Chapter 12 if your main interest is in RNA structures.

More on nucleic acid nomenclature

Don't panic if you get the impression that books, courses, and the technical literature all use many different words and abbreviations to designate the building blocks of nucleic acids: This is true. You will find "base," "base pair," "nucleoside," or "nucleotide." These different names designate slightly different chemical entities, but the differences are irrelevant for us now. Until now, we've used the term *nucleotide* — abbreviated *nt*, as in "a 400-nt long sequence." This way of labeling a sequence refers to the length of the DNA (or RNA) molecules in terms of the number of positions for nucleotides. For instance, the sequence in Figure 1-5 is 5-nt long.

Notice that we say "number of positions," rather than number of nucleotides. A 400-nt long DNA molecule has 400 positions for nucleotides, but it actually contains 800 nucleotides because every position contains a pair of nucleotides. To make this clearer, DNA sequence sizes are often given in *base pairs*, abbreviated *bp*. Thus, the DNA sequence in Figure 1-5 is 5-bp long. Larger units, such as kb (1000 bp) or Mb (mega bp) are used.

DNA Coding Regions: Pretending to Work with Protein Sequences

Only a small percentage of the hundreds of thousands of protein sequences that are found in current databases correspond to molecules that have actually been isolated by somebody or experimented upon. Determining the sequence of a protein is much more difficult than sequencing DNA. However, all the proteins that a given organism (a microbe, a human being) can synthesize are encoded in the DNA sequence of its genome. Thus, the smart shortcut that molecular biologists have been using is to read protein sequences directly at the information source: in the DNA sequence! This way, we can pretend to know the amino acid sequence of a protein that has never been isolated in a test tube.

Turning DNA into proteins: The genetic code

When you know a DNA sequence, you can translate it into the corresponding protein sequence by using the genetic code, the very same way the cell itself generates a protein sequence. The genetic code is universal (with some exceptions, otherwise life would be too simple!), and it is nature's solution to the problem of how one uniquely relates a 4-nucleotide sequence (A, T, G, C) to a suite of 20 amino-acid symbols. Understanding how the cell does this was one of the most brilliant achievements of the biologists of the 1960s. Yet, the final answer can be contained in a (miraculously small) table, as shown in Figure 1-9. Please, be quiet, as you enter the most sacred monument of modern biology.

Here is how to use the table shown in Figure 1-9: From a given starting point in your DNA sequence, start reading the sequence 3 nucleotides (one triplet) at a time. You then consult the genetic code table to read which amino acid corresponds to the current triplet (technically referred to as *codons*). For instance, the following DNA (or messenger RNA) sequence is decoded as follows:

1. **Read the DNA sequence:**

 ATGGAAGTATTTAAAGCGCCACCTATTGGGATATAAG

2. **Decompose it into successive triplets:**

 ATG GAA GTA TTT AAA GCG CCA CCT ATT GGG ATA TAA G . . .

3. **Translate each triplet into the corresponding amino acid:**

 M E V F K A P P I G I STOP

If your DNA sequence is correctly listed in the 5' to 3' orientation, you generate the protein sequence in the conventional N- to C-terminus as well. This way you don't have to think about these orientation details ever again.

Thus, if you know where a protein-coding region starts in a DNA sequence, your computer can pretend to be a cell and generate the corresponding amino-acid sequence! This simple computer translation exercise is at the origin of most of the so-called *protein sequences* that you can find in databases. Many sequence analysis programs acknowledge this fact by offering on-the-fly translation, so you can process DNA sequences as virtual protein sequences with a simple mouse click.

Table of Standard Genetic Code

		T	C	A	G
T		TTT Phe (F)	TCT Ser (S)	TAT Tyr (Y)	TGT Cys (C)
		TTC Phe (F)	TCC Ser (S)	TAC Tyr (Y)	TGC Cys (C)
		TTA Leu (L)	TCA Ser (S)	TAA Stop	TGA Stop
		TTG Leu (L)	TCG Ser (S)	TAG Stop	TGG Trp (W)
C		CTT Leu (L)	CCT Pro (P)	CAT His (H)	CGT Arg (R)
		CTC Leu (L)	CCC Pro (P)	CAC His (H)	CGC Arg (R)
		CTA Leu (L)	CCA Pro (P)	CAA Gln (Q)	CGA Arg (R)
		CTG Leu (L)	CCG Pro (P)	CAG Gln (Q)	CGG Arg (R)
A		ATT Ile (I)	ACT Thr (T)	AAT Asn (N)	AGT Ser (S)
		ATC Ile (I)	ACC Thr (T)	AAC Asn (N)	AGC Ser (S)
		ATA Ile (I)	ACA Thr (T)	AAA Lys (K)	AGA Arg (R)
		ATG Met (M)	ACG Thr (T)	AAG Lys (K)	AGG Arg (R)
G		GTT Val (V)	GCT Ala (A)	GAT Asp (D)	GGT Gly (G)
		GTC Val (V)	GCC Ala (A)	GAC Asp (D)	GGC Gly (G)
		GTA Val (V)	GCA Ala (A)	GAA Glu (E)	GGA Gly (G)
		GTG Val (V)	GCG Ala (A)	GAG Glu (E)	GGG Gly (G)

Figure 1-9:
The universal genetic code.

More with coding DNA sequences

Using the example in the first paragraphs of the section "DNA Coding Regions: Pretending to Work with Protein Sequences," you can see that the resulting protein sequence depends entirely on the way you converted your DNA sequence into triplets before using the genetic code. For instance, using the second position as starting point leads to

```
1-  ATGGAAGTATTTAAAGCGCCACCTATTGGGATATAAG
```

```
2-  A  TGG  AAG  TAT  TTA  AAG  CGC  CAC  CTA  TTG  GGA  TAT  AAG
```

```
3-     W    K    Y    L    K    R    H    L    L    G    Y    K
```

Beginning with the third position (GGA-AGT-. . .) again leads to an entirely different translation.

Because of the triplet-based genetic code, a given DNA interval, *on a given strand*, can theoretically be translated in three different ways — basically three perspectives known in the field as *reading frames*. Because the DNA can be used from both strands, there are a total of six possible reading frames to

translate a DNA sequence into proteins. With very few exceptions (found in exotic viruses), only one of these six frames is used for any given DNA coding region. An interval of DNA sequence that remains free of STOP (the translation of TAA, TGA, or TAG) is called an *open reading frame* (ORF).

Additional complications arise from the fact that some DNA sequences are not encoding proteins at all, and that higher organisms have large pieces of non-coding DNA inserted within their genes. A large part of bioinformatics is devoted to the development of methods to locate protein-coding regions in DNA sequences, to precisely delineate where genes start and end, or where they are interrupted by noncoding intervals called *introns*.

DNA/RNA bioinformatics covered in this book

Need a road map to the bioinformatic analyses relevant to DNA/RNA sequences covered in this book? Here it is:

- ✔ Retrieving DNA sequences from databases (Chapters 2 and 3)
- ✔ Computing nucleotide compositions (Chapter 5)
- ✔ Identifying restriction sites (Chapter 5)
- ✔ Designing polymerase chain reaction (PCR) primers (Chapter 5)
- ✔ Identifying open reading frames (ORFs) (Chapter 5)
- ✔ Predicting elements of DNA/RNA secondary structure (Chapter 12)
- ✔ Finding repeats (Chapter 5)
- ✔ Computing the optimal alignment between two or more DNA sequences (Chapters 7, 8, and 9)
- ✔ Finding polymorphic sites in genes (single nucleotide polymorphisms, SNPs) (Chapter 3)
- ✔ Assembling sequence fragments (Chapter 5)

Working with Entire Genomes

The first truly efficient technique to sequence DNA was discovered in 1977. In 1995, the first sequence of an entire genome (from the microbe *Hemophilus influenzae*) was determined. Between these two dates, DNA sequencing technologies improved steadily, but such technologies still tended to concentrate

on mining individual genes for information. During this period, biologists were mostly sequencing DNA fragments a few thousand nucleotides in length, simply because they were interested in specific genes that they had started working on years before. Most of the bioinformatics tools available today were created during that period. They include

- All basic sequence alignments programs
- Phylogenetic and classification methods
- Various display tools adapted to relatively small sequence objects (such as protein sequences of, at most, a few thousand characters long)

Genomics: Getting all the genes at once

The determination of the first complete genome sequence terminated the gene-by-gene routine and initiated the era of *genomics*, the genetic mapping, physical mapping, and sequencing of entire genomes. As a consequence, the DNA sequences we have to work with now are much longer — close to a million-bp in length for microbes and up to several billion-bp in length for animals and humans. This revolution called for the design of new bioinformatic tools and databases capable to store, query, analyze, and display these huge objects in a user-friendly manner. Chapters 3, 5, and 7 present some of the questions that biologists address at the genome scale, and show the relevant bioinformatic tools in action.

In contrast to the early days of the gene-by-gene approach, DNA sequences are now often obtained (and the putative protein sequences derived from said DNA sequences) without any prior knowledge of what is actually there. In essence, genes are both sequenced *and discovered* at the same time. This prompted the emergence of an entirely new branch of bioinformatics devoted to the *parsing* of large DNA sequences into their components (genes, transcription units, protein coding regions, regulatory elements, and so forth). This first pass is then followed by a longer phase of genome *annotation*, where the biological functions of these various elements are (more or less tentatively) predicted. Part IV of this book presents you with some of these most advanced techniques.

Figure 1-10, representing the whole genome of the bacterium *Rickettsia conorii*, illustrates this new level of complexity. This circular DNA molecule is 1.3 million-bp long, a size on the small side among bacteria. Each little rectangle in the two most external circles of features corresponds to a protein-coding gene (one circle per strand) in the circular genome. Each rectangle corresponds to approximately 1000 bp. Nobody knew which genes — or which proteins — were in that bacterium before the sequencing started.

Almost everything we know now about this bacterium (and many others we can describe as fairly inaccessible, such as those thriving on the ocean floor near volcanic vents at 100°C) has been derived from bioinformatic analyses.

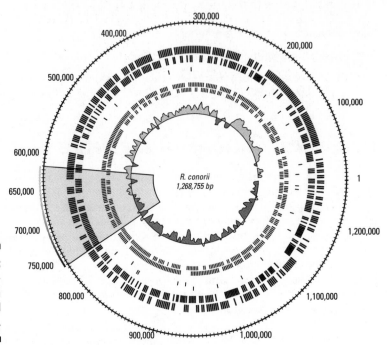

Figure 1-10:
Represen-
tation of a
bacterial
genome.

Genome bioinformatics covered in this book

The following list lets you know where in this book you'll find more in-depth coverage of specific topics related to genome bioinformatics:

- Finding which genomes are available (Chapters 3)
- Analyzing sequences in relation to specific genomes (Chapters 3 and 7)
- Locating gene homologues in a genome (Chapters 3 and 7)
- Displaying genomes (Chapter 3)
- Parsing a microbial genome sequence: ORFing (Chapter 5)

> ✔ Parsing a eukaryotic genome sequence: GenScan (Chapter 5)
>
> ✔ Finding orthologous and paralogous genes (Chapter 3)
>
> ✔ Discovering "unknown" genes (Chapter 7)
>
> ✔ Finding repeats (Chapter 5)

A Few Words on DNA Chips

A recent technological advance in genomics has been the use of *DNA chips* (more generally referred to as *DNA arrays*). In Figures 1-6, 1-7, and 1-8, you can see the unique property of "strand complementarity" exhibited by DNA or RNA molecules. This property is at the origin of most of the detection techniques routinely used in molecular biology. All such techniques work as follows:

1. The biologists prepare a set of target DNA molecules that are separated and fixed on a gel, a membrane, or a glass side.

 Each type of target molecule becomes individually addressable.

2. The biologists then prepare a mixture of other RNA or DNA molecules — known as *probes* — among which they want to detect those identical to some of the target defined in Step 1.

3. Using the proper combination of heating and solvent, they manage to make both the target and probe molecules single-stranded ("sticky").

4. They put the molecules in contact (for instance, by dropping the probe mixture on the target glass slide) and allow them to spontaneously reconstitute double-stranded molecules — a process known as *hybridization* or *annealing*.

5. After a while, all the corresponding complementary strands are zipped back up, causing some of the probe molecules to be retained on the spots of the corresponding targets.

6. Because the smart biologists didn't forget to flag the probe molecules with a fluorescent or a radioactive marker (symbolized by a black star in Figure 1-11), they can identify the target spots where probes are retained and even measure in which quantity.

These techniques are remarkably specific because a DNA sequence of about 20 bp is long enough to generate an adequate signal.

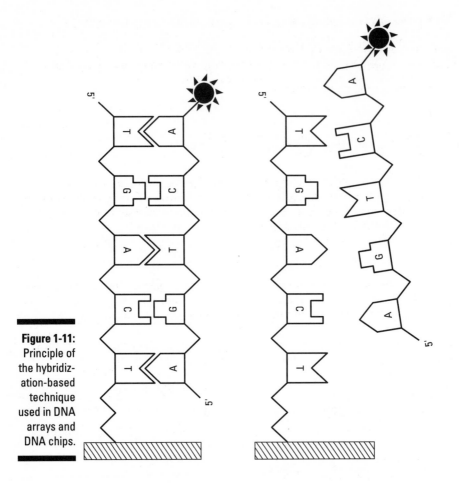

Figure 1-11:
Principle of
the hybridiz-
ation-based
technique
used in DNA
arrays and
DNA chips.

In the most recent implementations, the target DNA samples are so finely "printed" that all genes of a given genome (10- to 50- thousand) can be represented on a single microscope glass slide, hence the term *micro-array* used to designate them. In other techniques, the target molecules are attached to silicon chips at an even higher density, hence the term *DNA chips.*

These microarrays or DNA chips are mostly used to measure the level of activity for all genes at once (in parallel), in a high throughput manner. By varying the individuals, the organs, and the conditions (disease, stress) from which the probe molecules (messenger RNAs) are obtained, biologists can get a picture of the cellular activity (the technical term is *gene expression profiling*) from which they can eventually identify a specific pattern or signature. These signatures can then be used to discover the genes that are associated with some diseases, or serve as molecular diagnosis.

These new ways of playing with DNA are producing a new kind of data (images of arrays and matrices of DNA spot intensities) that has nothing in common with the usual sequence or molecular structure data. The use of DNA chips thus prompted the emergence of specialized bioinformatics concepts and tools that are much beyond the scope of this book.

If you have a particular interest in these topics — and you want to see what DNA chips and arrays look like, learn the basic vocabulary, follow cutting-edge DNA chip technology, and download a wealth of good software for free — check out the following Internet sites:

- ✔ cmgm.stanford.edu/pbrown/: This site from a leading laboratory in the field offers a complete guide to microarraying for the molecular biologist. (Build your own equipment!) It also provides access to the Stanford Microarray Database and forum.

- ✔ www.dev.nhgri.nih.gov/DIR/Microarray/main.html: This site provides detailed protocols, analysis tools, and related resources.

- ✔ www.ebi.ac.uk/microarray/index.html: Based at the European Bioinformatics Institute, this site provides access to Web-based tools for microarray analysis, and to a public repository for microarray data. (Your data submission is welcome!)

For commercial solutions, visit these sites:

- ✔ www.affymetrix.com/: The leading company in DNA chips

- ✔ www.axon.com/: The commercial provider of a popular DNA array reader and analysis software that features very nice pictures and animations

Chapter 2

How Most People Use Bioinformatics

In This Chapter

▶ Using PubMed to become quickly knowledgeable on any biological subject

▶ Retrieving protein and DNA sequences relevant to your work

▶ Running your first sequence/database comparison with BLAST

▶ Performing your first multiple sequence alignment with ClustalW

Personally, I'm always ready to learn, although I do not always like being taught.

— Sir Winston Churchill

*I*n this chapter, we show you enough bioinformatics to perform the routine — but nonetheless incredibly useful — tasks that experienced molecular biologists once spent months mastering by wandering helplessly on the Net. Chances are that you can fulfill 90 percent of your professional needs without knowing more than the content of this chapter. If you've already got a pretty good chunk of bioinformatics under your belt — enough to qualify as a power user — you may still find useful reminders and tricks here. If your background is in computer science, we show you the kinds of tools that biologists demand. So get on your favorite PC, follow the guide, and enjoy the ride!

Becoming an Instant Expert with PubMed/Medline

Think about the following situation: You just got the sequence of the gene you have been working on for years, and you gave it to the grumpy local specialist in bioinformatics to analyze it for you. (This is, of course, before you

had an opportunity to read this book!) The specialist diagnosis comes back, and here it is: "Jim, your sequence looks pretty much like a dUTPase."

Finding out about a protein by its name

The good news is, your gene actually resembles *something*, and you may have the beginning of a *story*. The bad news is that you've never heard the term dUTPase in your entire life. Where do you go from here? How do you find out more about this seemingly obscure term with the strange spelling? How do you quickly decide whether this story makes any sense in the context of what you already know about your gene?

Searching PubMed

The answer is to stay in your chair and go to PubMed! To do this, follow these steps:

1. **Using your favorite Internet browser, navigate to**
 www.ncbi.nlm.nih.gov/entrez/ **on the World Wide Web.**

 Your screen now looks like Figure 2-1, with a Search window already preset to PubMed and a For text box next to it, ready for you to type in a few words relevant to your question.

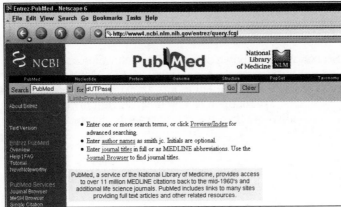

Figure 2-1:
The initial
PubMed
search
window.

2. **Type in** dUTPase **in the For window, and click the Go button.**

 Soon after, a Results list, like the one in Figure 2-2, appears on-screen.

 For our dUTPase example, we now have more than 200 references at our fingertips, more than enough to start unraveling the mysteries of dUTPase — including its relevance to your gene. But don't feel you have to rush to the library with your printed list of references before it closes: You may not even have to go.

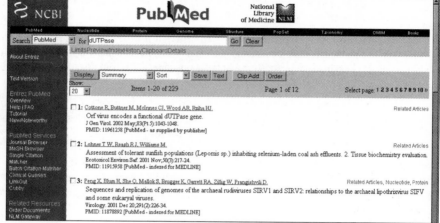

Figure 2-2:
The initial
result of a
standard
PubMed
search for
dUTPase.

3. **For any entry in the Results list, click the associated author names.**

 The author names stand out because they appear in a blue font and are underlined — two signs of a clickable hyperlink.

 Clicking the link calls up a page containing a rather detailed summary of the paper you chose. If you have a reasonable background in biology and a little flair, reading a few of these abstracts should be enough for you to get an idea of what dUTPase is all about.

4. **Save what you like to your hard drive by choosing your browser's File⇨Save As option.**

 Alternatively, you can transform the display into a more simpler (printer-friendly) format by first clicking the Text button on the right of the line with Display, and then choosing File⇨Save As.

Saving multiple summaries

When your search yields many references, the best move is to start scanning a few pages and select the most promising papers for future use by checking the corresponding boxes. To move through the Results pages, just use the Select Page button in the top right to select the next page.

When you think you have selected enough references, just do the following:

1. **Choose Abstract from the pull-down menu to the right of the Display button.**

2. **Click the Display button.**

 The abstracts of all the papers you selected in the various pages appear for you to read.

3. If you want to print this display as is, choose File⇨Print from your browser menu.

4. If you want to print this display in a more printer-friendly format, first click the Text button (on the same menu bar as the Display button) to display this information in a non-HTML test format and then choose File⇨Print from your browser menu.

5. If you'd like to save the file in the format of your choice, choose File⇨Save As from you browser menu, enter a new filename for the file, and then choose the file format you want to save to.

Although PubMed provides its own Save button on the same menu as the Display button, we believe the procedure involved in using this button is unnecessarily complex for the task at hand. We advise you to stay away from it and use the File⇨Save As command on your browser menu.

Searching PubMed using author's names

Although you can certainly search PubMed by using topic keywords such as protein names, you do have other search options available to you. For example, we're sure you've found yourself in the situation where one of your colleagues (or perhaps your advisor) has told you something like the following: "I read a paper by so and so, published not too long ago, and I think it was on dUTPase; you should check it out." In the world of card catalogs and stacks of periodicals, this isn't much of a lead. But on the World Wide Web, with PubMed at your side, you're almost home free. Suppose that you only remember one of these author's names, such as Abergel. Using PubMed, take the following steps to track down the elusive reference:

1. **Point your browser to** www.ncbi.nlm.nih.gov/entrez/.

2. **Type** Abergel — **the name of your prospective author — in the For window, and then click the Go button.**

 The Results list appears, as shown in Figure 2-3.

 We now have a list of many papers, all of them with "Abergel" as an author. Again, you can browse through this list, select some of them, and bring out their abstracts. However, the bad news is that none of them appear to have anything to do with dUTPase. To solve this problem, you simply need to combine the protein name and the author information in your query — and you don't even need to memorize a complicated syntax or Boolean symbols.

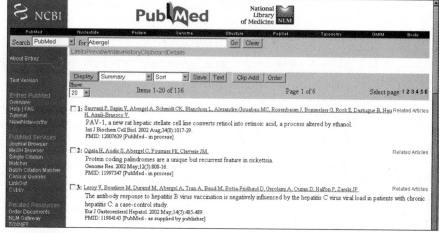

Figure 2-3:
The results
of a
standard
PubMed
search for
Abergel.

3. **Type dUTPase next to Abergel in the For window — don't forget to put a single space between the search terms — and click Go.**

By default, PubMed assumes that we want to use both search terms — Abergel AND dUTPase — when looking for the appropriate research papers. By default, PubMed also assumes that we want to look for these words in the title OR the abstract of the papers. Figure 2-4 shows the results of your first combined search.

Click here for the abstract.

Figure 2-4:
The results
of a
standard
combined
search for
Abergel and
dUTPase.

Only one paper appears in the list. This must be the paper your boss was talking about. You can now proceed.

4. **Click the underlined names of the authors to access the page containing a summary of the paper.**

The top-left Display pull-down menu automatically switched to the Abstract mode, and you can now read the summary (Figure 2-5).

Click here for full text.

Figure 2-5:
An abstract
of the
unique
paper
selected
with
Abergel and
dUTPase.

But wait! There's more! In fact, this is your lucky day: You've hit upon a paper from a journal that offers free access to the full text, indicated by the large colored rectangle that appears above the title.

5. **Click the free text rectangle**.

 This brings you directly to the publisher's site, where you have the option of reading an interactive on-line copy (HTML) or downloading reprints in pdf format, among other choices (Figure 2-6). An increasing number of scientific journals — including leading journals in their respective fields — are offering full-text access to articles. However, most of them will still ask you to pay a fee, to be a registered client (for instance, through your university), or to hold a personal subscription. For some journals, such as the one in Figure 2-6, full-text access becomes free six months after print publication.

 Thus, if you're lucky, you can go through the entire bibliographic process and quickly find out all you need about some mysterious biological subject without having to leave the comfort of your chair, thanks to PubMed and the cooperating publishers.

PubMed searches are *case insensitive*, meaning that the results of the searches that appear in Figures 2-1 through 2-4 do not depend on the capitalization of query terms when you type them: *DUTPASE*, *dUTPase*, or *dutpase*, as well as *Abergel* or *aBERGEL* all produce the same output. Be aware that this is NOT the case with all databases and all search programs used in Bioinformatics.

Where does PubMed come from?

PubMed is the public and freely available version of Medline, the most comprehensive database of primary scientific literature in the biomedical area. Maintained by the U.S. National Library of Medicine (physically located on the campus of the National Institutes of Health in Bethesda, MD), PubMed is the brainchild of the NCBI (The National Center for Biotechnology Information), a world-leading bioinformatic center. The nucleotide sequence database GenBank and the immensely popular BLAST software are two other famous productions of NCBI.

Click here for a PDF reprint.

Figure 2-6: The Journal of Virology grants free full-text access.

Searching PubMed using fields

In addition to searching by author and topic, PubMed offers you even more ways to narrow down bibliographic searches, thus making them more specific. To get a handle on these different ways, however, we need to know more about the internal structure of Medline entries — the pool of entries from which PubMed gathers its search results.

To gain some insight into this internal structure, refer to Figure 2-4. We had just used the PubMed site (www.ncbi.nlm.nih.gov/entrez/) to track down a single article after inputting Abergel and dUTPase as a query. With that Results list on your computer screen, you can now do the following:

1. **Click the small arrow of the pull-down menu, just to the right of the Display button.**

 The contents of the pull-down menu appear, displaying a lot of options that are different ways of displaying information related to the current article or are links to related information.

2. **Select the MEDLINE option (Figure 2-7) and click the Display button.**

 The Medline page appears, as shown in Figure 2-8.

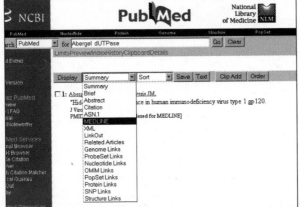

Figure 2-7:
Changing
the default
abstract
format into
MEDLINE.

In Figure 2-8 you can see the internal structure of a Medline database record. The information is spread out over separate sections, called *fields*, each one preceded by a specific abbreviation — TI for title fields, AB for abstracts, AD for the laboratory address, AU for the authors, SO for the journal abbreviation, and so forth. This structure applies to all Medline records.

Putting fields to good use

Unfortunately, not all PubMed searches are as easy and productive as the one we used in the dUTPase example. You can come up with an overwhelming number of *hits* — the standard term for search results — if you formulate queries that contain common names (such as Smith or Cohen) or use search terms (such as Down) that can occur in different contexts: titles (for example, Down syndrome), abstracts, author name, or can be part of an address (955 Down Street).

```
UI     - 99102635
PMID - 9847382
DA     - 19990128
DCOM- 19990128
LR     - 20011113
IS     - 0022-538X
VI     - 73
IP     - 1
DP     - 1999 Jan
TI     - "Hidden" dUTPase sequence in human immunodeficiency virus type 1 gp120.
PG     - 751-3
AB     - A coding region homologous to the sequence for essential eukaryotic enzyme
         dUTPase has been identified in different genomic regions of several viral
         lineages. ------------------------------------------------------------------
         --------------------------------------------------------------------------
         -------------- our results strongly suggest that an ancestral dUTPase
         gene has evolved into the present primate lentivirus CD4 and cytokine
         receptor interacting region of gp120.
AD     - Laboratory of Structural and Genetic Information, CNRS EP-91, Marseille
         F-13402, France. chantal@igs.cnrs-mrs.fr
FAU    - Abergel, C
AU     - Abergel C
FAU    - Robertson, D L
AU     - Robertson DL
FAU    - Claverie, J M
AU     - Claverie JM
LA     - eng
PT     - Journal Article
CY     - UNITED STATES
TA     - J Virol
JID    - 0113724
RN     - 0 (HIV Envelope Protein gp120)
RN     - EC 3.6.1.- (Pyrophosphatases)
RN     - EC 3.6.1.23 (dUTP pyrophosphatase)
SB     - IM
SB     - X
MH     - Amino Acid Sequence
MH     - HIV Envelope Protein gp120/*chemistry
MH     - Human
MH     - Molecular Sequence Data
MH     - Pyrophosphatases/*chemistry/genetics
MH     - Support, Non-U.S. Gov't
EDAT - 1998/12/16
MHDA- 1998/12/16 00:01
URLS - http://jvi.asm.org/cgi/content/full/73/1/751
PST    - ppublish
SO     - J Virol 1999 Jan;73(1):751-3.
```

Figure 2-8:
The internal
structure of
a MEDLINE
record.

To alleviate this problem, you can query PubMed by restricting the search for each word of your query to a given field. Thus, papers containing the word at the wrong place are not selected. To do so, you simply have to follow each term by the code of the field (in brackets) in which you want to find it. Changing the field can totally modify the result of the searches. For instance, try entering three different queries — Down [AU], Down [AB], and Down [AD] into the For text box at the PubMed site.

When we did it, we got 217, 81,011, and 724 totally unrelated references, respectively.

Using fields to find experts near you

Field-restricted queries are great for dealing with author's names or subject terms that can be found in different fields and thus inflate and confuse search results. But they have many other applications, too. For instance, we find field-restricted queries very useful when we want to locate experts in a particular research field in a given location.

Suppose that you live in Chicago and you'd like to know if anybody around can give you some insight into the mysteries of dUTPase. To get your personalized list of resident experts, just do the following:

1. **Point your browser to the PubMed site**
 (www.ncbi.nlm.nih.gov/entrez/).

2. **In the For window, enter** dUTPase [AB] Chicago [AD]**, and click Go.**

 By specifying the [AB] field, you'll be getting abstracts of potential articles; the [AD] field specifies the laboratory address.

 A couple of papers show up (Figure 2-9).

3. **Click on the list of authors (in blue, underlined)**

 You get the abstract of the corresponding article, together with the main laboratory address. Alternatively, to get all the abstracts at once,

4. **Go back to the previous URL (after Step 2) using the Go Back button of your browser.**

 1. Check all the little boxes on the left side of the paper numbers,

 2. Change the display option from *Summary* to *Abstract*,

 3. Click on the *Display* button.

On the top of each abstract, you get the name of the relevant laboratory in Chicago.

Check these boxes to select papers.

Change from summary to abstract here.

Restricted fields

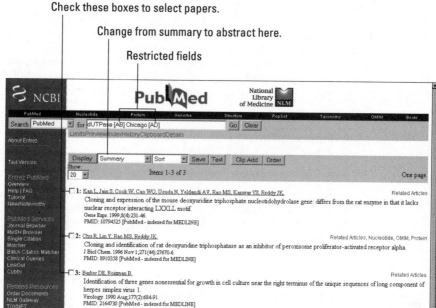

Figure 2-9:
Searching
for experts
on dUTPase
in Chicago.

With the information contained in the Medline records you have names, street addresses, and sometimes e-mail addresses. If needed, you can use a telephone book (or a Web search engine) to supplement this information and find out how to contact these experts the same day. But, please, don't call these people on our behalf!

Every time you change the Display option (or the Sort option next to it), don't forget to hit the Display button again, otherwise nothing will happen. The same is true with the Go button if you change your target database (in the top-left Search window) or your query terms (in the For window). In contrast, the Text button simultaneously causes a change of format and a re-display.

Searching PubMed using limits

PubMed offers yet another way of fine tuning your queries: You can predefine ranges for different attributes in different fields before running your search. Said like this, this sounds awfully technical and complex. But we show you here how to do it, again using our dUTPase example.

When you want to quickly find out the basics about a subject you know nothing about, it's best not to read any single highly specialized article. A better bet is to stick with review articles, where one expert summarizes the state-of-the-art for you. Unless you're an historian, you also have no need to know about the state of the art as it was 20 years ago (which is more like three centuries ago in the field of molecular biology). Thus, we want to limit our PubMed search to recent review articles about dUTPase. Here's how it's done:

1. **Point your browser to the PubMed site at**
 `www.ncbi.nlm.nih.gov/entrez/`.

2. **Type** dUTPase **in the For window.**

 If you clicked the Go button at this point, you'd get a list of more than 200 articles — a bit too much reading for a nonspecialist! You need to restrict this list to the most recent review articles written in English.

3. **Click the Limits button, just beneath the little arrow for the pull-down menu of the Search window.**

 The Limits screen appears, as shown in Figure 2-10. Here, below the Limited To line, you now have plenty of fields and attributes you can use for setting limits. You can go back later to this page and explore these various options.

4. **Choose Review from the Publication Types pull-down menu.**

5. **Choose English from the Language pull-down menu.**

 Unless you're fluent in French, *bien sûr*.

6. **Type in a suitable year in the Publication Date From box, such as 1997.**

 Filling in the To date field is optional (default is today).

7. **Select Title/Abstract in the pull-down menu above Publication Types.**

 The default here is All Fields. Choosing Title/Abstract (at the bottom of the menu) restricts the occurrence of dUTPase to the Title or Abstract fields in Medline records.

8. **Finally, click the Go button.**

 Your (new and more concise) Results page appears.

In our hands, searching for dUTPase using these limits resulted in three articles — much better than the 200 hits we would have gotten without the limits. More important, we've limited the search to review articles, which hopefully contain — in concise form — all we'd ever want to know about this subject.

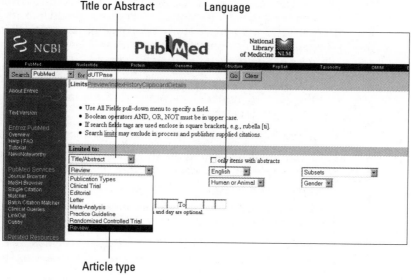

Figure 2-10:
Limiting a
search for
dUTPase to
recent
review
articles in
English.

A few more tips about PubMed

Finding out how to use PubMed to its full extent is certainly worth your time
because PubMed can be useful in many different ways. Remember that you
cannot break anything, and you risk nothing by exploring what all these
buttons — especially the ones we haven't talked about — might do. Here are
a few last-minute tips:

✔ **How to get the most out of your query:**

Quoted queries (for example, "down syndrome") behave as a single
word, and are a great way to improve the relevance of your search.

Adding initials to proper names (for example, "Abergel C") can
greatly reduce the number of hits.

Write down the PubMed Identifier (the number in the PMID field)
of that interesting paper you just found. It can be very useful in any
subsequent searches for related items, such as associated gene
and protein sequences.

Don't forget to deselect the Limit box when starting a new search.

Don't put too much initial faith in a search that produces no
results. Spelling mistakes, wrong field restrictions, or improper
limits settings can all throw off an effective search.

As a beginner researching a new subject, read through a couple of abstracts to enlarge your initial vocabulary and look for synonyms. For example, if you don't know that some papers on dUTPase might use the term "dUTP pyrophosphatase" instead, you may miss out on some interesting papers.

Try the Related Papers button — the one to the extreme right of the PubMed output — to enlarge a search that isn't giving you enough references.

✔ **Things you (unfortunately) won't find in PubMed:**

Names ranking beyond the 10th place in the author's list aren't retrievable for older papers (before 1995). This is a significant problem for many pioneering articles in genomics.

NO papers recorded before 1965 are in PubMed. Don't rely on PubMed to write an historical article on your field.

Most references recorded before 1976 have no abstract. Don't expect great results from PubMed searches involving them.

Retrieving Protein Sequences

If putting PubMed to use for bibliographic searches is the most common bioinformatics thing biologists want to do, then the next most popular thing is to retrieve relevant protein sequences from the Web in order to find out more about the subject at the molecular level. Although PubMed provides you with a direct link to the protein world, we find that it's sometimes confusing to use. So we guide you to another site, where finding protein sequences is really easy.

Expasy: A prime Internet site for protein information

Created and managed by one of the pioneers of protein bioinformatics, Prof. Amos Bairoch, the Expasy server is a world-leading resource for protein information. In addition to the SWISS-PROT protein sequence database, the Expasy site provides numerous analysis tools that we use throughout this book. It also provides a wealth of outside links for more specialized analyses on other servers.

After using PubMed to acquire some preliminary information about a particular function that you're interested in — dUTPase, for example — go ahead and find out more about it by retrieving a few examples of protein sequences that perform this function. (Don't worry; we show you how.) Because *Escherichia coli* (abbreviated E. coli) is one of the most studied organisms, we use it in our example to show you how to retrieve the sequence of the protein performing the dUTPase function in this bacterium.

1. **Point your browser to** www.expasy.org/sprot/**, the SWISS-PROT database home page (Figure 2-11).**

Click here for Quick Search.

Figure 2-11:
The Quick Search window at the top of the SWISS-PROT home page.

Click here for Advanced Search.

2. **Type** dUTPase coli **in the Quick Search window, and then click the Quick Search button.**

A full page of information about the dUTPase protein *of E. coli appears* on-screen, as shown in Figures 2-12 and 2-13. (No single browser window can hold such a wealth of information, so we've broken up the Results page into two figures.)

A SWISS-PROT entry can be roughly described as made up of four parts. For our example of dUTPase

- The top of the entry contains the entry name DUT_ECOLI on the first line and the unique identifier P06968 (called a primary accession number) on the second line. Like PubMed identifiers, these codes are worth writing down because they're used to cross-reference related entries in other databases.

- The top-middle section offers a biochemical description of the protein, including its standard name, its international Enzyme Committee number (E.C. does not mean E. Coli), as well as a couple of synonyms. These synonyms can be very useful to broaden your search for relevant articles in Medline. Then comes a list of bibliographic references relevant to the sequencing of the protein or some functional studies. Note that the PubMed ID is given for all cited articles to allow for easy cross-referencing.

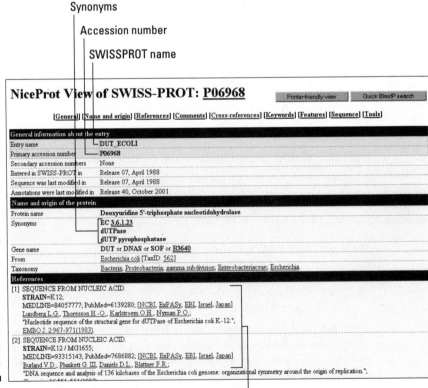

Figure 2-12:
SWISS-
PROT entry
for the
E. coli
dUTPase
protein
(upper half).

- The bottom-middle section offers a whole series of links to various functional classification schemes including relevant protein domains and functional signatures. We describe this section in more detail in chapter 4.

Links to domains and signatures.

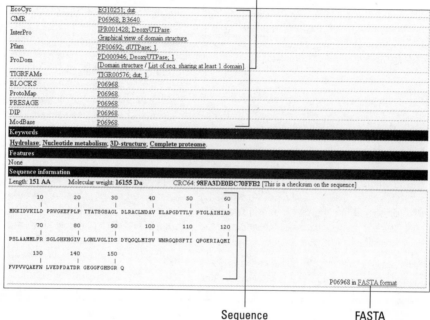

Figure 2-13: Amino-acid sequence of the *E. coli* dUTPase protein (bottom half).

Sequence FASTA

- The last section — the *sequence* section — provides you with the actual amino-acid sequence of the protein. In Figure 2-13, you can see that the E. coli dUTPase is made up of 151 amino acids, corresponding to a predicted molecular weight of 16,155 Da. (The sequence shown in Figure 2-13 uses the one-letter code format for describing amino acids. For more on the different formats used, see Chapter 1.)

For further studies, you need this amino-acid sequence in the FASTA format — a much simpler format that doesn't bother with punctuation. See the sidebar "The FASTA (and RAW) format" elsewhere in this chapter.

3. **To get the FASTA format, click the FASTA format button, on the extreme right of the very bottom of the entry.**

 The FASTA format for your sequence appears on a new page, as shown in Figure 2-14.

The FASTA (and RAW) format

FASTA is the name of a popular sequence alignment/database scanning program created by WR Pearson and DJ Lipman in 1988 (you can use your brand new PubMed skill to find the original article). The sequences used by FASTA have to obey the following format:

>My_Sequence_Name

ARCGTCRGCKINTANDRGCKINTAND

CKINTANDARCGTCRGCKINTANDRG

CKINTAND

The line starting with > (the *definition line*) contains a unique identifier followed by an optional short definition. The following lines contain the DNA or protein sequence (in one-letter code) until the next > character in the file indicates the beginning of a new sequence.

Because FASTA is easy to parse, this format has become hugely popular and is now the default input format for much sequence analysis software, including BLAST and CLUSTALW. Be aware, though, that programs using FASTA-formatted sequences as input are sometimes case sensitive. Always use CAPITAL letters for the one-letter codes. When using FASTA-formatted sequences on a PC, always use the TEXT option of your preferred Word processing software (no formatting, nothing else but ASCII characters). When displaying these sequences as a word processing document, use the Courier font for easy alignment.

Some programs analyzing one sequence at a time work with the *RAW* format. This is simply the sequence part of the FASTA format, without the definition line. Using the FASTA format when the RAW format is required may cause an error — or some of the definition line may end up included in the protein or DNA sequence (!).

In just three easy steps, you've gone all the way from an informal definition like dUTPase to the complete description of the protein performing this function in our favorite bacterium. And you thought bioinformatics was complicated?

Figure 2-14:
E. coli dUTPase protein sequence in FASTA format.

Definition line Sequence

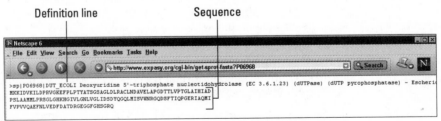

More advanced ways to retrieve protein sequences

Identifying the right protein in SWISS-PROT isn't always so easy because the information you start with may not be specific enough. For instance, the Quick Search feature may not work particularly well if the same term is found in different sections of the SWISS-PROT entry. To get better results, you have to use some field restriction (which we describe in the context of PubMed searches).

This type of search always depends on the organization of each particular database as well as on the idiosyncrasies of a particular piece of search software. For SWISS-PROT, the steps are as follows:

1. **Point your browser to** `www.expasy.org/sprot/`.

 You'll see the by-now-familiar SWISS-PROT database home page.

2. **Click the <u>Advanced Search</u> link (located near the bottom of the opening screen, below the Access to SWISS-PROT and TrEMBL banner).**

 The Advanced Search page appears (Figure 2-15). Using this electronic form you can retrieve protein sequences using keywords associated with 3 specific fields: gene definition, gene name, and organism.

Description

Figure 2-15: Using SWISS-PROT Advanced search.

Organism

3. **For the purposes of our example, type** dUTPase **in the description field, select Human in the Organism pull-down menu, and click the Submit Query button.**

Your results appear on a new page. If you use the dUTPase example, you'll successfully retrieve the human mitochondrial dUTPase protein, no ifs, ands, or buts — that's to say exactly the protein you were looking for. Its SWISS-PROT entry name is DUT_HUMAN, and its accession number is P33316.

Retrieving a list of related protein sequences

Many questions in molecular biology (your dissertation topic, your own research, or your personal interests) require downloading not only one sequence, but a large collection of similar protein sequences, all related to the same function. These biological questions typically include the detection of conserved functional motifs, the simultaneous alignment of multiple sequences, the assessment of their variability, or phylogenic studies.

Building your own specialized sequence data file, right on your PC, is easy and enables you to use it with other analysis programs accessible from other bioinformatic sites. To continue with our previous example, we now show you how to build a file containing all well-documented dUTPase protein sequences. To make it usable by other programs, you generate this file in FASTA format.

First, go back to the Advanced Search page of the Expasy server:

1. **Point your browser to** www.expasy.org/sprot/ **and click the Advanced Search link.**

2. **Type** dUTPase **in the Description window.**

Be sure that you don't type in anything else. In particular, don't put anything in the Organism window.

3. **Deselect the TrEMBL and TrEMBL-New boxes.**

Whereas TrEMBL is made from unsupervised computer translations of new DNA sequences, SWISS-PROT only includes entries validated by expert curators. Restricting the search to SWISS-PROT thus ensures — to the best of our knowledge — that all returned proteins are actual dUTPases.

4. **Click the Submit Query button.**

A Results page appears, as shown in Figure 2-16. This particular search (at the time of writing) yielded 78 SWISS-PROT entries. (You may get

more entries when you get around to doing your own search.). Scrolling down the list, you can see that all look like fine and upstanding dUTPase protein sequences. At this point, you can click each of the individual names (such as DUT_CANAL) to access a complete ID card. (For a sense of what you can find on such an ID card, refer to Figures 2-12 and 2-13.)

But to complete your assignment on dUTPases, for instance, you now have to gather all these sequences into a single file. This is easy, just do the following:

1. **Perform Steps 1 through 4 in the preceding list, and then click the Perform button at the bottom left of the Results screen. (See Figure 2-17.)**

 Clicking the Perform button sends your query to another site, the Sequence Retrieval System (or SRS) server. (We cover SRS in greater detail in the next section.) If everything goes well, your screen is now displaying something similar to Figure 2-18.

2. **Don't waste any time here and don't change anything; simply click the Save button.**

 The Download page appears, as shown in Figure 2-19. Here you can choose the format in which you want these sequences to be delivered to your hard drive.

3. **Save your sequences in FASTA format by selecting Fasta from the Sequence Format pull-down menu.**

Be careful with gene names

In the example shown in Figure 2-15 of an advanced (that is, field-based) search, we ignored the Gene Name window on purpose. Gene nomenclature can be very confusing, and gene names do not immediately relate to gene function. In other words, you cannot guess a gene name from its function; if you don't know the name, you're pretty much stuck. A further difficulty is that genes with an identical function may have totally different gene names in other organisms. For instance, the gene coding for the dUTPase protein is called dut in E. coli, dut1 in Baker's yeast, F2L for the vaccinia virus, UL50 in Herpes virus, and so forth.

However, knowing an exact gene name can be very useful because an international committee of some standing hands out such names and makes sure that they're unique within a given species.

For most situations, however, using the definition field — rather than the gene name field — is still best when exploring a database. You can then switch to the gene name when you know exactly what you're looking for or as a cross-reference between different databases.

4. Select FastaSeqs from the Use View pull-down menu.

5. Click the Save button.

Your sequences are ready to be saved to your hard drive (Figure 2-20).

Figure 2-16:
A partial
list of all
dUTPase
proteins
found in
SWISS-
PROT.

Figure 2-17:
Using the
Perform
button to
compile
results.

Figure 2-18:
dUTPase
sequences
ready for
down-
loading
from SRS.

You can save the content of this page in various ways, depending on the Internet browser version you're using. We recommend the following proce- dure because it always works — and it doesn't add mysterious extra charac- ters to the resulting file!

1. **Choose the Select All feature from your browser menu (in Internet Explorer, choose Edit⇨Select All) and then choose Copy. (Again, in Internet Explorer you choose Edit⇨Copy.)**

2. **Open a new document in your word processing program — Microsoft Word, for example.**

3. **Choose Paste from the Edit menu of your word processing program**

4. **Reformat the whole document with a Courier font (8 or 10) to realign the sequences.**

5. **Finally, save your document as** `dUTPaseDB.txt`, **using the Save As type option** *text only* **(*.txt).**

You now have, on your own personal computer, all the well-characterized dUTPase protein sequences known to man. Keep this file on your hard drive; you need it later in this chapter.

Figure 2-19:
Setting up
the form to
get your
sequences
in FASTA
format.

```
>DUT_ADEG1
MDPFGSSSVPPCSTSDLPEPKLYFVRLSPHAVPPVRATHGAAGYDLFSAYDIKVPARGRA
LVPTDLVFQFPPGCYGRIAPRSGLAAKFFIDVGAGVIDPDYRGNVSVVLFNFSESSFNIR
RGDRVAQLILERIMVPELSELTQLGETDRGASGFGSTGMGAVDRNQRSVLEWLTPGSR
>DUT_ADEG8
MSFDSGCPPTPPVKLLFKKHSPFAVTPQRATSGAAGYDLCSSADVVVPPKSRSLIPTDLS
FQFPRGVYGRIAPRSGLAVKFFIDVGAGVIDSDYRGIVSVLLFNFSDHNFNVRRGDRIAQ
LILERHLTPDLEERSGLDETARGAAGFGSTGGFDTGVCPSSFS
>DUT_AGRTS
MTVQNDNRPLLRLVRLANGADLELPSYETRGAAGMDLRAAVPADEPLNLQPGERALVPTG
FIFEVPQGYEAQIRPRSGLAIKNGITCLNSPGTVDSDYRGEVKVILANLGQDDFTIERGM
RIAQMVIAPVTQVTVSEVTETSETARGAGGFGSTGV
>DUT_AQUAE
MSKVILKIKRLPHAQDLPLPSYATPHSSGLDLRAAIEKPLKIKPFERVLIPTGLILEIPE
GYEGQVRPRSGLAWKKGLTVLNAPGTIDADYRGEVKVILVNLGNEEVVIERGERIAQLVI
APVQRVEVVEVEEVSQTQRGEGGFGSTGTK
>DUT_BPT5
MIKIKLTHPDCMPKIGSEDAAGMDLRAFFGTNPAADLRAIAPGKSLMIDTGVAVEIPRGW
FGLVVPRSSLGKRHLMIANTAGVIDSDYRGTIKMNLYNYGSEMQTLENFERLCQLVVLPH
YSTHNFKIVDELEETIRGEGGFGSSGSK
>DUT_BRAJA
MSTKVTVELQRLPHAEGLPLPAYQTAEAAGLDLMAAVPEDAPLTLASGRYALVPTGLAIV
LPPGHEAQVRPRSGLAAKHGVTVLNSPGTIDADYRGEIKVILINHGAAVLRDQARRADRA
KMVIAPVVQAALVPVATLSATDRGAGGFGSTGR
>DUT_BRUME
MTAASSSAPTLGIIRLEHAKGLDLPAYETAGSAGMDLRAAVAEDRQIVLLPGRRTLVPTG
LILEIPQGYEVQIRPRSGLAFKNGITCLNTPGTIDSDYRGEVKVILINLGDDDFRIERGM
RIAQAVFAPVIQPKIEERAKISETARGAGGFGSTGTA
>DUT_BUCAI
MSNIEIKILDSRMKNNFSLPSYATLGSSGLDLRACLDETVKLKAHKTILIPTGIAIYIAN
PNITALILPRSGLGHKKGIVLGNLVGLIDSDYQGQLMISLWNRSDQDFYVNPHDRVAQII
FVPIIRPCFLLVKNFNETSRSKKGFGHSGVSGVI
>DUT_CANAL
MTSEDQSLKKQKLESTQSLKVYLRSPKGKVPTKGSALAAGYDLYSAEAATIPAHGQGLVS
TDISIIVPIGTYGRVAPRSGLAVKHGISTGAGVIDADYRGVKVVLFNHSEKDFEIKEGD
RIAQLVLEQIVNADIKEISLEELDNTERGEGGFGSTGKN
>DUT_CAUCR
MSAIPAGLAIRFKRWEGNADLPVPAYATVGAAGFDLRAHVPDDAPIVLKPGARCMAPTGF
SVAIPDGYEMQVRPRSGLALKNGVTVVNAPGTVDSDYRGQVCVLLINLGEEDFTIRRGDR
IAQGVIAAAPQWPLVEVEDLDATERGAGGFGSTGV
>DUT_CHLMU
MKFFCKLESGSSLPEYATSGASGADVRANISEPIAILPGQRALIPTGISVDIPHGYEIQV
```

Figure 2-20:
FASTA
sequences
ready to be
saved to
your hard
drive.

Retrieving DNA Sequences

Protein sequences are simple objects with a relatively narrow distribution in size (300 amino-acids ±200, with the exception of a few giant ones), clearly defined boundaries, and specific functional attributes. Furthermore, proteins of microbes or higher eukaryotes (animal and plants) have roughly the same properties.

This is not true of the corresponding gene (DNA) sequences. Gene sizes in humans may vary from microbe-like lengths (a few thousands bp) to several hundred thousands bp.

Beware of parasite characters in downloaded sequence files

In the dUTPAse example, we didn't use the File⇨Save As command on the Internet browser main menu to download the content of the window because files saved using the File⇨Save As command

✔ May contain some hidden *parasite charac-ters* — Control/Alt/Shift + something — often displayed as **<PRE>** at the beginning of the file, thus corrupting the FASTA format.

✔ Are trickier to reopen by your PC word-processing software, because they don't have the right extension (for instance .doc), or ask you to choose an encoding scheme (about which we know nothing) to load the file.

It is our experience that the File⇨Save As option produces unpredictable results, depend-ing on the navigator type, version, or implemen-tation. In general, inspecting the sequence data files that you download from the Internet for unexpected leading or trailing signs is a good practice. For most sequence analysis programs, FASTA formatted sequence files have to begin by a definition line, such as:

```
>P0343456 My_Sequence_definition
```

and nothing else! Any leading character (even blank ones) different from > may produce an error For instance, the definition line might be considered part of the protein sequence. Sequence data files also have to end with a final <New Line> character (showing as a blank line).

The good news is that usually no constraints restrict the length of the definition and sequence lines (yet, use reasonable numbers, such as 60 to 100 characters, to be safe). Except for the constraint of having > as the first char-acter of any definition line, you can freely use blank characters, such as <Space>, <Tab>, and line delimiters without interfering with the parsing of the sequences.

Finally, do not use characters that are not among the standard amino-acid codes such as <-> or <*>. They aren't treated in a consistent way by different analysis programs. They're skipped (deleting a position), replaced by X, or may simply cause an error. (For more on stan-dard amino-acid codes, see Chapter 1.)

Not all DNA is coding for protein

Various types of DNA sequences are involved in defining a gene:

✔ Regulatory regions (usually preceding the coding region)

✔ Untranslated regions that precede and follow the coding regions

✔ The protein-coding region

In eukaryotes (yeast, plants, animals), the protein-coding region is divided by a variable number of exons that contribute to the final protein, interspersed with introns that do not.

As a consequence, working with DNA sequences is always trickier than working with protein sequences. Go to Chapters 3 and 5 to find out more about the architecture of genes.

Going from protein to DNA sequences

In databases, the correspondence between protein and DNA sequences is *not* one to one. Many different — even nonoverlapping — DNA sequences can be linked to the same protein or gene name, as the following list makes clear:

- ✔ The *primary transcript* — the sequence encompassing the full genome interval that is transcribed (the hnRNA, including exons + introns)

- ✔ the *mature transcript* — the mRNA, without the introns)

- ✔ The strict protein coding region — the open reading frame or ORF

- ✔ Numerous types of partial sequences (such as Expressed Sequence Tags, also known as ESTs)

Life is a lot simpler if you can stick to working with just protein sequences, but we realize that situations will arise that will require that you go back to the DNA sequence. The most common situation of this kind involves the design of PCR primers, when you try to clone your protein of interest or just check it for the presence of mutations. Your problem, succinctly stated, is this: Given a protein sequence, how can I retrieve the DNA sequence encompassing its coding region?

One protein, many DNA entries

Figure 2-21 shows that retrieving relevant DNA sequences is not as easy as working with proteins. There are four DNA links to the simple *E. coli* dUTPase protein: X01714 is a 1609-bp sequence, V01578 is a 2568-bp sequence, L10328 a 136,254-bp one, and AE000441 is 10562-bp long. They all contain the dUTPase ORF, but are linked to various articles and authors, and represent successive stages in the sequencing of the *E. coli* genome. The protein-DNA correspondence is even more complex for higher organisms due to the existence of introns, and the presence of database entries related to mRNA and genomic sequences for the same protein.

Retrieving the DNA sequence relevant to my protein

Imagine that you want to retrieve enough DNA sequence to clone the dUTPase gene of *E. coli.* Sounds tricky, right? We show you how it's done in the following steps:

1. **Point your browser to** www.expasy.org/sprot/.

2. **To access the *E. coli* dUTPase entry quickly, simply enter the accession number (P06968) in the Quick Search window and then click the Quick Search button.**

 Again, we have the information page devoted to protein P06968 at hand.

3. **Click the Cross-References button near the top of the form.**

 Your browser jumps to the relevant section, as shown in Figure 2-21. The Cross-References section, consists of successive groups of lines introduced by keywords such as *EMBL*, *PIR*, *PDB*, and so on. All these keywords correspond to databases of various kinds that can give you additional information about your protein sequence. By clicking on the different underlined words, you jump right to the corresponding entry in those different databases. This web of links between different sites concerning the same object are the "cross-references." Take the EMBL group of lines, for instance. There are four of them starting with obscure numbers and followed by a set of four underlined links to — EMBL, GenBank, DDBJ (those are databases), and CoDingSequence (this is the precise DNA segment coding for the protein). Let's see what happens when we use them.

Click here to get the DNA sequence.

Figure 2-21: Cross-Reference section of the P06968 information page.

4. **Click the GenBank link in the EMBL row's first line.**

 The GenBank record (Accession # X01714) corresponding to protein P06968 appears, as shown in Figure 2-22. It is several pages long.

Roughly speaking, a GenBank entry consists of four parts. For our dUTPase

- The *locus name* (ECDUT) — (an arbitrary identifier) — followed by a short definition line, and a unique accession number (X01714). This number is the most stable identifier for the entry; we recommend that you write it down because it's used to cross-reference related entries in other databases — as we just did with SWISS-PROT. A few lines below, the SOURCE and ORGANISM fields describe the biological origin of the DNA sequence.

- The *Reference section* lists article(s) relevant to the sequence determination. This list can be quite long for large sequences.

- The *Features section* lists the definitions and exact ranges of multiple types of elements that have been recognized in the sequence. Our example — the X01714 entry — includes promoter elements, ribosome binding sites (RBS), and protein coding segments (CDS). Figure 2-23 shows — right next to the keyword CDS — the limits of the dUTPase ORF as 343..798. The CDS range is followed by the name and the amino-acid translation of the corresponding protein.

- The *Sequence section* rounds out the GenBank entry, where the nucleotides are listed between the Origin keyword and the final // that signals the very end of the entry. Numbering is provided to help relate the location of the dUTPase ORF (343-798) to the actual nucleotide sequence. (See Figure 2-24.)

Now that you've read through the entry, you can save the whole thing to your hard drive by using the steps we defined in the preceding list for saving SWISS-PROT sequences.

However, to use the nucleotide sequence as input for other programs (for instance, to design primers), you may need to isolate it from the text part of the entry and convert it in FASTA format. Here's how that's done:

1. **Scroll back to the top of the page for the ECDUT/X01714 entry.**

 Refer to Figure 2-22 for what your screen should look like.

2. **Choose FASTA from the Display pull-down menu, as shown in Figure 2-25.**

3. **Transform the content of this window into plain text by clicking the Text button.**

4. **Save the FASTA sequence by using the following protocol:**

 a. In the Edit menu of your Web browser, click Select All and then click Copy.

 b. Open a default Word document and, in the Edit menu of Word, click Paste. Then select a Courier font (8 or 10).

 c. Finally, save your document as dUTPaseDNA.txt by choosing the Save as type option text only (*.txt).

The unified DNA sequence databases

The SWISS-PROT Cross-Reference section offers links to three databases: EMBL, GenBank, and DDBJ: What are these? These databases are in fact different views of the same data set. They are the products of a consortium including the European Molecular Biology Laboratory (EMBL), the DNA Data Base of Japan (DDBJ), and the American GenBank operation at the National Center of Biotechnology Information (NCBI). All three entities basically share all

sequences entered by each member of the consortium. Although the primary DNA sequences are the same, each site has its own way of recording and presenting the data. Thus, the formats of GenBank, EMBL, and DDBJ entries differ in a few details. The CoDingSequence button on the protein information page (refer to Figure 2-21) can be used to extract the nucleotide sequence strictly corresponding to the ORF.

Accession number

GenBank name

Figure 2-22:
Top of
GenBank
entry
ECDUT/
X01714.

Bibliography

Features

Range of the dUTPase ORF

```
FEATURES             Location/Qualifiers
     source          1..1609
                     /organism="Escherichia coli"
                     /db_xref="taxon:562"
     promoter        286..291
                     /note="-35 region"
     promoter        310..316
                     /note="-10 region"
     misc_feature    322..324
                     /note="put. transcription start region"
     RBS             330..333
                     /note="put. rRNA binding site"
     CDS             343..798
                     /note="dUTP-ase (aa 1-151)"
                     /codon_start=1
                     /transl_table=11
                     /protein_id="CAA25859.1"
                     /db_xref="GI:41297"
                     /db_xref="SWISS-PROT:P06968"
                     /translation="MKKIDVKILDPRVGKEFPLPTYATSGSAGLDLRACLNDAVELAP
                     GDTTLVPTGLAIHIADPSLAAMMLPRSGLGHKHGIVLGNLVGLIDSDYQGQLMISVWN
                     RGQDSFTIQPGERIAQMIFVPVVQAEFNLVEDFDATDRGEGGFGHSGRQ"
     misc_feature    831..851
                     /note="put.stem-loop structure"
     repeat_unit     831..838
                     /note="inverted repeat A"
     repeat_unit     844..851
                     /note="inverted repeat A'"
     misc_feature    866..893
                     /note="put. stem-loop structure"
     repeat_unit     866..872
                     /note="imp. inverted repeat B"
     repeat_unit     888..893
                     /note="imp. inverted repeat B'"
     RBS             889..895
                     /note="pot. rRNA binding site"
     CDS             905..1540
```

Figure 2-23:
The Features section of GenBank entry ECDUT/ X01714.

ORF translation

Start of the dUTPase ORF

```
BASE COUNT      395 a     421 c     413 g     380 t
ORIGIN
        1 cagagaaaat caaaaagcag gccacgcagg gtgatgaatt aacaataaaa atggttaaaa
       61 accccgatat cgtcgcaggc gttgccgcac taaaagacca tcgaccctac gtcgttggat
      121 ttgccgccga aacaaataat gtggaagaat agcccggca aaaacgtatc cgtaaaaacc
      181 ttgatctgat ctgcgcgaac gatgtttccc agccaactca aggaatttaac agcgacaaca
      241 acgcattaca cctttctgg caggacggag ataaagtctt accgcttgag cgcaaagagc
      301 tccttggcca attattactc gacgagatcg tgacccgtta tgatgaaaaa aatcgacgtt
      361 aagattctgg acccgcgcgt tgggaaggaa tttccgctcc cgacttatgc cacctctggc
      421 tctgccggac ttgacctgcg tgcctgctctc aacgacgccg tagaactggc tccgggtgac
      481 actacgctgg ttccgaccgg gctggcgatt catattgccg atccttcact ggcggcaatg
      541 atgctgccgc gctccggatt gggacataag cacggtatcg tgcttggtaa cctggtagga
      601 ttgatcgatt ctgactatca gggccagttg atgatttccg tgtggaaccg tggtcaggac
      661 agcttcacca ttcaacctgg cgaaccgcatc gcccagatga tttttgttcc ggtagtacag
      721 gctgaattta atctggtgga agatttcgac gacaccgacc gcggtgaagg cggctttggt
      781 cactctggtc gtcagtaaca catacgcatc cgaataacgt cataacatag ccgcaaacat
      841 ttcgtttgcg gtcatagcgt gggtgccgcc tggcaagtgc ttattttcag gggtattttg
      901 taacatggca gaaaaacaaa ctgcgaaaag gaaccgtcgc gaggaaatac ttcagtctct
      961 ggcgctgatg ctggaatcca gcgatggaag ccaacgtatc acgacggcaa aactggccgc
     1021 ctctgtcggc gtttccgaag cggcactgta tcgccacttc ccagtaagga cccgcatgtt
     1081 cgatagcctg attgagttta tcgaagatag cctgattact cgcatcaacc tgattctgaa
     1141 agatgagaaa gacaccacag cgcgcctgcg tctgattgtg ttgctgcttc tcggttttgg
     1201 tgagcgtaat cctggcctga cccgcatcct cactggtcat gcgctaatgt ttgaacagga
     1261 tcgcctgcaa gggcgcatca accagctgtt cgagcgtatt gaagcgcagc tgcgccaggt
     1321 attgcgtgaa aagagaatgc gtgagggtga aggttacacc accgatgaaa ccctgctggc
     1381 aagccagatc ctggccttct gtgaaggtat gctgtcacgt tttgtccgca gcgaatttaa
     1441 ataccgcccg acggatgatt ttgacgcccg ctggcccgcta attgcggcca gttgcagtaa
     1501 tatgcagccg gatgactttt catccggcga gtttctttaa acgccaaact cttcgcgata
     1561 ggccttaacc gccgccagat gttccgccat ttccggcttc tcttccagg
//-
```

Figure 2-24:
The Sequence section of GenBank entry ECDUT/ X01714.

End of the dUTPase ORF

End of the X01714 entry

Select FASTA format for the X01714 entry.

Text button for printer friendly display

Figure 2-25:
Nucleotide
sequence of
GenBank
entry
ECDUT/
X01714 in
FASTA
format

Using BLAST to Compare My Protein Sequence to Other Protein Sequences

After you know the basics of retrieving a protein sequence from a database (see the earlier, aptly named section "Retrieving Protein Sequences"), you're ready to perform your first analysis with it. For most people, the next step is to perform a BLAST search.

BLAST (short for *Basic Local Alignment Search Tool* — with a name like that, no wonder they shortened it to BLAST) is a great sequence comparison tool that quickly tells you which of the other known proteins out there has a sequence similar to yours. You can then use this information for a variety of purposes, including the prediction of protein function, 3-D structure and domain organization, or the identification of homologues in other organisms. In Chapter 7, we show you how to use BLAST in great detail. At this point, we only want to get you started and show you how easy it is to do.

BLAST comes to us under the auspices of the National Center for Biotechnology Information (NCBI), already familiar to us as the hosts for the PubMed bibliographic database we cover at the beginning of this chapter. You'll soon discover that using BLAST is just as easy as using PubMed. To start off:

1. **Point your favorite Internet browser to** `www.ncbi.nlm.nih.gov/BLAST/`.

 The BLAST home page — probably the most frequented bioinformatic Web page in the world — appears, as shown in Figure 2-26. Because this is your first time here, keeping things simple is best.

2. **Click the <u>Standard Protein-Protein BLAST</u> link.**

 A Query screen appears, as shown in Figure 2-27.

3. **Open the dUTPase FASTA-formatted protein sequence file.**

 This is the file that you (hopefully) created on your PC by using the steps in the "Retrieving a list of related protein sequences" section, earlier in this chapter.

4. **Using your browser's Edit menu, copy and paste ONE of the protein sequences (with its definition line) into the BLAST Search window.**

 We used DUT_ADEG1. If you don't have the protein sequence ready, open a new navigator window and go get this sequence on the Expasy server:

 a. Point your browser to `www.expasy.org/sprot/`.

 b. Enter *DUT_ADEG1* in the Search Window, click on Quick Search.

 c. At the very bottom of the Nice Prot form, click on **Q89662 in FASTA format**.

 d. Copy the protein sequence with its header line.

 e. Close that navigator window, and paste the sequence in the BLAST Search window.

 At this point, you need a FASTA-formatted protein sequence.

5. **Deselect the Do CD-Search box, but don't change the Choose Database setting.**

 The *nr* (for nonredundant) default database includes all protein sequences known to man, so you don't have to bother about selecting a specific organism or subset.

6. **Click the BLAST! button.**

 You just launched your first BLAST search. Welcome to the club!

 The next form, entitled Formatting BLAST, allows you to change a number of output features. (refer to Figure 2-28). For now, keep it as it is.

 c. At the very bottom of the Nice Prot form, click on **Q89662 in FASTA format**.

Standard protien BLAST

Figure 2-26:
NCBI
BLAST
entry page.

Place your query sequence here.

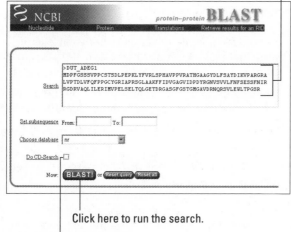

Figure 2-27:
NCBI
BLAST
query form.

Click here to run the search.

Uncheck the CD-search box.

Figure 2-28:
NCBI
BLAST
"format-
while-you-
wait" form.

Click here once and wait.

7. **Click the Format! button.**

 BLAST now gives you an estimated running time for your query. If the time of the day is right, this shouldn't be more than a few seconds.

8. **When the Results page appears, scroll down the page until you reach a long list of sequences (Figure 2-29).**

 What you have here are all the sequences that have a significant similarity with your query sequence, ranked by decreasing score values. If you used DUT_ADEG1 or another protein sequence taken from a database, the best matching protein is probably the one that you started with. This is what happened in Figure 2-29, with the Fowl adenovirus dUTPase. The highest similarity score is 338. The score value depends on the length of the most similar segments between 2 sequences, and the quality of the match. It is not normalized to 100 percent. See Chapter 7 to find out more about the significance of BLAST scores.

 In this list of sequences, you would simply click an underlined identifier to link to the corresponding *nr* (nonredundant) database entry.

9. **Click one of the underlined scores in the second column from the right.**

 This brings you further down in the page (Figure 2-30), to see the alignment between your query sequence and the matching nr sequence of the protein corresponding to this score.

In our example, the best match is an identical one. By going down the list, you can see less perfect matches, slowly degrading as the corresponding score decreases and the E-value increases. The *E-value* is an assessment of the statistical significance of the score. E-values close to 1 are a warning that the conclusion you might draw from the alignments is not reliable. Go to Chapter 7 to find out more about these issues.

Click here to see the nr entry. Click here to see an alignment.

```
                                                        Score   E
Sequences producing significant alignments:            (bits)  Value

gi|9628836|ref|NP_043869.1|  (NC_001720) dUTP pyrophosphatas...   338   2e-92
gi|281211|pir||S26429  probable dUTP pyrophosphatase (EC 3.6...   336   6e-92
gi|9633174|ref|NP_050279.1|  (NC_000899) dUTPase homolog [Fo...   182   2e-45
gi|1071889|pir||A46256  dUTP pyrophosphatase (EC 3.6.1.23) -...   177   4e-44
gi|16758374|ref|NP_446044.1|  (NM_053592) dUTPase [Rattus no...   176   8e-44
gi|7513966|pir||T10819  dUTP pyrophosphatase (EC 3.6.1.23) -...   176   8e-44
gi|3046554|gb|AAD03407.1|  (AF056304) dUTPase homolog [orf v...   176   1e-43
gi|4503423|ref|NP_001939.1|  (NM_001948) dUTP pyrophosphatas...   174   4e-43
gi|13435139|ref|NP_076084.1|  (NM_023595) 5031412I06Rik [Mus...   173   9e-43
gi|12833509|dbj|BAB22551.1|  (AK003076) DUTPASE-data source:...   172   1e-42
gi|18043437|gb|AAH19979.1|AAH19979  (BC019979) Similar to dU...   171   4e-42
gi|139356|sp|P14597|DUT_ORFN2  DEOXYURIDINE 5'-TRIPHOSPHATE ...   171   4e-42
gi|3041664|sp|P33316|DUT_HUMAN  Deoxyuridine 5'-triphosphate...   169   1e-41
gi|19921126|ref|NP_609479.1|  (NM_135635) BcDNA:LD08534 gene...   168   3e-41
gi|18583771|ref|XP_056825.3|  (XM_056825) dUTP pyrophosphata...   167   5e-41
gi|12856347|dbj|BAB30639.1|  (AK017216) DUTPASE-data source:...   167   5e-41
gi|18767711|ref|NP_572002.1|  (NC_003407) dUTPase-like prote...   164   3e-40
gi|9634708|ref|NP_039001.1|  (NC_002188) ORF FPV038 dUTP pyr...   158   2e-38
gi|416922|sp|P32518|DUT_LYCES  DEOXYURIDINE 5'-TRIPHOSPHATE ...   156   1e-37
gi|15232681|ref|NP_190278.1|  (NM_114561) dUTP pyrophosphata...   154   4e-37
gi|9964316|ref|NP_064784.1|  (NC_002520) putative dUTPase [A...   152   1e-36
gi|20178416|ref|NP_619837.1|  (NC_003663) V047 [Cowpox virus...   152   2e-36
gi|12642133|gb|AAK00135.1|AF221544_2  (AF221544) ORF2 [porci...   151   3e-36
gi|1808634|emb|CAA64116.1|  (X94355) G2L [Cowpox virus] >gi|...   150   4e-36
gi|17508109|ref|NP_492542.1|  (NM_060141) deoxyuridine 5'-tr...   150   4e-35
gi|17974947|ref|NP_536461.1|  (NC_003310) C8L [Monkeypox vir...   149   2e-35
gi|6969680|gb|AAF33892.1|  (AF095689) TF2L [Vaccinia virus (...   148   2e-35
gi|13876670|gb|AAK43558.1|  (AF336131) dUTPase [lumpy skin d...   148   2e-35
gi|18640099|ref|NP_570173.1|  (NC_003389) SPV013 dUTPase [Sw...   148   2e-35
gi|15150457|ref|NP_150452.1|  (NC_003027) LSDV018 dUTPase [l...   148   2e-35
gi|9790951|ref|NP_063678.1|  (NC_001559) dUTP pyrophosphatas...   147   5e-35
gi|418297|sp|P32208|DUT_SPVKA  DEOXYURIDINE 5'-TRIPHOSPHATE ...   146   1e-34
gi|9627547|ref|NP_042070.1|  (NC_001611) C6L [Variola virus]...   145   2e-34
gi|335626|gb|AAA48246.1|  (M34368) F2L protein [Vaccinia vir...   144   4e-34
gi|9633648|ref|NP_051726.1|  (NC_001132) deoxyuridine 5'-trip...   144   5e-34
gi|19114785|ref|NP_593873.1|  (NC_003424) deoxyuridine 5'-tr...   144   5e-34
```

Figure 2-29: NCBI BLAST output form: best match listing.

Matching residues

Query sequence Database sequence

```
                              Alignments

>gi|9628836|ref|NP_043869.1|  (NC_001720) dUTP pyrophosphatase [Fowl adenovirus A]
  gi|7531092|sp|Q89662|DUT_ADEG1  Deoxyuridine 5'-triphosphate nucleotidohydrolase (dUTPase) (dUTP
     pyrophosphatase)
  gi|58424|emb|CAA78921.1|  (Z17216) open reading frame [Fowl adenovirus 1]
  gi|1314433|gb|AAC54895.1|  (U46933) ORF1 [Fowl adenovirus 1]
          Length = 178

 Score =  338 bits (866), Expect = 2e-92
 Identities = 178/178 (100%), Positives = 178/178 (100%)

Query: 1    MDPFGSSSVPPCSTSDLPEPKLYFVRLSPHAVPPVRATHGAAGYDLFSAYDIKVPARGRA 60
            MDPFGSSSVPPCSTSDLPEPKLYFVRLSPHAVPPVRATHGAAGYDLFSAYDIKVPARGRA
Sbjct: 1    MDPFGSSSVPPCSTSDLPEPKLYFVRLSPHAVPPVRATHGAAGYDLFSAYDIKVPARGRA 60

Query: 61   LVPTDLVFQFPPGCYGRIAPRSGLAAKFFIDVGAGVIDPDYRGNVSVVLFNFSESSFNIR 120
            LVPTDLVFQFPPGCYGRIAPRSGLAAKFFIDVGAGVIDPDYRGNVSVVLFNFSESSFNIR
Sbjct: 61   LVPTDLVFQFPPGCYGRIAPRSGLAAKFFIDVGAGVIDPDYRGNVSVVLFNFSESSFNIR 120

Query: 121  RGDRVAQLILERIMVPELSELTQLGETDRGASGFGSTGMGAVDRNQRSVLEWLTPGSR 178
            RGDRVAQLILERIMVPELSELTQLGETDRGASGFGSTGMGAVDRNQRSVLEWLTPGSR
Sbjct: 121  RGDRVAQLILERIMVPELSELTQLGETDRGASGFGSTGMGAVDRNQRSVLEWLTPGSR 178
```

Figure 2-30: NCBI BLAST output form: local sequence alignments.

Making a Multiple Protein Sequence Alignment with ClustalW

Besides running database searches to identify similar proteins in a pairwise manner, the second most common bioinformatic task that biologists like to perform with protein sequences is a multiple alignment. *Multiple alignments* consist in lining up many similar proteins side by side for the sake of comparison. Multiple alignments are used to

- Identify sequence positions where specific amino acids really matter for the structural integrity or the function of a given protein
- Define specific sequence signatures for protein families
- Classify sequences and build evolutionary trees

We detail all the intricacies of making good multiple alignments in Chapter 9. At this point, however, we simply want to show you that performing a multiple alignment is really easy, especially when you have the pleasure of using some nice Internet server, such as the one maintained by the Protein Information Resource (PIR) people at Washington, D.C.'s Georgetown University.

The PIR actually originated from the *Atlas of Protein Sequences*, the first protein sequence collection built by the late Prof. M. Dayhoff in the late 1970s. The PIR site offers some useful protein analysis tools and databases that we invite you to explore by yourself. Among these tools, it offers a multiple alignment server (running the standard ClustalW program) that is really easy to use for beginners.

Do the following to get your feet wet using by ClustalW:

1. **Point your browser to** `pir.georgetown.edu`.

 The PIR home page appears, as shown in Figure 2-31.

2. **Under the Search and Retrieval heading, click the <u>ClustalW</u> or <u>Alignment</u> link in the scroll-down menu to display the input form.**

 The input form appears. At this point, you need a few FASTA-formatted protein sequences.

3. **Open the dUTPase FASTA-formatted sequence file that you created on your PC in the previous section.**

 Alternatively, go to the SWISS-PROT home page at `www.expasy.org/ sprot/` and grab a few dUTPAse sequences (such as DUT_AQUAE, DUT_BRAJA, DUT_BPT5, DUT_CANAL, and DUT_BUCAI), as we spell out in the section "Expasy: A prime Internet site for protein information," earlier in this chapter.

4. **Copy/paste five of these sequences (all at once) in the input window, (Figure 2-32).**

 Be careful to copy the definition line and the entire amino-acid sequence for each of them.

5. **Click the Submit button.**

 Within a few seconds, your screen is proudly displaying its first successful phylogenetic tree and multiple alignment (Figure 2-33).

This output (if you used the same sequences as we did here) illustrates the power of multiple alignments. It clearly delineates a region of high versus low sequence conservation in these related proteins. Positions 100 percent conserved (highlighted by < * >) or nearly conserved (highlighted by <:> or <.>) tend to occur in a clustered fashion along the sequence, defining regions that are probably directly involved in shaping the catalytic site or performing the pyrophosphatase biochemical reaction. Various types of *functional signatures* can be extracted from such a multiple alignment. Finally, evolutionary relationships between proteins are inferred from phylogenetic trees. In Chapters 9, 11, and 13, we tell you more about these key applications of bioinformatics.

Figure 2-31: Another useful protein analysis center: PIR.

Figure 2-32:
Filling up the
multiple
alignment
input
window at
PIR.

Figure 2-33:
Phylo-
genetic
tree and
multiple
alignment
of five
bacterial
dUTPase
sequences

Part II
A Survival Guide to Bioinformatics

The 5th Wave By Rich Tennant

Andy was beginning to feel there must be
an easier way to earn $50.

In this part . . .

Bioinformatics is about finding and interpreting biological data. In this part, we show you how to go shopping in the main bioinformatics databases. We've got lots of great tips for picking up the freshest, ripest data (and not just the low hanging fruits!). Because all these databases are linked with one another, we also show you how you can travel across all of them and gather every piece of information that you need. This provides a fascinating journey across human knowledge and a compulsory starting point for any research project.

With the right data in the fridge, it's time to start cooking. In this part, we show simple recipes for working with DNA and protein sequences in order to predict their most basic properties.

Chapter 3

Using Nucleotide Sequence Databases

..

In This Chapter

▶ A quick refresher on the structure of genes and genomes

▶ Making use and sense of GenBank

▶ Finding out about LocusLink

▶ Working with complete microbial genomes

▶ Browsing the human and other animal genomes

▶ Submitting your own sequence to GenBank

..

The secret of success is to know something nobody else knows.

— Aristotle Onassis (1906-1975)

Sequence databases are great tools because they offer a unique window on the past. They make it possible to answer today's biological questions by enabling us to analyze sequences that may have been determined as many as 20 years ago, when the whole technology emerged. By doing this, they connect past and present molecular biology.

The first databases were in fact created as some sort of sequence museum, where sequences could be preserved for all eternity in pristine form, just as they were determined, interpreted, and published by their original authors. This *historical* (time capsule!) perspective pretty much remains in GenBank, the leading nucleotide sequence repository maintained as a consortium between the U.S. National Center for Biotechnology Information (NCBI), the European Molecular Biology Laboratory (EMBL), and the DNA Data Bank of Japan (DDBJ). In this chapter, we show you how to use GenBank and decipher its entries.

Repository-type databases are great tools when you want to come up with a bibliography for a particular sequence, but they do not provide easy access to sequence data when your query deals with broader issues related to a gene or function rather than with a specific paper. For this reason, the second-generation nucleotide sequence databases have adopted a more *gene-centric* perspective, where all the sequence information relevant to a given gene is made accessible at once. In addition to unraveling the mysteries of the more traditional GenBank, then, we also show you how to use NCBI's LocusLink, a good example of a gene-centric database.

These days, nucleotide sequences are routinely determined at the whole genome or chromosome scale — at least for microorganisms. We now have information not only about individual gene sequences, but also about their relative positions, strand orientation, and the presence or absence of bio-chemical functions within an entire organism. To take advantage of this more global information, researchers have had to design state-of-the-art *genome-centric* sequence information management systems that are able to connect specialized sequence collections and browsing tools. As an added bonus, this chapter presents some of the great genome-centric resources dealing with viruses, bacteria, or (you guessed it) human beings.

However, before you start delving into these information management systems in earnest, you may find it useful to read the next section. There, we quickly summarize the fundamentals of genes and genomes and introduce the vocabulary you need to read GenBank fluently. (If you're an experienced biologist, you can skip this section completely or read it word for word in hopes of finding some embarrassing mistake. Go ahead. We dare you!)

Reading into Genes and Genomes

True, nucleotide sequences are universal, but the structure of the genes they encode is markedly different between *prokaryotic* organisms (organisms lacking a true nucleus) and *eukaryotic* organisms (the kind lucky enough to have a true nucleus). You have to know the basic architecture of both types of genomes and genes to make sense of a simple GenBank entry. Remember that we need to define a gene as the contiguous genome segment encompassing all the nucleotide sequence information necessary to its successful expression — successful expression here meaning the production of protein or RNA. This handy definition includes both the coding and regulatory parts of a gene.

Prokaryotes: Small bugs, simple genes

The three most basic classes of living organisms are the prokaryotes (usually bacteria), the archea (bacteria-like organisms living in extreme conditions), and the eukaryotes. Eukaryotes go all the way from microscopic yeast to humans, animals, and plants.

For bioinformatic purposes, prokaryotes and archea are very much the same. With few exceptions — far beyond the realm of this book — they have the following properties in common:

- ✔ They are microscopic organisms.

- ✔ Their genome is a single, circular DNA molecule.

- ✔ Their genome size is in the order of a few million base pairs [0.6-8].

- ✔ Their *gene density* — the number of genes per base pairs in the genome — is approximately one gene per 1,000 base pairs.

- ✔ Their genome is lean and mean, containing few useless parts (70 percent is coding for proteins).

- ✔ Their genes do not overlap.

- ✔ Their genes are transcribed (copied into messenger RNA) right after a control region called a *promoter*.

- ✔ These messenger RNA (mRNA) are collinear with the genome sequence. In other words, genes are in a single piece, not interrupted by noncoding patches (called introns).

- ✔ Protein sequences are derived by translating the longest open reading frame (from ATG to STOP) spanning the gene transcript sequence.

Figure 3-1 illustrates the simple collinear relationship between a bacterial genomic (gene) sequence, the transcript (mRNA), the open reading frame (ORF), and the final protein. (For a peek at a typical (small) bacterial genome, flip to Figure 1-10 in Chapter 1.) The mRNA sequence gets translated into a protein right after a special signal, called the Ribosome Binding Site (RBS in Figure 3-1). The ribosome is the main piece of machinery of the cell's protein translation apparatus.

Accordingly, database entries describing a coding prokaryotic sequence should include three important features:

- ✔ The coordinates of some promoter elements

- ✔ The coordinates of the RBS

- ✔ The coordinates of the ORF boundaries

That's about all there is to say regarding the simple architecture of prokaryotic genes. Not all genes encode proteins; for some of them, the function is directly carried out by the transcribed RNA molecule. This includes transfer RNA (tRNA), ribosomal RNA (rRNA), and a few other fancy RNAs that shall remain nameless.

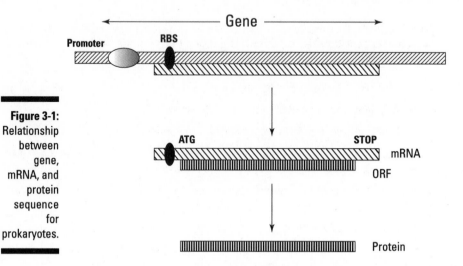

Figure 3-1: Relationship between gene, mRNA, and protein sequence for prokaryotes.

Eukaryotes: Bigger bugs, complex genes

Eukaryotes encompass a wide variety of organisms, from microscopic yeasts and fungi, to elephants, plants, and humans. Yet eukaryotes, too, have basic properties in common, properties that end up making them a bit more challenging when it comes to genomic and bioinformatics analysis:

✔ Their genome consists of multiple linear pieces of DNA called *chromosomes* (up to a hundred million base pairs long).

✔ Their genome size (10-3,000 million base pairs), especially for animal and plants, is *much bigger* than in prokaryotes.

✔ Their gene density is much lower than that for prokaryotes (one human gene per 100,000 base pairs).

✔ Their genome is no model of efficiency, containing many useless parts (less than 5 percent of the human genome code for proteins).

✔ Genes on opposite DNA strands might overlap — although that's a relatively rare occurrence.

✔ Their genes are transcribed right after a control region called a *promoter*, but sequence elements located far away can have a strong influence on this process.

✔ Gene sequences are not collinear with the final messenger RNA (mRNA) and protein sequences. Only small bits (the exons) are retained in the mature mRNA that encodes the final product.

✔ Genes often exhibit more than one mRNA (and protein) form.

Figure 3-2 summarizes the complex relationship between a genomic (gene) sequence, the mature mRNA, and the protein. Using the same drawing convention as in Figure 3-1, Figure 3-2 shows that a given eukaryotic gene often produces several different forms of mRNA and proteins, depending on the selection of exons.

Genes of higher eukaryotes (animals) may span up to millions of base pairs — the human dystrophin gene, for example, is 2.2 million base pairs long. The relationships between a gene DNA sequence, its primary transcript, the various forms of mature mRNA, and the final protein sequence can be very complex. This is why understanding a database entry that describes all these things may take a bit of work. In addition, a lot of database entries correspond to partial gene sequences, or different gene-related objects such as promoter regions, mRNA, or genome fragments. Throughout the rest of this chapter, we show you several examples so that, after a bit of study, you'll be able to decipher the most intricate GenBank entries all by yourself.

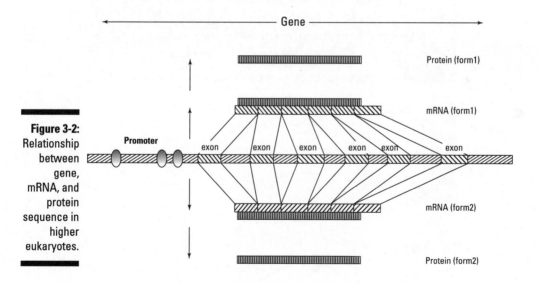

Figure 3-2:
Relationship between gene, mRNA, and protein sequence in higher eukaryotes.

Making Use (And Sense) of GenBank

For prokaryotes, the limited size of the genes involved as well as the simple (linear) relationship between the gene DNA sequence, the mRNA, the ORFs, and the final protein sequence make it relatively easy to annotate and store all that information in database records. This is why database entries corresponding to bacterial genes are easy to read and understand. In this section, we take you through the entry of the *Escherichia coli* dUTPase gene step by step (handy GenBank ID X01714).

Making sense of the GenBank entry of a prokaryotic gene

First things first, we have to fetch our sample GenBank entry, as follows:

1. **Point your browser to** www.ncbi.nlm.nih.gov/entrez/.

 The NCBI PubMed home page appears.

2. **From the Search pull-down menu, choose Nucleotide (Figure 3-3).**

Choose Nucleotide.

Figure 3-3: Searching for nucleotide sequence X01714 at NCBI PubMed.

3. **Type the GenBank ID X01714 in the For window, and click Go.**

 A Results page appears, displaying a short definition for the nucleotide sequence entry, preceded by its ID X01714 (underlined in blue).

4. **Click the X01714 hyperlink. (Alternatively, you can change the display option from Summary to GenBank, and then click the Display button.)**

 The X01714 GenBank entry appears in the default GenBank format, as shown in Figure 3-4.

Use this menu for FASTA format.

Click here to get .txt format.

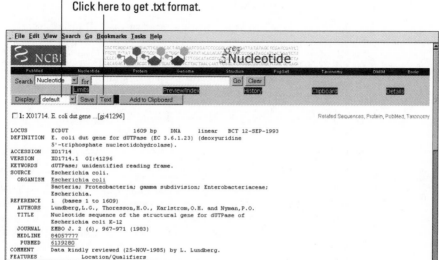

Figure 3-4:
GenBank
entry
X01714.

Gurus refer to this as the GenBank *flat file format* because you can read it in a linear fashion, and it doesn't involve indexes, pointers, or accessory information (which are customary in the structure of more sophisticated databases).

In fact, what you have here isn't totally flat because it contains a few hyperlinks.

5. **Click the Text button on the menu bar above the X01714 entry to generate a true flat file (text) format of the entry.**

6. **Save your entry as a text document on your hard drive by choosing File⇨Save As from your browser's main menu.**

Reading the header of a GenBank prokaryotic entry

The entry text you just saved is divided into sections, with keywords introducing each section. Typical keywords include *LOCUS, DEFINITION, ACCESSION, VERSION, KEYWORDS,* and *SOURCE*. Refer to Figure 3-4 as we go through these keywords step by step:

- ✔ **LOCUS** gives us the locus name (an arbitrary name; here it's ECDUT), the size of the nucleotide sequence in base pairs, the nature of the molecule (here it's DNA), and its topology (*linear* or *circular* molecule). In this particular case, it's *linear*.

- ✔ **DEFINITION** provides a short definition of the gene that corresponds to the entry sequence. Here it's the *E. coli dUTPase gene*.

- ✔ **ACCESSION** lists the accession number, a unique identifier within and across various databases (such as the protein sequence databases). Here, the accession number is X01714.

- ✔ **VERSION** fills you in on synonymous or past ID numbers.

- ✔ **KEYWORDS** introduces a list of terms that broadly characterize the entry. You can use these terms as keywords for certain database searches. (For more on using keywords in database searches, check out Chapter 2 where we discuss using Entrez with limits.)

- ✔ **SOURCE** divulges the common name of the relevant organism to which the sequence belongs.

- ✔ **ORGANISM** gives a more complete identification of the organism, complete with its technical taxonomic classification.

 Notice that ORGANISM is set in a bit from the left margin of the page — more so at least than the other keywords. This isn't a mistake. Computer scientists call this *indentation*. The precise position where a keyword begins on the various lines denotes its subordination to other keywords. Keywords starting at the same position are said to be at the same level. The left-most position delimits sections of the GenBank entry that are independent, such as the preceding six sections.

 In contrast, ORGANISM is indented to indicate that it's part of the SOURCE section. (The same thing happens in the REFERENCE section.)

- ✔ **REFERENCE** introduces a section where the credits for the sequence determination are given (different parts of the sequences can be credited to different authors). The REFERENCE section contains multiple parts as denoted by the indentation of the following self-explanatory keywords: AUTHORS, TITLE, JOURNAL, and MEDLINE. The PUBMED keyword is subordinated to MEDLINE and introduces the unique PubMed ID in the Medline database, useful for cross-referencing.

- ✔ **COMMENT** introduces the last line of the top of the X01714 GenBank entry. It contains free-formatted text, such as acknowledgments or info that doesn't fit in the previous sections.

The next (middle) section of the entry is the *FEATURES table* (see Figure 3-5). It describes precisely the gene regions and the associated biological properties that have been identified in the nucleotide sequence. This entire section

is under the control of the keyword *FEATURES*. A large variety of biologically related keywords are subordinated to *FEATURES*, as we make clear in the following list:

- ✓ **source** indicates the origin of specific regions of the sequence. This is useful when you want to distinguish cloning vectors from host sequences. In X01714, the whole sequence comes from E. coli genomic DNA.

- ✓ **promoter** shows the precise coordinates of a promoter element. In X01714, a *-35 box* is indicated from position 286 to 291 (indicated by *286..291*) in the nucleotide sequence.

 promoter introduces another line that contains the coordinates of another promoter element, this time a *-10 box* at positions 310 to 316.

- ✓ **misc feature** (miscellaneous feature) indicates the putative location of the transcription start (mRNA synthesis). For X01714, this is at position 322 to 324.

- ✓ **RBS** (Ribosome Binding Site) indicates the location of the last upstream element. For X01714, this is at position 330 to 333.

Protein sequence

```
JOURNAL     EMBO J. 2 (6), 967-971 (1983)
MEDLINE     84057777
 PUBMED     6139280
COMMENT     Data kindly reviewed (25-NOV-1985) by L. Lundberg.
FEATURES             Location/Qualifiers
     source          1..1609
                     /organism="Escherichia coli"
                     /db_xref="taxon:562"
     promoter        286..291
                     /note="-35 region"
     promoter        310..316
                     /note="-10 region"
     misc feature    322..324
                     /note="put. transcription start region"
     RBS             330..333
                     /note="put. rRNA binding site"
     CDS             343..798
                     /note="dUTP-ase (aa 1-151)"
                     /codon_start=1
                     /transl_table=11
                     /protein_id="CAA25859.1"
                     /db_xref="GI:41297"
                     /db_xref="SWISS-PROT:P06968"
                     /translation="MKKIDVKILDPRVGKEFPLPTYATSGSAGLDLRACLNDAVELAP
                     GDTTLVPTGLAIHIADPSLAANMLPRSGLGHKHGIVLGNLVGLIDSDYQGQLMISVWN
                     RGQDSFTIQPGERIAQMIFVPVVQAEFNLVEDFDATDRGEGGFGHSGRQ"
     misc feature    831..851
                     /note="put.stem-loop structure"
     repeat unit     831..838
                     /note="inverted repeat A"
     repeat unit     844..851
                     /note="inverted repeat A'"
     misc feature    866..893
                     /note="put. stem-loop structure"
     repeat unit     866..872
                     /note="imp. inverted repeat B"
     repeat unit     888..893
                     /note="imp. inverted repeat B'"
     RBS             889..895
                     /note="pot. rRNA binding site"
     CDS             905..1540
                     /note="unidentified reading frame"
                     /codon_start=1
```

Figure 3-5: GenBank entry X01714: the FEATURES table part.

- ✔ **CDS** (CoDing Segment) introduces a complex section that describes the gene open reading frame (ORF).

 The first line indicates the coordinates of the ORF from its initial ATG to the last nucleotide of the first stop codon TAA. (For X01714, it's 343 to 798.)

 Each of the following lines (indented at the same level) gives the name of a protein product, indicates the reading frame to use (here 343 is the first base of the first codon), the genetic code to apply, and a number of IDs for the protein sequence.

 /translation, the final keyword of the CDS section, introduces the conceptual amino-acid sequence of the coding segment. This sequence is a computer translation that uses the coordinates, reading frame, and genetic code indicated in the preceding lines.

- ✔ Another **misc feature** follows the CDS section. It contains lines that point out putative stem-loop structures and repeats. These are potential regulatory elements of the dUTPase gene.

The FEATURES section of entry X01714 is typical of a simple, well-annotated bacterial nucleotide sequence, centered on a well-identified gene. However, this straightforward entry includes a complication, in the form of a putative additional reading frame. We selected this entry because we were interested in one gene: the dUTPase. However, the entry exhibits an extra putative gene, indicated by an additional RBS element and a second CDS section.

GenBank entries containing more than one gene are frequent.

Using the Sequence section of a prokaryotic entry

The last section of GenBank entry X01714 is the nucleotide sequence section. It starts with the *BASE COUNT* keyword and finishes with the end-of-entry line introduced by two slash marks (//). Each line of nucleotide sequence starts with the position number of the first nucleotide in that line. Each line contains 60 nucleotides.

Because it mixes numbers and nucleotides, you cannot directly use the sequence section of a GenBank entry for direct input (by cut-and-paste) on most similar search or alignments servers. For this, you need to generate a FASTA formatted file of the sequence, as follows:

1. **To the right of the Display button (at the top of the page), choose FASTA from the pull-down menu to change from the current default option (Figure 3-4).**

2. **Click the Text button to refresh the screen.**

 The nucleotide sequence appears now in a form suitable for most sequence analysis programs.

3. **Save this entry by choosing File⇨Save As from your browser's main menu or by copying/pasting it into an empty Word document.**

Making sense of the GenBank entry of an eukaryotic mRNA

We now continue with the dUTPase gene because it's quite ubiquitous and present in both prokaryotes and eukaryotes: Humans have it, too. Looking at the GenBank entries describing the human version of dUTPase clearly illustrates the increased complexity of higher eukaryotes compared to prokaryote genes. We start here with a simple one, GenBank entry U90223.

1. **Point your browser to** www.ncbi.nlm.nih.gov/entrez/.

 The NCBI Pub Med home page appears on cue.

2. **From the Search pull-down menu, choose Nucleotide.**

3. **Type in GenBank ID U90223 in the For window, and click Go.**

 A Results page appears, displaying a short definition for the nucleotide sequence entry, preceded by its ID U90223 (underlined in blue).

4. **Click the U90223 hyperlink. (Alternatively, you can change the display option from Summary to GenBank, and then click the Display button.)**

 You're now ready to read your first human GenBank entry.

5. **Click the Text button to generate a totally flat format.**

6. **Save your result by choosing File⇨Save As.**

Although it's related to a human gene, GenBank entry U90223 doesn't look very different from entry X01714, the one that describes its bacterial homologue. The top part of the entry follows the general information keywords order: LOCUS, ACCESSION, DEFINITION, and VERSION.

The KEYWORD line, which is supposed to list quickly relevant searchable terms, such as dUTPase, is empty for entry U90223. Unfortunately, this isn't a fluke. It illustrates a common problem in sequence databases — annotations may be incomplete. As a result, keyword-based database searches are usually not retrieving all the relevant sequences in GenBank. On a related note, the information slated for the SOURCE and REFERENCE sections may also be

missing or incomplete. The U90223 entry, for example, says that the sequence here comes from an unpublished source. However, a simple PubMed search (see Chapter 2) using the author's names — Ladner RD and Caradonna SJ — brings us the relevant article in the *Journal of Biological Chemistry*, 1997! A word to the wise: You should never expect GenBank (or any sequence database) annotations to be up-to-date. This explains why it is sometimes impossible to retrieve some sequences with PubMed searches.

The FEATURES section indicates a coding region (63-821) sequence that corresponds to the mitochondrial form of human dUTPase. Following the conceptual amino-acid translation of the ORF, the *sig peptide* keyword indicates the location of a mitochondrial targeting sequence, and the *mat peptide* keyword provides the exact boundaries of the mature peptide.

The final part of the U90223 entry is the sequence section, as shown in Figure 3-6.

This human GenBank entry isn't really more complex than its bacterial homologue because we carefully selected an entry describing an *mRNA* sequence, not a *genomic* one. At the level of the mature mRNA, the relationship between a eukaryotic protein and its encoding nucleotide sequences is as collinear as its prokaryotic counterpart (refer to Figures 3-1 and 3-2). As you're about to find out, genuine genomic data can be a bit more tricky.

Redundancy in GenBank

Because better technology has caused the average size of nucleotide sequences to increase with time, large GenBank entries containing multiple genes have become very frequent. This can confuse the link between a given gene, the relevant nucleotide regions, and the Genbank ID. This problem is at its worst for complete bacterial genome sequences, where all genes are regrouped within a single entry. Increasingly, large DNA sequences often overlap with shorter, previously determined ones, creating redundancy in the current GenBank. For instance, four GenBank entries (X01714, V01578, L10328, AE000441) encompass the E. coli dUTPase gene. These entries originate from different publications and their sizes are respectively 1609, 2568, 136,254, and 10,562 nucleotides. The latter is an artificially generated entry representing 1/400 of the complete E. coli genome sequence. You have to remember that many different GenBank entries are relevant to a specific gene, especially for human and popular laboratory organisms, such as E. coli, the yeast S. cerevisiae, the Drosophila fly, and the mouse. Not all these entries exhibit the same quality or level of annotation: Some are plain sequences with empty features tables, some emphasize coding regions, while others contain detailed information about regulatory features in the promoter. Some of these entries are correct while others can be obsolete. When your interest is set on a specific gene, do not stop at the first relevant GenBank entry, but browse among all of them in order to grab every bit of information. Unfortunately, you can't really avoid this redundancy because it's the price you pay for the historical perspective adopted in GenBank.

Protein sequence

```
JOURNAL     Submitted (19-FEB-1997) Dept. of Molecular Biology, Univ. of Med.
            and Dent. of NJ-School of Osteopathic Medicine, 2 Medical Center
            Drive, Stratford, NJ 08084, USA
FEATURES             Location/Qualifiers
     source          1..960
                     /organism="Homo sapiens"
                     /db_xref="taxon:9606"
     CDS             63..821
                     /note="mitochondrial dUTPase isoform; DUT-M"
                     /codon_start=1
                     /product="deoxyuridine triphosphate nucleotidohydrolase
                     precursor"
                     /protein_id="AAB94642.1"
                     /db_xref="GI:2735292"
                     /translation="MTPLCPRPALCYHFLTSLLRSAMQNARGTAEGRSRGTLRARPAP
                     RPPAAQHGIPRPLSSAGRLSQGCRGASTVGAAGWKGELPKAGGSPAPGPETPAISPSK
                     RARPAEVGGMQLRFARLSEHATAPTRGSARAAGYDLYSAYDYTIPPMEKAVVKTDIQI
                     ALPSGCYGRVAPRSGLAAKHFIDVGAGVIDEDYRGNVGVVLFNFGKEKFEVKKGDRIA
                     QLICERIFYPEIEEVQALDDTERGSGGFGSTGKN"
     sig_peptide     63..269
                     /note="mitochondrial targeting presequence"
     mat_peptide     270..818
                     /product="deoxyuridine triphosphate nucleotidohydrolase"
BASE COUNT      221 a      258 c      272 g      209 t
ORIGIN
        1 ggtggaagcc tggcgcacgt ccggaggtgc cgaggaccca accagcccaa actctggggg
       61 aaatgactcc cctctgccct cgccccgcgc tctgctacca tttccttacg tctctgcttc
      121 gctcagcgat gcaaaacgcg cgagggcacgg cagagggccg aagccgcggt actctccggg
      181 ccaggcccgc ccctcggccg ccggcggcgc agcacgggat tccccggccg ctgtccagcg
      241 ctggccgcct gagccaaggc tgccgcggag ccagtacagt cggggcccgct ggctggaagg
      301 gcgagcttcc taaggcgggg ggaagcccgg cgccggggcc ggagacaccc gccatttcac
      361 ccagtaacgg ggcccggcct gcggaggtgg gcggcatgca gctccgcttt gcccgggctct
      421 ccgagcacgc cacggcccc accgggggct ccgcgcgcgc cgcggggctac gacctgtaca
      481 gtgcctatga ttacacaata ccacctatgg agaaagctgt tgtgaaaacg gacattcaga
      541 tagcgctccc ttctgggtgt tatggaagag tggctccacg gtcaggcttg gctgcaaaac
      601 actttattga tgtaggagct ggtgtcatag atgaagatta tagaggaaat gttggtgttg
      661 tactgtttaa ttttggcaaa gaaaagtttg aagtcaaaaa aggtgatcga attgcacagc
      721 tcatttgcga acggattttt tatccagaaa tagaagaagt tcaggctttg gatgacaccg
      781 aaaggggttc aggaaggtttt ggttccactg gaaagaatta aaatttatgc caagaacaga
      841 aaacaagaag tcatacctttt ttcttaaaaa aaaaaaagt ttttgcttca agtgttttgg
      901 tgttttgcac ttctgtaaac ttactagctt taccttctaa aagtactgca tttttttactt
//
```

Figure 3-6: GenBank entry U90223: the sequence part.

Making sense of a GenBank eukaryotic genomic entry

The previous section described the GenBank entry corresponding to the mRNA sequence of the human dUTPase gene. Here, we now want to have a look at the GenBank entry AF018430 related to the *gene sequence* (as it is on the chromosome) from which this mRNA originated.

We first have to fetch the entry by using the same protocol as before:

1. **Point your browser to** www.ncbi.nlm.nih.gov/entrez/.

 The NCBI PubMed home page appears.

2. **From the Search pull-downmenu, choose Nucleotide.**

3. **Type in AF018430 in the For window, and click Go.**

4. **Click the <u>AF018430</u> link.**

 The AF018430 entry appears, as shown in Figure 3-7.

You're now ready to read your first human genomic GenBank entry. We've selected a simple one for you to start with, but it contains some of the specificities only encountered in eukaryotic entries.

The top part contains general information introduced by the usual keywords:

- **LOCUS:** The locus name is HSDUT2. The rest of the line tells us that we're dealing with 1177 bp of linear DNA.

- **DEFINITION:** This line indicates that the name of the gene is *DUT* and specifies that the entry encompasses *exon 3* of the gene. This reminds us that human genes generally spread their protein coding regions across many disjointed pieces, the exons.

- The **ACCESSION, VERSION,** and **KEYWORDS** lines are standard. Note that for AF018430, the latter is empty, meaning that you cannot retrieve this entry via a keyword field-restricted search.

- **SEGMENT:** This field relates to the mosaic structure of eukaryotic genes. It indicates that this current GenBank entry is the second segment of a super entry made of four. You need all four entries to reconstruct the complete mRNA sequence used as a template for producing the protein.

- The **ORGANISM, SOURCE,** and **REFERENCE** lines are standard.

The FEATURES table is what really makes a eukaryotic genomic entry special and, as such, is much longer than the ones for prokaryotic organisms. It contains the following elements (see Figure 3-8):

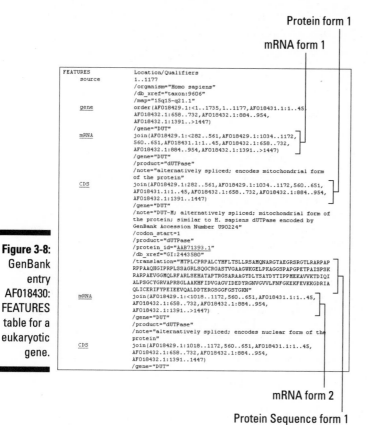

Figure 3-8:
GenBank
entry
AF018430:
FEATURES
table for a
eukaryotic
gene.

✔ The **source** section contains a */map* section. For AF018430, it indicates that the sequence belongs to chromosome 15, and was more precisely mapped on the long arm (q) of this chromosome, between cytogenetic bands q15 and q21.

✔ The **gene** keyword introduces complex looking formulas. Their purpose is to precisely describe the reconstruction of the various mRNAs spread over several separate entries.

Concentrate on the first *gene order* formula to understand how this works:

```
AF018429.1:<1..1735,1..1177,AF018431.1:1..45,
AF018432.1:658..732,AF018432.1:884..954,
AF018432.1:1391..>1447
```

This is nothing more than a recipe telling us how to reconstruct the mRNA sequence. All it says is:

1. Take nucleotides from positions 1 to 1735 from entry AF018429.

2. Add nucleotides from positions 1 to 1177 from the current entry (AF018430).

3. Add nucleotides 1 to 45 from entry AF018431.

4. Add nucleotides 658 to 732 from entry AF018432.

5. Add nucleotides 884 to 954 from the same entry (AF018432).

6. Add nucleotides 1391 to 1447 from the same entry again.

The < (less than sign) at the beginning of the formula indicates that the gene might actually start before the indicated position, while the > (greater than sign) at the end of the formula indicates that the gene might actually continue beyond the indicated position.

✔ **The mRNA:** The AF018430 entry has two mRNA fields, which are interesting because they illustrate the way GenBank represents alternative splicing patterns. Alternative splicing is a common property of higher eukaryotic gene expression. If you use the same parsing logic as gene order, which we describe in the preceding bullet, you can produce the results that we summarize in Table 3-1.

Table 3-1 makes it easy to understand what's going on exactly:

- There are two alternative mRNA: type 1 (for the mitochondria) and type 2 (for the nucleus).

- The nuclear mRNA does not contain the first exon (stored in entry AF018429) although the mitochondrial mRNA uses this exon.

- The nuclear mRNA uses a part of the second exon (stored in the AF018429) in a different reading frame from its mitochondrial counterpart.

- The two protein sequences become identical on the third exon. You can see this in the two */translation* fields (*ETPAISPSK* and so on). See Figure 3-9 for the sequence of the nuclear form of the protein.

Table 3-1 Mitochondrial or Nuclear dUTPase mRNAs in AF018430

mRNA	AF018429	AF018430	AF018431	AF018432
Type 1 (Mitochondria)	282-56 1034-1172	560-651	1-45	658-732 884-954 1391-1447 ->
Type 2 (Nuclear)	<1018-1172	560-651	1-45	658-732 884-954 1391-1447 ->

✔ The **exon** keyword indicates the position of the sole exon present in this sequence. This is near the end of the FEATURES part of the AF018430 entry (Figure 3-9, positions 560-651). Look at AF018432 if you want to see multiple occurrences of the exon field in a single entry.

The sequence section, located in the bottom part of the entry, is formatted as usual and appears in Figure 3-9.

TIP

Understanding how to splice back the various nucleotide sequence fragments to form an mRNA and the associated coding regions from segmented GenBank entries was the main difficulty of this chapter. You did it! Now you can sit back and relax while we continue our tour of GenBank.

Protein sequence form 2

```
    CDS           join(AF018429.1:1018..1172,560..651,AF018431.1:1..45,
                  AF018432.1:658..732,AF018432.1:884..954,
                  AF018432.1:1391..1447)
                  /gene="DUT"
                  /note="DUT-N; alternatively spliced; nuclear form of the
                  protein; similar to H. sapiens dUTPase encoded by GenBank
                  Accession Number U90224"
                  /codon_start=1
                  /product="dUTPase"
                  /protein_id="AAB71394.1"
                  /db_xref="GI:2443581"
                  /translation="MPCSEETPAISPSKRARPAEVGGMQLRFARLSEHATAPTRGSAR
                  AAGYDLYSAYDYTIPPMEKAVVKTDIQIALPSGCYGRVAPRSGLAAKHFIDVGAGVID
                  EDYRGNVGVVLFNFGKEKFEVKKGDRIAQLICERIFYPEIEEVQALDDTERGSGGFGS
                  TGKN"
    exon          560..651
                  /gene="DUT"
                  /number=3
BASE COUNT      334 a    207 c    221 g    415 t
ORIGIN
        1 tccctaaatc aacacagatc atgtggagga ataaaatggg gttaatatat gtaaaaccaa
       61 ttaggaaact gtttctgggg caacacagta aagggcttat tcaatggata ggctagtatt
      121 attagttagt aattgggccc tttttttctt tgttttctttt cttcattttt ttccttttca
      181 aactatgggt tgtaaagcat ccaccttttg aaagtttgcc tttctgccct ttcacgctga
      241 taagtacctc agtttccaat aaacttttgt tcaggggcaa acatttacaa tgttgacatc
      301 tcttcacacc accaaaaata ttcatggaga attatttttat ctaaagctgt cttttttaata
      361 ataaaatagc cacctctacc ttcttcataa acttttaaga tgaattggta attcatcata
      421 gcaaggttga ttttagaaac taaagttgca ttaattcatt aaatacactg aaagtaantt
      481 tgtatgcttg gtcacaaaga aaatataaaa acaattttat aaatagattt gcagtttattt
      541 tctttcaata ttttcttagt gcctatgatt acacaatacc acctatggag aaagctgttg
      601 tgaaaacgga cattcagata gcgctcccctt ctgggtgtta tggaagagtg ggtaagtcat
      661 ttaagaaaca ggtaactatt tgtcaagttc tcctttgtga tagattcttc atgtttcatt
      721 tggggtaata agcaggcaat attgcttggg ctgtgtccta aaagaagcac catttgtgat
      781 agcaaatgca ctctttgaaa ggctttattt acatcctgc tttgcctctt tttgaccctt
      841 ttattttttct ccttcctcac tggagctttt aggctcacac tggcctagaa ggctgttctc
      901 agaacatggc attttatatt atgagagtaa aacttctgac ctgttggtcc cagaatgtgt
      961 aagcctactt aacctttttct tgtttggcca tggggtttag ggtaagggat actcttcagt
     1021 gttttgtagag gcactgggag gaagctagga caaaatggag ttacacgtca acaggtttga
     1081 ttttttcctgg aagcgaattc agtgtttacc agacagtttcc tttgcagagc gttagttcct
     1141 ttttgactac ttccaagtta acttaaggag gcatgga
//
```

Figure 3-9:
GenBank
entry
AF018430:
Bottom and
sequence
part.

Genomic sequence

Working with related GenBank entries

In every example that we've used so far, we knew the accession numbers for the entries we wanted to look up — X01714, U90223, or AF018430, for example. You normally get these accession numbers by reading articles that explicitly mention them when reporting about the corresponding sequence.

That's a good strategy to start off with, but after you've accessed the first GenBank entry relevant to your work, you can retrieve other related genes

by using the <u>Related Sequences</u> link. This link appears in a pull-down menu on the right side of the screen for each GenBank entry retrieved by its accession number. If you click this link for entry U90223, for example, it returns 14 human dUTPase-related GenBank entries. They include various mRNA forms and partial sequences, partial genomic sequences (around exons), and two large (154kb and 192kb) sequences of the 15q21.1 genomic region.

Retrieving GenBank entries without accession numbers

Although GenBank isn't the best database for keyword-based searches (see "Using a Gene-Centric Database," a bit later in this chapter, for a better method), querying GenBank by using gene or protein keywords rather than accession numbers is still possible. Suppose that you want to find the nucleotide sequence encoding the human dUTPase, but you don't have any handy accession numbers lying around. Following is how you should proceed, step by step:

1. **Point your browser to** `www.ncbi.nlm.nih.gov/entrez/`.

 The NCBI PubMed home page appears.

2. **From the Search pull-down menu, choose Nucleotide.**

3. **Type** human [organism] AND dUTPase [Protein name] **in the Search window, and then click Go.**

 This search retrieves four GenBank entries: AH005568, AF018430, AF018431, AF018432. They encompass exon 1 and 2 (AH005568), exon 3 (AF018430), exon 4 (AF018431), and exons 5, 6, and 7 (AF018432). The three last entries indicate the full amino acid sequence of the two forms (nuclear and mitochondrial) of the dUTPase protein, as well as the alternative exon usage pattern. Not a bad start!

4. **Click the <u>Links</u> link to the right of the AF018432 entry ID line, and then choose Related Sequences from the pull-down menu that appears.**

 This retrieves a total of 12 entries. Among these entries, some contain mRNA sequences such as U90223.

 Note: We do not yet have all the entries related to human dUTPase in GenBank.

5. **Type** human [organism] AND dUTPase [Title] **in the Search window, and then click the Go button.**

 Notice that we use dUTPase as a title here — not as a protein name, which is what we did in Step 3.

 This search retrieves a whopping 190 GenBank entries! Most of them, however, are short, single-pass, partial mRNA sequences called *Expressed Sequence Tags* (ESTs).

To remove these ESTs from the list, you can use the *Search-within-Limits* trick that we use with Entrez/PubMed in Chapter 2:

1. **Open the Limits setting menu by clicking the <u>Limits</u> link (below the Search window).**

2. **Select the Exclude ESTs check box, as shown in Figure 3-10.**

Check box here to exclude ESTs. Query

Figure 3-10:
Searching
GenBank by
using Limits.

3. **Scroll to the top of the form, and click the Go button.**

 Only nine GenBank entries survive this particular limits setting.

Feel free to use your new knowledge of the *Search-within-Limits* protocol to try other field-restricted searches in GenBank.

Using a Gene-Centric Database

Besides the traditional GenBank, NCBI recently developed other databases (or database interfaces) that are more adapted to gene-centric queries.

Making a gene-centered query involves asking a question that relates directly to a specific gene, rather than going through all known pieces of sequences related to that gene. The main advantage of gene-centric databases is that they return results that are more synthetic than a long list of GenBank entries, and make much more sense to the biologists. Basically, you get the whole story at once.

In this section, we take you through the LocusLink/RefSeq resource available on the NCBI server. This resource makes it possible to gather important

information related to a genetic *locus*, a specific place on a chromosome where a given gene has been identified. Thanks to this service, you can rapidly find out everything that is known about your favorite gene, or its genomic surrounding.

1. **Point your browser to** www.ncbi.nlm.nih.gov/.

 The NCBI home page appears, eager to serve.

 This is not the same URL as Entrez/PubMed!

2. **From the Search pull-down menu, choose LocusLink.**

3. **Type** DUT **(the gene name for dUTPase) in the For window, and then click Go.**

 Your browser returns a list of three Locus ID (genes), as shown in Figure 3-11:

 - **Hs DUT** (*Homo sapiens*, human beings)

 - **Rn Dut** (*Rattus norvegicus*, the Norway rat)

 - **Mm Dutp** (*Mus musculus*, the house mouse)

 The short definitions indicate that the LocusIDs retrieved are indeed the dUTPase genes in each of these three organisms.

 Each locus line finishes with a set of small, colored squares. These are links to various information sets (when available) concerning this locus. Here is a brief summary of each square's meaning:

 P (red): Lists selected publications in Medline.

 O: Links to the relevant entry in the Online Mendelian Inheritance in Man (OMIM) database, a database created by the great geneticist Victor A. McKusick and his colleagues at Johns Hopkins University. OMIM entries contain a description of the locus and neighboring genetic loci. These entries also summarize in plain text the relevance of the gene to human diseases. They are usually fascinating — and highly recommended — readings!

 R: Lists the Reference Sequences (RefSeq) for that gene.

 G: Lists selected relevant sequences in GenBank.

 P (green): Lists a core set of relevant protein sequences in GenBank, including alternative forms.

 H: Provides a link to potentially homologous genes in other higher eukaryotic genomes.

 This is not a complete homology set.

Check box View loci

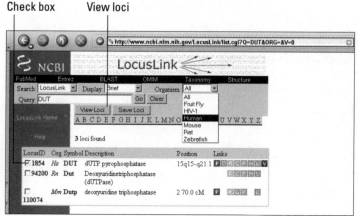

Figure 3-11:
LocusLink
entry page
for genes
named DUT.

U: Provides a link to the corresponding UniGene database entry. UniGene focuses on mRNA and Expressed Sequence Tag information.

V: Provides a link to the single nucleotide polymorphisms (SNP) reported for that gene.

Each of these links provides in turn other links by which you can reach a large spectrum of information, going beyond the sequences themselves. You can explore them at your leisure.

Keeping our focus on the gene sequence, we now turn to a detailed picture of this particular gene as well as its genomic environment.

1. **Click to select the Hs DUT (ID:1854) check box on the Results page from the preceding search, and then click the View Loci button (refer to Figure 3-11).**

 You get a long page of information, the top of which you can see in Figure 3-12.

 Figure 3-12 displays a schematic view (in red) of the *Hs DUT* gene structure, with its seven exons spread over 11,909 base pairs of genomic DNA on chromosome 15. (Here, the schematic view doesn't look like much; it's just a straight line with some blocks on it, set down immediately beneath the menu bar.)

Click here for Evidence Viewer.

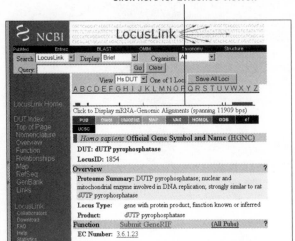

Figure 3-12:
LocusLink
report for
human gene
Hs DUT.

This long entry is a complete report on the *Hs DUT* gene. As you scroll down the page, you see that it contains the following sections (indicated in purple rectangles):

- Overview
- Function
- Relationships
- Map Information
- NCBI Reference Sequences (RefSeq)
- GenBank Sequences
- Additional Links

Each section provides multiple links that help you get a complete (and sometimes redundant picture) of what's known about your gene. This one-stop shopping capacity illustrates the concept of a gene-centric database.

2. Click the top red schematic structure of the gene.

A separate window opens, displaying the Evidence Viewer (Figure 3-13). This window contains a nucleotide level description of our gene structure.

What you have here are all the types of mRNA sequences that have been observed and recorded in GenBank for this gene, and are thus supporting its exon architecture (s). You can see that variations involve the two first exons, and the last one. These variants (alternative transcripts) include the mitochondrial and nuclear forms of dUTPase (Table 3-1).

Figure 3-13:
Detailed
alternative
structures
of mRNA for
the human
gene Hs
DUT.

Exons

3. **Click the <u>Go to full display with alignments</u> link at the top-left of the screen.**

 You can review for yourself every evidence (experimental argument in favor), down to the nucleotide level that supports the various mRNA models.

Working with Whole Genome Databases

The most recent genome-centric databases are the modern bioinformatic response to the proliferation of complete genome sequencing projects. The goal of these new types of resources is to gather all the information you need on a given organism, clearly separated from all the others, so that you can more easily target your analyses on all genes from a specific genome. These new resources also promote comparisons of whole genomes with whole genomes, a new field of endeavor called *comparative genomics*.

These new databases still provide sequence information on isolated genes, but now bring in the additional capacity of showing them in relation to each other within the global framework of an entire genome.

The best way to interact with these databases is to zoom in and out, in order to reach the level of detail you need for the question you're interested in. Genome-centric databases also provide the easiest way to gather every gene/ protein sequence of a given organism at once, with only a few mouse clicks.

We start our exploration of whole genome databases by taking a peek at the nice little Viral Genome section available on the NCBI server.

Working with complete viral genomes

Viruses are fascinating objects, on the edge of the living world. They function as minimal molecular bits of machinery, cleverly designed to ensure the multiplication of nucleic acid molecules (the *viral genome*) at the expense of cellular hosts (eukaryotic, bacterial, or archaebacterial). While going about their business, viruses might go unnoticed or trigger dreadful diseases and epidemics, such as smallpox, poliomyelitis, or AIDS. Viral genomes come in all varieties of biochemical forms (RNA/DNA, circular/linear, single/double stranded) and size (a few kb to half a million).

Visiting the NCBI Viral Genome resource is a great way to find out more about them. To demonstrate, we show you what you can find out about the type 2 human immunodeficiency virus, or HIV-2.

1. **Point your browser to** www.ncbi.nlm.nih.gov/entrez/.

 You've probably memorized this address by now. The NCBI Pub Med home page appears.

2. **On the black menu bar at the top of the form, click Genome.**

 This takes you to the Entrez Genome page.

3. **Click the <u>Viruses</u> link (on the right side of the form, under the Related Resources section).**

 The Viral Genomes reference page appears.

4. **Scroll down the Viral Reference Genomes page until you reach the table of available viral genome sequences grouped by class (Deltavirus, Retroid viruses, and so on), as shown in Figure 3-14.**

5. **Type HIV2 in the Search window, and then click the Find button.**

 Your browser returns a nice global picture of the HIV-2 genome, as shown in Figure 3-15. This picture indicates the identity and respective positions of all the genes.

 Similar pictures are available for all viruses, regardless of their sizes.

 These are live maps, and if you click the blue bar that symbolizes the genomic DNA (below the scale), you can zoom in on the corresponding region, down to the nucleotide sequence level, as shown in Figure 3-16.

 Figure 3-16 shows a position in the genome where the end of the Pol gene slightly overlaps with the beginning of the Vif gene. Viruses commonly have the same nucleotide sequence involved in the making of two different amino-acid sequences.

6. **Click your browser's Back button.**

 This takes you back to the HIV-2 genome entry page. (Refer to Figure 3-15.)

Virus classes

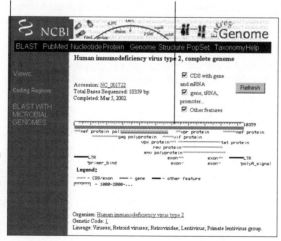

Figure 3-14:
The table of
all available
viral
genome
sequences.

Enter virus name.

Download sequences. Click here to zoom in.

Figure 3-15:
Structure of
the HIV-2
genome.

Figure 3-16:
Zooming in
on the HIV-2
genome: the
Pol/Vif
overlap.

7. **Click the** <u>Coding Regions</u> **link, on the left side of the page (blue background).**

The Protein List page appears, as shown in Figure 3-17. Here, you can retrieve the genes and proteins sequences in different formats — GenBank record, flat file format, or FASTA format — by clicking one of the three lozenges in the left column next to each protein.

If you want, you can download the complete set of sequences by clicking the Save button, after choosing your preferred format from the pull-down menu.

Export all sequences. Choose format.

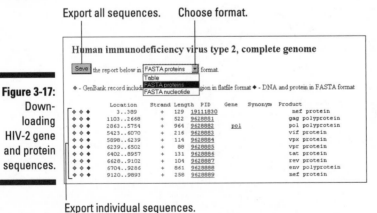

Figure 3-17:
Down-
loading
HIV-2 gene
and protein
sequences.

Export individual sequences.

Working with complete bacterial genomes

NCBI Entrez offers a nice interface to all publicly available complete bacterial genome sequences. For a quick tour, do the following:

1. **Point your browser to** www.ncbi.nlm.nih.gov/entrez/.

 The NCBI PubMed home page dutifully appears.

2. **Click Genome on the black menu bar near the top of the form.**

 This takes you to the Entrez Genome page.

3. **Click the <u>Microbial</u> link, near the top of the Related Resources section (right column).**

 This returns a long and complex-looking form, listing every bacterial and archebacterial genome available on the site. You can see part of the form in Figure 3-18.

 A set of lettered buttons accompanies each bacterial genome. They work as follows:

 > **G**: Takes you to the complete genome sequence in the classical GenBank flat file format. It contains a long, long, feature table!

 > **F**: Returns the complete genome sequence in FASTA format. There is no feature table at all.

 > **T**: Returns the taxonomical distribution of the homologues throughout the whole genome.

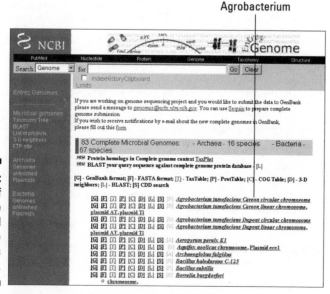

Figure 3-18: The table of all available bacterial genome sequences (top part).

This taxonomical information is great to identify potential lateral gene transfers.

P: Returns a table of all protein coding regions, in a format similar to Figure 3-17.

This table is the ultimate source to download gene sequences from a given bacteria, individually or in bunches.

C: Returns a classification (and the distribution across the genome) of all the genes according to a repertoire of recurrent protein families and domains known as COG (Clusters of Orthologous Groups).

D: Returns a list of the genes that exhibit significant sequence similarities with proteins of known 3-D structures. Precomputed alignments and immediate links to relevant protein structures are also provided — making it a great resource.

L: Provides a specific form to run BLAST similarity searches of your query against the current bacterial gene/protein set. All combinations of searches are possible: DNA vs. DNA, protein vs. protein, DNA vs. protein.

BLAST is the perfect tool to find out whether your favorite gene has a homologue in this bacterium.

For more information on BLAST and searching techniques, see Chapter 7.

S: Returns a precomputed search of all the genes against the protein domain database CDD (Conserved Domain Database), with the relevant links. We explain CDD in detail in Chapter 6.

Each line ends with the underlined name of the bacterial species. This name is a hyperlink to the entry page of the bacterial genome.

4. Click the <u>Agrobacterium cereon tumefaciens circular chromosome</u> link.

Your browser returns a nice circular map of the bacterial genome, as shown in Figure 3-19. On this map, you can click to zoom into a particular region of the genome. Try it, you'll like it!

5. Click in the circle. (We selected an area right around half past one o'clock.)

You get a local gene map similar to the one shown in Figure 3-20. On this map, each gene appears as a colored rectangle. The color code depends on a broad functional classification. (You can find a precise description of this color code at www.ncbi.nlm.nih.gov/cgi-bin/COG/palox?fun=all.)

Click the <u>Coding Regions</u> link (in the left margin) to obtain an individual sequence downloading form, similar to what you see in Figure 3-17.

Click here.

Figure 3-19:
Home page
for the
Agrobac-
terium
cereon
tumefaciens
circular
chromosome.

Click here.

Figure 3-20:
Local gene
map of the
Agrobac-
terium
cereon
tumefaciens.

6. Click one of the individual rectangles.

In Figure 3-20, we selected the gene labeled *AGR_C_632*. This returns the precomputed results of a comprehensive BLAST similarity search between this precise protein sequence and a large list of species. Figure 3-21 shows you what this display looks like. Who said you needed powerful supercomputers to do serious bioinformatics?

Finding out about organelle genomes

If you have a special interest in the genomes of organelles such as *Mitochondria* and *chloroplasts* — the miniature power plants of the eukaryotic cells — take the time to visit the organelle section on the Entrez Genome page. It is organized like the bacterial section we just described.

Figure 3-21: Precomputed BLAST search results for gene AGR_C_632.

More bacterial genomics at TIGR

The Institute for Genome Research (better known by its acronym TIGR) is home to a team of scientists that pioneered the field of bacterial genomics. In 1995, these scientists produced the two first complete bacterial genomes: *Hemophilus influenzae* and *Mycoplasma genitalium*. Since then, they have contributed to 17 complete bacterial genomes, with more on the way.

TIGR offers a site that is quite complementary to the NCBI resource because it keeps track of all ongoing bacterial genome sequencing projects (not only of the completed ones), and offers the possibility to run similarity searches on its genome sequence data much before the completion of the projects. In addition, TIGR's scientists offer a specific view on the genomes they have sequenced through their own genome viewing software.

To pay them a quick visit, do the following:

1. **Point your browser to** www.tigr.org/tdb/.

 The TIGR home page appears, as shown in Figure 3-22.

Get a list of all past and current projects.

Genome Projects - The Institute for Genomic Research - Netscape 6

File Edit View Search Go Bookmarks Tasks Help

http://www.tigr.org/tdb/

HTIGR THE INSTITUTE FOR GENOMIC RESEARCH

Home | What's New | Search | About TIGR | Careers | Site Map | Contact Us | FAQ

Comprehensive
Microbial Resource

Unfinished
Microbial Genomes

Eukaryotic Resources

Gene Indices

Parasites Databases

TIGRFAMS

Fungal Databases

Human Sequencing
Projects

Genome Projects

TIGR's Genome Projects are a collection of curated databases containing DNA and protein sequence, gene expression, cellular role, protein family, and taxonomic data for microbes, plants and humans. Anonymous FTP access to sequence data is also provided. Please read the disclaimer regarding use of data. The TIGR clone distribution policy is available for viewing.

Comprehensive Microbial Resource (CMR)

Containing analysis on world-wide completed microbial genome sequencing, including all TIGR's published organisms:
Archaeoglobus fulgidus
Borrelia burgdorferi
Caulobacter crescentus
Chlamydia pneumoniae
Chlamydia muridarum
Deinococcus radiodurans
Haemophilus influenzae
Helicobacter pylori

Eukaryotic Projects

Containing analysis on TIGR's Eukaryotic Projects.

The TIGR *Arabidopsis thaliana* Database

Provides access to genomic sequence data and annotation generated at TIGR and assemblies of *Arabidopsis* ESTs from world-wide sequencing projects.

Archeoglobus

Figure 3-22:
The TIGR
home page.

2. **Click the worldwide completed microbial genome sequencing link (just beneath the Comprehensive Microbial Resource (CMR) link).**

 The TIGR Microbial Database page appears, containing a very long list (usually quite up-to-date) of all the microorganisms (including bacteria, archebacteria, yeast, and fungi) whose genomes have been sequenced and made publicly available.

 This list provides links to the corresponding laboratories, funding agencies, and the relevant publications.

3. **Click the Microbial Genomes in Progress link near the bottom of this list.**

 Your browser returns a list of all the sequencing projects currently going on throughout the world. For some of them, links are provided to the project sequence databases.

 Some laboratories make their partial data available for similarity searches or downloading long before they release it officially via the central database GenBank. Knowing this can make all the difference if you live amidst fierce competition.

To see what a particular genome might look like, do the following:

1. **Go to the TIGR home page by pointing your browser to** `www.tigr.org/tdb/`.

 The TIGR home page reappears.

2. **Click the first genome listed (*Archaeoglobus fulgidus*) beneath the <u>Comprehensive Microbial Resource (CMR)</u> link; refer to Figure 3-22.**

 Your browser displays a simple-looking genome page (shown in Figure 3-23) that contains all you'd ever want to know about *Archaeoglobus fulgidus* and its genome. This page includes links from which you can freely download sequences and functional annotations.

3. **Choose Main Archaeoglobus fulgidus from the pull-down menu in the lower-right corner of the page.**

 As soon as you release your mouse button, your browser takes you to a spectacular genome viewer, as shown in Figure 3-24.

It would take pages to demonstrate all you can do by mousing around the TIGR genome browser. Just pretend this is a video game and discover the rules while you play (with an infinite number of lives). Bear in mind that top scores can win Nobel prizes!

Figure 3-23: TIGR's Archaeoglobus fulgidus genome page.

Click here to zoom in.

Click here to zoom in.

Figure 3-24: Browsing through the Archaeo-globus fulgidus genome at TIGR.

Exploring the Human Genome

The sequencing of the human genome is probably one of the greatest scientific accomplishments of modern times. With three thousand million (and then some) nucleotides spread over 23 chromosomes, this genome is definitely a complex object to deal with. The major task ahead for bioinformatics is to integrate all the past, present, and future information that human genes contain in a maintainable, user-friendly resource.

If you want to make sense of human data, you must have clear ideas on the current state of the data:

✔ The complete nucleotide sequence of the human genome is now at hand — except for a few thousand holes that are still being filled up.

✔ This sequence was obtained in raw format and the next challenge is the annotation of the raw data; creating a detailed and accurate features table of the human genome.

✔ Throughout the world, new information is generated daily on human gene properties and functions, using a wide array of techniques. Ideally, someone has to gather it all, package it nicely, and offer it to the whole research community for free!

Having doubts that the last goal will ever be attainable in this less-than-perfect world? Well, think again, because this is where we're taking you right now.

Finding out about the Ensembl project

The Internet home page of Ensembl (www.ensembl.org) says it all: Ensembl is a joint project between the European Bioinformatics Institute (EBI) and the Sanger Institute, both located near Cambridge (UK). Together they've developed a software system to produce and maintain automatic annotation on eukaryotic genomes. This project, like the others that we describe in this chapter, relies heavily on the collaboration of a large number of individual laboratories. It all started as part of the International Human Genome Project, and it is now continuing for the Mouse Genome project and other animals. Data and software are also freely flowing between numerous national database and bioinformatics centers from all over the world.

Getting started on the Ensembl site

Considering how complex the human genome is, you may not be surprised to find that you can attack the Ensembl resources from many different angles. To quickly find out about your options, the best thing to do is to jump on the guided tour that Ensembl proposes on its home page. Here's how it's done:

1. **Point your browser to** www.ensembl.org/.

 The Project Ensembl home page appears, as shown in Figure 3-25.

2. **Click the <u>Ensembl tour</u> link, under the Help and Documentation section.**

 A page entitled A Guided Tour of Ensembl appears. It proposes a slide show controlled by a set of forward/backward buttons located at the bottom.

3. **Scroll down to the bottom of the page and click the right-pointing arrow on the control panel you see. (The arrow is the one *right next* to the little Home icon.)**

 You can also click a specific slide number at the top of the page.

4. **When you're done with the tour, go back to the Ensembl home page by clicking the Home icon.**

 By now, you know that Ensembl lets you browse the human genome, starting from a chromosome location down to the level of a single gene, an individual exon, or a nucleotide. Ensembl lets you grab individual genes along with their models, review the evidence that supports all this, and much more.

Click here to start browsing the human genome.

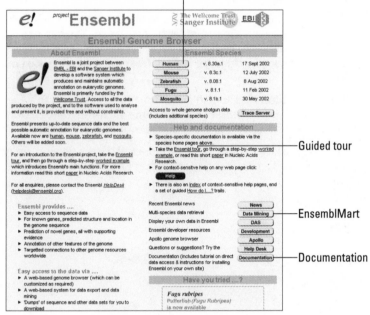

Figure 3-25:
Starting
your journey
through the
Human
Genome at
Ensembl.

5. **Click the Documentation button on the lower-right side of the Ensembl home page (in the Help and Documentation section).**

 The Documentation page provides what you'd expect: a list of documents. They're all you need to turn yourself into the local Ensembl guru.

6. **On the Documentation page, click the <u>Sitemap</u> link near the bottom-left corner.**

 A neat Sitemap page appears, explaining all the entry points that you can use as well as the various analyses you can carry out. It's a live map, so you can directly try the analyses from here.

Zooming in and out along human chromosomes

Chromosome scanning and browsing are some of the most spectacular tools on this site. They let you explore the human genome from any arbitrary location. Here's how it's done:

1. **Point your browser to** www.ensembl.org.

 The Project Ensembl home page appears.

2. Click the Human button under the Ensembl Species heading in the right column.

The Homo Sapiens page appears, as shown in Figure 3-26, with a Human Genome Browser form prominently displayed. You can use this form's pull-down menus to search for many types of objects by name. (Figure 3-26 shows us searching for the DUT gene. You can try this out by yourself later.)

Random exploration is also good fun, and you can do this by clicking on any of the chromosomes (chromosome 15, for instance). Doing so produces a nice diagram, like the one shown in Figure 3-27.

Because we know that DUT, the human dUTPase gene, is on chromosome 15, we can check to see whether Ensembl has this gene at the right position.

Click here for chromosome 15.

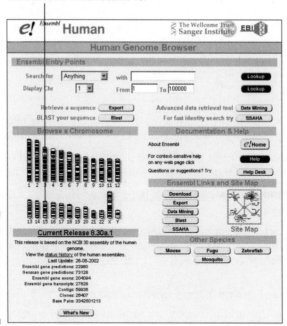

Figure 3-26:
Ensembl general genome browsing entry point.

3. Click chromosome 15, right in the middle of the grey band annotated as q21.1. (Refer to Figure 3-27.)

The Contig View page — a fairly complex page, if truth be known — appears, as shown in Figure 3-28. Here you can see three linear diagrams:

> The top one is the whole chromosome. The little red rectangle indicates your current position.

> The middle one is an overview of the part in the little red rectangle. In this graph, a bigger rectangle shows your current position.

The detailed view, just below the overview, represents the gene content of the rectangle in the overview.

At the top of Figure 3-28, try clicking exactly on the period in Q21.1 in the Chromosome 15 diagram. This zooms you onto the DUT gene location. Then click the DUT gene name as soon as you see it. If you did it right, you can see that Ensembl returns a DUT gene structure similar to the one reported by LocusLink (refer to Figure 3-12).

Figure 3-27:
A global view of chromosome 15.

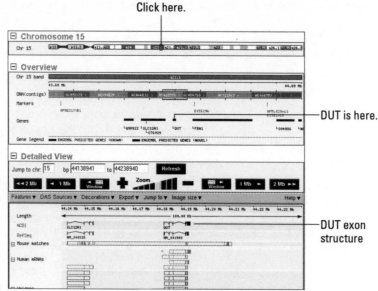

Figure 3-28: Zooming in on band q21.1 on chromosome 15.

Given that you have so many objects to click and things to do in these diagrams, we recommend that you explore this new universe by yourself — when you have lots of time.

Finding genes with coding SNPs

There's more to Ensembl fun than just random browsing. If you're a serious scientist, it can answer meaningful questions very easily, very precisely, and very quickly, even if the output isn't always as pretty as what we show you in the preceding section.

Imagine that, as a competent human geneticist, you ask yourself the following questions:

> *Which genes on chromosome 15 have been linked to a disease and which SNPs (Single Nucleotide Polymorphisms) have been identified?*

If you're just dying to find a candidate gene for a disease you've just genetically mapped on chromosome 15, this question makes a lot of sense.

To make things even more interesting, we can look for *coding SNPs*: Nucleotide changes that may alter the sequence and possibly the shape and/or function of the corresponding protein. Now, *that's* a doozy of a question. Believe it or not, you can answer it in less than a minute with Ensembl! Follow these steps to see how:

1. **Point your browser to** www.ensembl.org/.

 The Project Ensembl home page appears.

2. **Click the Data Mining button on the right side of the page, in the Help and Documentation section. (Refer to Figure 3-25.)**

 Your browser displays the Ensembl Human MartView page, as shown in Figure 3-29. This form is the place where you can precisely describe your question (query) to the Ensembl database.

3. **Click the Next button.**

 Don't change anything; just keep the default *Homo sapiens* and *Ensembl Genes* settings.

 The browser takes you to the *REGION* part of the form.

4. **Select the Chromosome radio button, choose 15 from the accompanying pull-down menu, and then click Next.**

5. **The browser takes you to the *FILTER* part of the form, where you need to make selections based on your interest in disease genes:**

 • Select the Disease Genes check box.

 • Further down the page, select the Entries Associated with SNPs of Type check box and the Coding Only radio button.

6. **Ignore all other options and click Next.**

 The browser takes you to the *OUTPUT* part of the form, where you indicate what information you want back from the search.

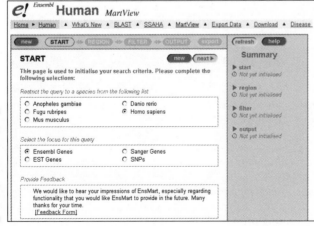

Figure 3-29:
First page
of the
Ensembl-
Mart query
form.

7. **In the Gene Group settings, keep the default Ensembl Gene ID.**

8. **In the Chromosome Group settings, select the Chromosome Name check box and the Band check box.**

9. **Ignore the Protein Group settings and move right along to the Disease Group settings. Select the Disease OMIM ID check box.**

10. **Ignore the Mouse Homology Group settings, skip to the SNP Group settings, and select these options: SNP Type, Allele, and CDS Location (bp).**

11. **Ignore all remaining settings and click the blue Export button at the top-right or bottom-right of the page.**

In no time at all, Ensembl returns a table that looks like Figure 3-30.

This table contains every disease gene you asked for, along with the coding SNP nucleotide changes and their precise locations the coding region of each affected protein. In Figure 3-30, each Ensembl Gene ID, Band ID, and OMIM ID is an active hyperlink.

Click here for a gene report. Click here for a disease description.

Figure 3-30:
Output of
your first
Ensembl-
Mart query.

For instance, if you click the third Ensembl ID (ENSG00000128739), your browser takes you to a comprehensive gene card, shown in Figure 3-31. This card includes information on the structure of the gene and on the functional attributes of the corresponding protein sequence. The other links direct you toward a chromosome display or the OMIM disease description.

With such a one-stop shopping design, you have everything at hand to decide whether these genes make any sense in the context of your research.

The range and complexity of the questions you can address through the Ensembl MartView resource is truly impressive. We really encourage you to spend some time playing with it, even though it isn't as visual as the zooming tools we show you in previous sections.

More human genome browsing tools

Several other laboratories have developed browsing resources for the human, the mouse, and other animal genomes. These are worth looking into because you may find that some of these resources are better suited to your personal needs (or taste). The Genome Bioinformatics Laboratory at the University of California in Santa Cruz has developed another popular human genome viewer. It has the advantage of being much more direct to interact with than its counterparts at the EBI or NCBI.

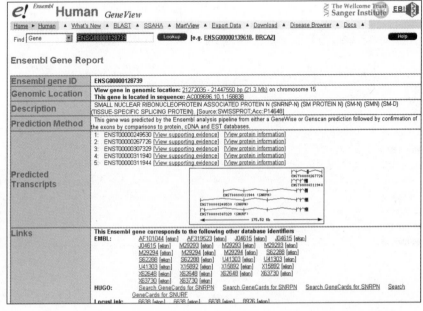

Figure 3-31: Ensembl Gene card for gene ENSG000001 28739 (top part).

In this section, we show you how fast you can visualize the genomic location of a GenBank entry with the Santa Cruz browser. For this purpose, we use our old friend from GenBank: U90223, the human dUTPase mRNA.

1. **Point your browser to** `genome.cse.ucsc.edu/`.

 The UCSC Genome Bioinformatics home page appears.

2. **Choose Human from the Organism pull-down menu, and then click the Browser link (just below the Organism window).**

 This takes you to the Human Genome Browser Gateway, as shown in Figure 3-32.

 On this genome browser, you can identify the region you want to look at by using a wide range of criteria, a simple syntax, and a minimal number of mouse clicks.

3. **Type** U90223 **in the Position text box, and then click the Submit button.**

 It takes a couple of seconds for the corresponding genome region to be found and displayed in a highly tunable fashion, but pretty soon something like Figure 3-33 appears. It is quick, efficient, and nice!

Enter U90223.

Figure 3-32: The UCSC human genome browser gateway.

Figure 3-33:
The Human
dUTPase
gene region,
as displayed
in Santa
Cruz.

The other big player in the Human (and Mouse) Genome Browser category is NCBI. Powerful browsing tools — sometimes similar, sometimes complementary — to the ones Ensembl and UCSC offer are waiting for you there. See for yourself by checking out `www.ncbi.nlm.nih.gov/genome/guide/human/`.

Chapter 4

Using Protein and Specialized Sequence Databases

Mankind is a catalyzing enzyme for the transition from a carbon-based to a silicon-based intelligence.

— Gerard Bricogne

*I*n this chapter, we concentrate on finding out information about protein sequences. The databases on proteins don't limit themselves to providing sequences. In this chapter, we show you that by mastering the — sometimes tangled — network of Web links available to you on the Internet, you can relate proteins to their gene sequences, to their function, and even to their 3-D structures.

Don't let that abundance and diversity of information fool you. As it is for the nucleotide sequence world, where everything revolves around one central resource (that is, GenBank), most of the links between genes, proteins, and functions rely on SWISS-PROT, the central resource on annotated protein sequences cofunded by the Swiss Institute of Bioinformatics (SIB) and the European Bioinformatics Institute (EBI). In this chapter, we show you how to extract every bit of information you need from this database.

SWISS-PROT: A personal vision of the protein world

Although GenBank and SWISS-PROT occupy similarly central roles for the international community of biologists, their overall philosophies aren't quite the same. GenBank, as a primary sequence repository, obeys a relatively strict historical point of view. In GenBank, the authors have full authority over the content of the entries they submit. GenBank annotators are only responsible for the recently introduced RefSeq entries that they derive from their own expert analysis of the community-submitted entries. However, the RefSeq entries never replace the original GenBank entries, and all these layers of information are maintained side by side.

In contrast, SWISS-PROT is not a repository database but a derived information resource, conveying the vision of its head and founder, Amos Bairoch (helped out by a group of experts). As a consequence, the only truly current SWISS-PROT version is the one on Amos' portable computer. This idea of "personal vision" also means that Amos Bairoch doesn't need anybody's permission to correct or change a SWISS-PROT entry on the spot. To do this, Amos simply needs to be convinced by an expert, or by his own evaluation of the literature, that a change is necessary. Believe us, we have seen this happening on our own kitchen table!

The benefit of such a philosophy is obviously great when it comes to flexibility; blatant errors can be removed very quickly while hot discoveries are incorporated immediately. This is the reason why SWISS-PROT is considered the best-annotated protein database, and why it occupies such a pivotal role in molecular biology and genomics. The down side is that the limit to what the best man in the world can do (or supervise), even when helped by a large team of annotators — at the EBI in Hinxton as well as the SIB in Geneva — and a circle of experts. These days, SWISS-PROT has troubles coping with the present rate of new (nucleotide) sequence determination and is falling behind in terms of completeness.

To alleviate this problem, a SWISS-PROT buffer has been created that is called TrEMBL (for automatic TRanslation of EMBL nucleotide sequences). TrEMBL entries are generated at the EBI from GenBank submissions and annotated mostly automatically, using sequence similarity as a main criterion. Upon visual inspection, manual correction, and final approval by Amos Bairoch, TrEMBL entries are then converted into *bona fide* SWISS-PROT entries.

The idea that such a key worldwide information resource rests on a single's man shoulders may come as a shock to you. However, this has been quite common since the early days of bioinformatics. Among other famous examples of one-man shows, we can cite Elvin Kabat 's Immunoglobulin sequence database, Richard J. Roberts' Restriction Enzyme Database, or Victor McKusick's database of human genetic diseases (Mendelian Inheritance in Man, OMIM). To paraphrase Sir Winston Churchill: "In the field of bioinformatics, rarely have so many owed so much to so few!"

You have many ways of landing on the SWISS-PROT database. Most genomic databases, genome browsers, or sites running a sequence retrieval system (SRS) link you directly to the relevant SWISS-PROT entry. If you want to know how to query SWISS-PROT to find an entry by keyword, go to Chapter 2, where we explain how to do this in some detail.

The focus of this chapter is on giving you a good understanding of the content of a SWISS-PROT entry so that you know exactly what is what in the very exhaustive annotation of this database. We then show you how you can use this annotation as a starting point to visit more specialized databases. You can use these databases to further your investigations of the function or structure of your protein.

We start this chapter by giving you some general ideas on how the cell turns a native protein sequence into a mature protein. These post-translational modifications are very important for the protein function. Finding out about them is one of the main reasons you want to use a protein database. (See Chapter 6 if you want to find out how to predict these modifications.)

From Translated ORFs to Mature Proteins

Because sequencing DNA molecules is so much easier and cheaper than sequencing proteins, most of the amino-acid sequences we know do in fact correspond to computer translations of open reading frames (ORFs) detected through the analysis of genomic data. Most of the corresponding proteins have never actually been isolated by anybody. This is why protein specialists hate us molecular biologists and bioinformaticists when we keep talking about translated ORFs as if they were genuine proteins. For a while, at the beginning of the genomic era, the easy thing to do was to consider these protein specialists a bunch of grumpy old men and simply ignore them.

However, after genomes were sequenced, people's interest in proteins came back in a large-scale format known as proteomics. *Proteomics* is the scientific field dealing with the visualization and quantification of the set of protein molecules present in a given tissue or organism. The rapid accumulation of data from proteomics analysis quickly demonstrated that protein specialists had very good reasons to be angry with us.

ORFs: What you see is NOT what you get

When biologists perform a 2-D gel electrophoresis (a tricky experiment separating protein molecules according to their mass in one direction and their charge in the other), real proteins almost never behave as you'd expect given the computer translation. In the world of ORFs and proteins, *what you see is NOT what you get*.

The reason is that, when translated, the nascent amino-acid chain can be heavily modified on becoming a mature protein. Even simple physico-chemical properties of a mature protein like size, molecular weight, or

isoelectric point are hard to predict if you know only the computer-generated amino-acid sequence. This complex process of protein maturation is referred to as *post-translational modification*. Figure 4-1 summarizes it.

Protein maturation may include any combination of the following:

✔ Cuts within the amino-acid chain

✔ Removal of fragments of the amino-acid chain (this is the case for insulin)

✔ Chemical modifications of specific amino acids (methylation, for example)

✔ Addition of lipid molecules (myristylation, for example)

✔ Addition of glycosidic (sugar) molecules (glycosylation, for example)

A major role for a *protein* database (in contrast with an ORF sequence collection) is to display this type of information, when it is available from experimental data or is predicted using various computational techniques. (For more on such techniques, see Chapter 6). Hence, post-translational modifications make up a sizeable portion of the protein features recorded in SWISS-PROT entries.

Figure 4-1: Possible modifications encountered during protein maturation.

A personal final destination for each protein

In order to correctly accomplish their functions, proteins have to reach the right destination in the organism or within the cell. As it is translated, the peptidic chain may expose a variety of highly specific sequence signals — "zip codes," if you will — which the cell then uses to direct the protein to the appropriate compartment (in or out of the cell). This sorting always involves the transport of the protein across one or several membranes and is also referred to as *translocation*. The final protein destinations include

✔ Getting attached to the cell membrane

✔ Being secreted outside the cell

✔ Being transported into the periplasm (for bacteria)

✔ Being transported to the mitochondria or any other organelle

✔ Being transported into the cell nucleus

Because knowing the final compartment where a protein ends up is important in understanding its function, this information (proven or predicted) is one of the important features recorded in protein databases like SWISS-PROT.

A combinatorial diversity of folds and functions

The most important step in turning a newly synthesized peptide chain into a functional protein is the folding of this chain into a compact and stable 3-D structure. Except for small proteins (less than 100 amino acids), the final protein structure generally consists of several relatively independent domains. You can imagine these domains as a basic set of fairly rigid Lego bricks. Nature can assemble these bricks to produce the immense variety of existing proteins. For instance, given a basic set of three domains (A, B, C), you can end up with Protein AAA or AB or BCC or BAC, and so on and so forth.

Most natural proteins are made of combinations of one to ten domains picked from a set of a few thousands. Despite significant sequence variations, the domains are identifiable by their *scaffold sequence signatures* — the motifs in the protein amino-acid texts that remain recognizable despite zillions years of divergent evolution. The recognition and the definition of protein domains is a major research topic of bioinformatics. (See Chapters 6, 7, 9 and 11.) Knowing the domain architecture underlying a particular protein sequence is very important because it gives hints about its possible 3-D structure and suggests its potential biochemical or cellular function.

For this reason, a significant part of each SWISS-PROT entry is devoted to the description of the domain organization of proteins, as identified or predicted by all available methods.

Reading a SWISS-PROT Entry

Despite the complicated concepts that their descriptions rely on, proteins are much simpler objects than genes:

- ✔ Proteins correspond to relatively small sequences (350 amino acids long, on the average).
- ✔ Unlike genes, proteins have clear beginnings and clear ends.
- ✔ Proteins are defined on a single strand.
- ✔ Whatever modifications occur between the ORF sequence and the mature protein, the amino acids they contain remain in the same order.

Given all these reasons, you'd probably expect SWISS-PROT entries to be easier to read and understand than GenBank's — and you'd be right!

Like a GenBank entry — which we cover with great flair in Chapter 3 — a SWISS-PROT entry is divided into several main sections:

- ✔ The general information part
- ✔ The bibliographic information part
- ✔ The functional information part
- ✔ The feature table
- ✔ The sequence part

Using the human *epidermal growth factor receptor* (EGF-R) as an example, the following sections show you how to decipher a SWISS-PROT entry from top to bottom, step by step (or line by line).

Deciphering the EGF-R SWISS-PROT entry

The EGF receptor is a rather complex molecule. We use it here because it illustrates remarkably well the wealth of information you can glean from a SWISS-PROT entry.

Chapter 2 shows you more sophisticated ways of querying SWISS-PROT, but for our purposes here we can use the following basic search:

1. **Point your browser to** `www.expasy.ch/sprot/`.

 The SWISS-PROT page of the ExPASy server appears.

2. **Type the SWISS-PROT ID** P00533 **in the Quick Search window (top left of the page, under the SWISS-PROT logo).**

 This is the entry corresponding to the human receptor of the epidermal growth factor. We chose it because this protein is hot! It is an oncogene (a trigger for cancer), and it exhibits a fairly detailed annotation. Because more people work on them, hot proteins are usually better annotated than boring ones.

3. **Click the Quick Search button.**

 The NiceProt View of SWISS-PROT: P00533 page appears. If you want to print this page, click the Printer-friendly View button at the top. Because of the large amount of information available that deals with this particular protein, this entry is probably among the most complicated you'll ever come across when interacting with SWISS-PROT. Most other entries don't exhibit a third of the keywords and fields that you can find here.

General information about the entry

You're now ready to go through this fairly complex entry step by step. Blue banners separate the main sections across the screen. The first one is entitled General Information about the Entry and contains the following six fields:

- ✔ **Entry Name:** *EGFR_HUMAN.* This SWISS-PROT identifier gives a quick idea of what the entry is all about. Here, the name is quite explicit, but don't count on it because this isn't always the case! For instance, the fly lysozyme entry name is *LYSA_DROME*, certainly more ambiguous and more difficult to remember (DROME stands for *DRO*sophila *ME*lanogaster).

 As a general rule, the entry name is not supposed to be meaningful, nor stable. As a consequence, you're strongly advised not to use it for cross-reference purpose, or as a sequence identifier in an article. Instead, you should use the primary accession number that appears on the next line.

- ✔ **Primary Accession Number:** *P00533:* This is the truly unique and stable identifier of this sequence. You must use this number as reference in your work.

 Accession numbers are great because they allow you to retrace the database history of a given protein. For instance, you can imagine that the EGF-receptor protein, being an unusually long protein, was first

described as a partial sequence. Then people worked further to determine longer fragments of this protein, and so on. At each step, they published a paper and submitted a new version of the EGF-receptor protein (perhaps in the process correcting earlier mistakes). Unlike GenBank, SWISS-PROT keeps only one version of the EGF receptor and summarizes our knowledge up to this day.

Ancient accession numbers don't go away though; they simply get stored in the secondary accession number field.

✔ **Secondary Accession Numbers: *P06268, Q14225, et al*:** Secondary accession numbers make sure that when you read an old paper mentioning Q12225 or P06268 as accession numbers for EGF-R, you can still use the old reference to find the most recent one. See for yourself and try to use one of these older numbers in the Quick Search window.

Whatever secondary accession number you type in, you always fall back on the current one: P00533. This is a clean and simple way to remove old entries while keeping them unambiguously linked to the state-of-the-art version.

The next lines in the General information are self-explanatory.

✔ **Entered in SWISS-PROT: *Release 01, July 1986***

✔ **Sequence was last modified in: *Release 35, November 1997***

This suggests that it took ten years to get the whole sequence right!

✔ **Annotations were last modified in: *Release 41, June 2002***

This suggests that a lot of work is still going on about this important protein.

Notice that, for the whole entry, the field definition (left part of each line) is an active link directing you to the relevant part of the SWISS-PROT manual. Reading a few of those might earn you some respect from the local bioinformatics gurus!

Name and origin of the protein

At this point, we encounter a new section (introduced by the blue banner): Name and Origin of the Protein. It contains the following five fields:

✔ **Protein name: *Epidermal growth factor receptor [Precursor]*:** This field contains general descriptive information about the sequence. It is generally sufficient to identify the protein precisely. If the sequence is a fragment, this field indicates it.

Here, the [Precursor] attribute indicates that the sequence must be cleaved to form the mature form of the protein.

✔ **Synonyms:** *Receptor protein-tyrosine kinase ErbB-1*, *EC 2.7.1.112:*
The E.C. number (2.7.1.112) encodes the biochemical reaction (protein kinase) that this protein performs. E.C. stands for Enzyme Nomenclature Committee, and we give you a bit more information about it in Table 4-1, later in this chapter.

This little line doesn't look very impressive, but it can be the starting point of a fascinating journey toward a complete understanding of this protein enzymatic function. For example, click the 2.7.1.112 link to see the NiceZyme view of the enzyme, as shown in Figure 4-2. The links on this page lead to various types of information; you can

- Locate your protein within a chart of all cellular metabolic pathways

- Get a description and sequence profile of the family it belongs to

- Get the chemical formulas of the biochemical compounds it can interact with

✔ **Gene name:** *EGFR or ERBB1:* This field is useful when you need to interact with nucleotide sequence and genome databases. It can help keep queries clear and unambiguous. (For more on gene names, see Chapter 3.)

Figure 4-2: NiceZyme view of SWISS-PROT entry P00533.

✔ **From: _Homo sapiens (Human)_ _[TaxID:9606]_:** The From field defines precisely the origin of the protein. The species designation consists, in most cases, of the Latin genus and species designation followed by the English name (in parentheses). For viruses, only the common English name is given. Clicking the Taxonomic ID 9606 link gets you to the H.sapiens page of the NEWT taxonomy database maintained by the SWISS-PROT group (www.ebi.ac.uk/newt/index.html).

✔ **Taxonomy: Eukaryota; Metazoa; Chordata; Craniata; Vertebrata:** This field lists the taxonomic classification of the source organism. The taxonomic classification used is that maintained at the National Center for Biotechnology Information (NCBI). Each group name is a link to a list of all the SWISS-PROT entries from that group.

This field is very handy if you want to quickly gather all the protein sequences of a given species for your research projects.

The References

The next section of the P00533 entry is the References section, and it really is as simple as it looks. It is a list of the bibliographical references used to build the current entry. In this example, it includes the various studies that have contributed to the sequencing, to the definition of several isoforms of the protein as expressed in different tissues or tumors, and to the mapping of various post-translational modifications, disulphide bonds, or ligand-binding sites. This protein's importance can be measured by the number of references here — 19 is pretty impressive.

For each of the published citations, SWISS-PROT provides a direct link to the abstract as well as to the relevant NCBI/PubMed entry. References are underlined — as in Science 224:843-848(1984) — when the full text is available (for a fee or free of charge) from the publisher.

The Comments

The Comments section holds any useful information that just doesn't fit anywhere else. The comments are free texts grouped together in comment blocks; a block contains one or more comment lines and is introduced by a topic keyword. The topic keywords themselves are pretty wide ranging, as the listing in Figure 4-3 makes clear. (You can get to this listing — part of the SWISS-PROT manual — by clicking the Comments heading.)

Topic	Description
ALTERNATIVE PRODUCTS	Description of the existence of related protein sequence(s) produced by alternative splicing of the same gene or by the use of alternative initiation codons
BIOTECHNOLOGY	Description of the use of a specific protein in a biotechnological process
CATALYTIC ACTIVITY	Description of the reaction(s) catalyzed by an enzyme [1]
CAUTION	This topic warns you about possible errors and/or grounds for confusion
COFACTOR	Description of an enzyme cofactor
DATABASE	Description of a cross-reference to a network database/resource for a specific protein [2]
DEVELOPMENTAL STAGE	Description of the developmental specific expression of a protein
DISEASE	Description of the disease(s) associated with a deficiency of a protein
DOMAIN	Description of the domain structure of a protein
ENZYME REGULATION	Description of an enzyme regulatory mechanism
FUNCTION	General description of the function(s) of a protein
INDUCTION	Description of the compound(s) which stimulate the synthesis of a protein
MASS SPECTROMETRY	Reports the exact molecular weight of a protein or part of a protein as determined by mass spectrometric methods [3]
MISCELLANEOUS	Any comment which does not belong to any of the other defined topics
PATHWAY	Description of the metabolic pathway(s) with which a protein is associated
PHARMACEUTICAL	Description of the use of a specific protein as a pharmaceutical drug
POLYMORPHISM	Description of polymorphism(s)
PTM	Description of a posttranslational modification
SIMILARITY	Description of the similarities(s) (sequence or structural) of a protein with other proteins
SUBCELLULAR LOCATION	Description of the subcellular location of the mature protein
SUBUNIT	Description of the quaternary structure of a protein
TISSUE SPECIFICITY	Description of the tissue specificity of a protein

Figure 4-3:
Key topics
in the
Comments
section of
SWISS-
PROT.

The Comments section of entry P00533 is particularly rich, as it includes information about the following topics:

- ✔ **Function:** Receptor for EGF, triggering cell division
- ✔ **Catalytic Activity:** Protein kinase
- ✔ **Subcellular Location:** Cellular membrane protein
- ✔ **Alternative Products:** The isoforms from splice variants
- ✔ **Tissue Specificity:** Placenta, ovarian cancer
- ✔ **Miscellaneous:** Used here to describe the molecular mechanism (dimerization) underlying the function of the protein
- ✔ **Similarity:** With other sequences of the EGF receptor family

The content of the Comments section is often sufficient to decide whether a match (with BLAST or otherwise) of your query with a particular SWISS-PROT entry makes sense or not.

Let's skip the Copyright notice — the stuff for the lawyers out there — and jump into the big Cross-References section that comes next.

The Cross-References

The Cross-References section contains links to entries in other databases that contain some information about our protein.

Most of the links in this section take you to new databases. With all this Web jumping about from database to database, it can be easy to get lost and lose your original SWISS-PROT entry. To avoid this, when you browse the Cross-References section, open the links in a new window: Click with the *right* button of your mouse on the link (not the left, as you usually do) and choose Open in a New Window from the context menu that appears.

There are bunches of fields in the Cross-References section. The following list will help you keep at least the major ones straight:

✔ **EMBL:** This field contains all necessary links with the nucleotide sequences world. As we point out in Chapter 3, numerous GenBank, EMBL, and DDBJ entries can be related to a single protein sequence.

Click the CoDingSequence link to send a query to the EBI SRS server for finding the CDS (*CoD*ing *S*egment, the part of the nucleotide sequence that precisely encodes the protein from start to stop) of these entries.

✔ **PIR:** This field contains the accession numbers of the corresponding entries in the Protein Information Resource (PIR) that Georgetown University (Washington, D.C., USA.) maintains. This independent resource offers some interesting tools that we present later in this chapter, in the section "Finding out more about modified amino acids."

✔ **HSSP:** This field contains a link to a sequence homologue of the current query for which 3-D structural information (X-ray crystal structures) is available. Here, this is SWISS-PROT entry P11362, the basic fibroblast growth factor 1, for which several crystal structures exist in the protein Databank (PDB). Click the PDB link to see the relevant page. (The PDB ID here is 1FGK.)

In Chapter 11, we explain all the great things you can do when homologous structural information is available for your protein. Whenever your SWISS-PROT entry has its own 3-D structure, a PDB field providing the cognate link replaces the HSSP field.

There is no PDB or HSSP field when no direct or indirect structural information exists. This is the case for a sizeable portion of the entries in SWISS-PROT.

✔ **SWISS-2DPAGE:** This field contains a link to the proteomics world. Clicking SWISS-2DPAGE brings you to a collection of 2-D gel electrophoreses for different human tissues or tumors. If your protein has been identified on a gel, the spots are highlighted (this is the case for

P00533). Otherwise, only a box of its expected location (drawn after its theoretical isoelectric point pI and predicted molecular weight) is shown. This is great if you're ready for real experimental work!

✔ **GENEW:** This field (for human proteins only) provides a link to the Genew database of official human gene names provided by the HUGO (*HU*man *G*enome *O*rganization) Gene Nomenclature Committee.

✔ **MIM:** This field (mostly for human/mouse proteins) provides a link to the relevant entry in OMIM, the human genetic disease database Online Mendelian Inheritance in Man. (For more on OMIM, see Chapter 3.)

✔ **GeneCards:** This field provides a link to the relevant entry in the GeneCards database (bioinformatics.weizmann.ac.il/cards/) maintained by the Weizmann Institute of Science, based in Israel.

GeneLynx: This field provides a link to the relevant entry of GeneLynx (www.genelynx.org). GeneLynx is a collection of hyperlinks to the publicly available resources associated with each human gene.

✔ **GlycoSuiteDb:** This field is only there when your protein has a known glycosylation pattern. It provides a link to the relevant chemical composition of the sugar moiety. This is a nice site, but the service is only provided to academics free of charge and requires a personal registration, a main drawback.

✔ **SOURCE:** This field is only here for human proteins. It lets you identify publicly available cDNA clones relevant to your protein. Stanford University maintains this service.

✔ **Ensembl:** This field is only here for human and mouse proteins. It provides a link to the relevant gene report in the Ensembl database, a major human genome resource. (For more on Ensembl, see Chapter 3.)

With the NCBI/PubMed and the PDB links, we believe this is one of the most useful hyperlinks provided in the SWISS-PROT entries for mammalian genes.

✔ **Profiles/domains:** The Cross-References section contains a series of fields that deal with the classification of the protein in families of related sequences based on their sharing of basic domains.

Many different concurrent classification systems exist. They all use different methods (algorithms) for the definition and detection of the recurrent structural/sequence domains, as well as for the clustering of the proteins in families. The principle behind these classifications and the use you can make of them are well described in Chapters 7, 9, and 11. The relevant keywords are

- InterPro (summarizing most of the following ones)

- Pfam

- Prodom

- SMART

- PROSITE

- BLOCKs

If you have time for only one click, use the InterPro link; it is a summary of all the others (except for BLOCKs). Clicking the Graphical view of domain structure link in the InterPro row shows you at once that many different recurrent domains have been identified in the EGF receptor protein P00533.

Below these keywords pointing to well-established domain databases, you can find links to other protein classification schemes, such as ProtoNet and ProtoMap. It is not clear to us how stable these links and services will be. To the nonexpert, they do appear quite redundant with the former gene report and classification sites.

Stability is also not guaranteed either for two 3-D structure modeling-related links introduced by the PRESAGE and ModBase keywords. These two services are apparently provided by two biotech companies, and their conditions of access may change without notice. We advise you to make sure that your work doesn't depend too much on these free services offered by for-profit companies. Of course, if you don't belong to the academic world, you aren't even allowed to look at them without explicit permission.

✔ **DIP:** Near the end of this huge Cross-References section, the keyword DIP provides us with our last interesting link. DIP stands for Database of Interacting Proteins, a database maintained by UCLA. This is an interesting resource in that it lists all the proteins that have been experimentally shown to interact with our EGF-receptor protein. Be aware that you can't use the DIP database for commercial purposes without a license from Protein Pathways, Inc (but you can at least look at it without a cumbersome login/registration procedure).

This ends the very long Cross-References section for entry P00533. If you've been clicking many of these links, you've pretty much found out everything there is to know about your protein.

The Keywords

The following section of the entry is the Keywords field. It is a simple list of terms relevant to your current protein, such as transmembrane, receptor, glycoprotein, and so on. Clicking any one of these keywords brings out a list of all SWISS-PROT entries that contain the same term. With the increasing size of the database, we're not sure that this type of query is truly useful anymore. The keyword *receptor*, for example, retrieved more than 1,570 entries!

Using the Advanced search and Full text search features of the ExPASy server's SRS search (see Chapter 2) allows you to query SWISS-PROT by using keywords in combination, thus performing more specific searches.

The Features

Still going down the P00533 entry, we now enter the Features section. This is where information on the protein is precisely mapped onto the sequence. We already went through this concept with GenBank in Chapter 3. It is much simpler for proteins.

Figure 4-4 shows the beginning of the Features section for P00533.

The whole section is organized in a table with the following self-explanatory headings:

```
Key    From   To    Length   Description
```

The Key field indicates the kind of feature you're looking at. The numbers in the From and To fields correspond to the N-terminus->C-terminus numbering of the sequence, whereas the Length field does the math for you and gives the length of the sequence. The Description field gives . . . well, a description.

Click a link to open an information window.

Figure 4-4: Top of the Features section for Entry P00533.

Entries in the Key can be long and varied. Going down the Key field column for the P00533 entry (refer to Figure 4-4), we can see numbers for the following key terms:

- ✔ **SIGNAL:** The numbers here indicate that residues 1 to 24 are a signal peptide, as expected for a transmembrane receptor protein.

- ✔ **CHAIN:** Here we see that the mature peptidic chain goes from residue 25 to 1210 (the C-terminus), meaning that the signal peptide is indeed excised from the final protein.

Notice that clicking the <u>CHAIN</u> link (like any other keyword link in this field) brings out the relevant part of the SWISS-PROT user manual (refer to Figure 4-4) where you can read about the precise meaning of this term. Clicking the underlined sequence range (for example, <u>1 24</u>) highlights this segment in the display of the sequence.

- ✔ **DOMAIN:** The numbers here state that residue 25 to 645 constitute the part of the mature protein that is exposed outside of the cell (the extracellular domain).

- ✔ **TRANSMEM:** Here we see that the following 23 residues (646 to 668) are the transmembrane segment of the protein.

- ✔ **DOMAIN:** These numbers indicate another domain. Residues 669 to 1210 are all inside the cell, forming the intracellular domain of the protein.

Notice that what SWISS-PROT calls DOMAIN here, corresponds to a broader topological definition than what specialists usually mean with this word in databases like InterPro or Pfam. For instance, the extracellular domain of our EGF-R protein contains several InterPro domains (click the link to see for yourself).

- ✔ **REPEAT:** This field points to a sequence segment that is repeated almost perfectly elsewhere in the protein. Here the [75-300] segment is similar to the [390-600] segment.

We find the DOMAIN keyword used again to indicate a serine-rich segment, and the region of the sequence responsible for the protein kinase activity. This confirms what we've just said about the loose usage of this term in SWISS-PROT. Let's keep moving, however, in search of newer terms:

- ✔ **NP-BIND:** The numbers here indicate the extent of the nucleotide phosphate binding region — from residue 718 to 726 — for binding ATP.

- ✔ **BINDING:** Here we can see the precise binding site for ATP — on lysine at position 745.

- ✔ **ACT_SITE:** The numbers here indicate amino acids involved in the activity of an enzyme.

To the nonspecialist, the last three Key field terms address fairly overlapping concepts. The last part of the Features table, shown in Figure 4-5, indicates residues that are chemically modified after translation. The corresponding keywords are

- **DISULPHID:** The numbers here indicate cysteine pairs forming a bond in the mature protein; there are plenty such pairs here. Note that they're all in the extracellular domain, as one would expect.

- **MOD_RES:** The numbers here indicate residues undergoing phosphorylation. Click MOD_RES to find out the list of chemical modifications associated with that keyword.

- **CARBOHYD:** Here we can see the residues on which carbohydrate molecules have been attached as well as the type of link (C-, N- or O-). Clicking the CARBOHYD keyword can tell you much more about the intricacies of glycosylation.

Finally, the Features section ends up with a series of specialized features:

- **VARSPLIC:** The numbers here indicate amino-acid changes from the translation of different mRNAs generated by alternative splicing.

- **CONFLICT:** This feature is used to indicate discrepancies between different sources of the same protein sequence — basically errors or unrecognized polymorphisms.

Click here.

Figure 4-5: The bottom of the Features section for entry P00533.

```
DISULFID    510    523
DISULFID    526    535
DISULFID    539    555
DISULFID    558    571
DISULFID    562    579
DISULFID    582    591
DISULFID    595    617
DISULFID    620    628
DISULFID    624    636
MOD_RES     678    678    PHOSPHORYLATION (BY PKC).
MOD_RES    1092   1092    PHOSPHORYLATION (AUTO-).
MOD_RES    1110   1110    PHOSPHORYLATION
MOD_RES    1172   1172    PHOSPHORYLATION
MOD_RES    1197   1197    PHOSPHORYLATION
CARBOHYD     56     56    N-LINKED (GLCNAC
                          /FTId=CAR_000227
CARBOHYD    128    128    N-LINKED (GLCNAC
CARBOHYD    175    175    N-LINKED (GLCNAC
CARBOHYD    352    352    N-LINKED (GLCNAC
CARBOHYD    361    361    N-LINKED (GLCNAC
CARBOHYD    413    413    N-LINKED (GLCNAC
CARBOHYD    444    444    N-LINKED (GLCNAC
CARBOHYD    528    528    N-LINKED (GLCNAC
CARBOHYD    568    568    N-LINKED (GLCNAC
CARBOHYD    603    603    N-LINKED (GLCNAC
VARSPLIC    404    405    FL -> LS (IN ISO
VARSPLIC    406   1210    MISSING (IN ISOFORM 2).
```

SWISS-PROT user manual : FT CARBOHYD - Netscape 6

SWISS-PROT user manual FT CARBOHYD

CARBOHYD - Glycosylation site.

This key describes the occurrence of the attachment of a glycan (mono- or polysaccharide) to a residue of the protein:

- The type of linkage (C-, N- or O-linked) to the protein is indicated.
- If the nature of the reducing terminal sugar is known, its abbreviation is shown between parenthesis. If three dots '...' follow the abbreviation this indicates extension of the carbohydrate chain. Conversely no dots means that a single monosaccharide is linked.

Before leaving this long Features section, look along the right side for the Feature Table Viewer icon (refer to Figure 4-5, where it's near the top right). Click the icon and scroll down the new page that appears to see a little sequence browser, as shown in Figure 4-6. It gives you a global picture of the protein primary structure. With P00533, this shows you at a glance that all the phosphorylation events take place in the intracellular domain, while all the other modifications (disulphide bridges and glycosylations) occur in the extracellular domain. If you've done a bit of biochemistry, you know that this makes sense and is as it should be!

Intracellular

Figure 4-6:
The Feature
table viewer
for entry
P00533.

Transmembrane

Finally, the sequence itself

There's nothing special to say about this straightforward section. Remember that you can get the sequence in the more convenient FASTA format version by clicking the P00533 in FASTA format link, located at the bottom right of the entry. Use the File➪Save As option of your browser to save it, or cut and paste it into a word processor.

Finding Out More about Your Protein

After carefully reading the SWISS-PROT entry on your protein of interest (or its homologue) and using all the hyperlinks it provides, you probably know 90 percent of what the whole world knows about this protein.

Your next problem is to make sense of, integrate, or eventually understand what you have read, in SWISS-PROT or through its numerous hyperlinks. For instance, perhaps you're not quite sure what myristilation is? Or you don't necessarily understand the biochemical reaction catalyzed by a protein? Or you'd like to know in which pathway this enzyme is involved?

In brief, there may come a time when you need additional information to better understand what you've read in SWISS-PROT. In the following sections, we point out some extremely valuable resources for our knowledge-craving readers!

Finding out more about modified amino acids

If you want to find out everything you always wanted to know about the modification of amino acids during protein maturation, have a look at PIR-RESID, the post-translational modification database maintained by the Protein Information Resource at Georgetown University (Washington, D.C.). For instance, you can use this resource to quickly find out what myristylation is exactly. Here's how you can access PIR-RESID:

1. **Point your browser to**
 `pir.georgetown.edu/pirwww/dbinfo/resid.html`.

 The PIR-RESID Database page appears.

2. **Click the <u>Search and Browse RESID Database</u> link at the top (and middle) of the page.**

 In the search form that appears,

3. **Enter** myristylation **in the leftmost string search window, and change RESID ID into Keyword in the pull-down menu next to it (refer to Figure 4-7).**

4. **Click the Submit button.**

 The query returns a table describing 3 entries: AA0059 (N-myristoyl-glycine), AA0078 (N6-myristoyl-L-lysine), and AA0307 (S-myristoyl-L-cysteine)

5. **Click the <u>AA0078</u> link in the RESID ID leftmost column.**

 You obtain a complete report on this modified amino acid, ending with its detailed chemical formula.

Not impressed with this one? Why don't you have a go at the modified cysteine: L-cysteinyl molybdopterin guanine dinucleotide (AA0281)? Reading this compound name aloud several times can only help the processing of your query!

While you're there, take the time to explore the other components of PIR resource; you may often find it complementary to the SWISS-PROT database.

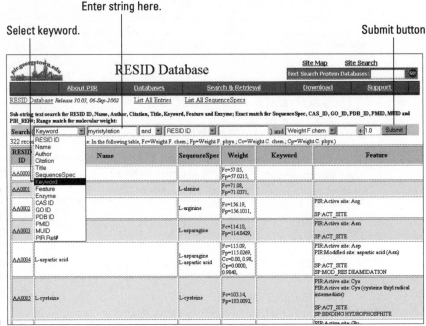

Figure 4-7:
Querying
the
PIR-RESID
database for
myristylation.

Some advanced biochemistry sites

If you want to brush up your biochemistry and chemistry skills, visiting the following sites could help:

✔ The Glycan Structure Database at www.glycosuite.com/

 This one is free for academics only and registration is required.

✔ The Lipid Bank at lipid.bio.m.u-tokyo.ac.jp/

✔ ChemIDplus at chem.sis.nlm.nih.gov/chemidplus/

 This one is fun; you can query their databanks by drawing the molecule of your dreams!

Be aware that using these databases is reserved to specialists. We only introduce them here so that you know they exist. There is some learning to do before you can search them in a sensible way.

Finding out more about biochemical pathways

Reading that your protein is involved in the biosynthesis of di-amino-pymelate probably rings no bell at all! To find out which biochemical pathway (a succession of elementary reactions leading to a compound central to the cell's well being) your protein belongs to, pay a visit to the Boehringer site.

If you've ever done any formal biochemical studies, we're pretty sure that you've seen this huge chart hanging on a wall in your biochemistry department. Did you know that you can search it automatically these days? Yep, this is the kind of era we're living in!

1. **Point your browser to** `www.expasy.ch/cgi-bin/search-biochem-index`.

2. **Type** pimelate **in the Keyword Search window and then click the Submit Query button.**

 Your browser displays the Pimelate page. Pimelate — a central molecule for the creation of bacterial cell walls — is just an example here. If you like, you can substitute the name of the compound you're working on at the moment.

3. **Click J3 below the 6-DIAMINOPIMELATE, the first entry in the list.**

 This brings up a local chart where you can see that this compound is central to the biosynthesis of bacterial cell walls.

Table 4-1 lists more great biochemical pathway resources.

Table 4-1	Biochemical Pathways Resources Available Online
Address	*Description*
`www.genome.ad.jp/kegg`	The famous Kyoto Encyclopedia of Genes and Genomes (KEGG). E.C. numbers or gene names are the best starting points for this resource.
`brenda.bc.uni-koeln.de`	The Comprehensive Enzyme Information System BRENDA is a must if you plan to get into real experiments!
`www.chem.qmul.ac.uk/iubmb`	The official site for Enzyme Nomenclature of the International Union of Biochemistry and Molecular Biology (IUBMB).
`www.ecocyc.org`	The Encyclopedia of E. coli Genes and Metabolism. Now progressively extending to other bacteria.

Finding out more about protein structures

If you have analyzed your protein at the sequence level by all available means (motif search in Chapter 6, database search in Chapter 7, multiple alignment in Chapter 9), your next question probably relates to the position of certain amino-acid residues in your protein 3-D structure. Is this amino acid buried or at the surface? Is it close or far from another residue? Is it in the active site? The list of questions goes on. All these questions require structural information. In Chapter 11, we show you how to relate sequences and structural information. For now, Table 4-2 gives you a list of the major resources dedicated to this topic.

Table 4-2	Databases Dedicated to Structural Information
Address	*Description*
www.rcsb.org/pdb	PDB, the official repository database for protein 3-D structures.
www.ncbi.nlm.nih.gov/Structure	MMDB, NCBI's database of macromolecular 3-D structures, with tools for their visualization and comparative analysis.
scop.mrc-lmb.cam.ac.uk/scop	SCOP, a Structural Classification Of Proteins, aims to provide a description of the structural and evolutionary relationships between all proteins whose structure is known.
www.biochem.ucl.ac.uk/bsm/cath_new	CATH (*C*lass, *A*rchitecture, *T*opology, *H*omologous superfamily) also provides a hierarchical classification of protein domain structures.
www.expasy.ch/swissmod/SWISS-MODEL.html	Swiss-Model is a fully automated protein structure homology-modeling server, accessible via the ExPASy server.

Finding out more about major protein families

Some protein families are so hot that they have become a whole world of their own. There is always a time when the amount of detailed information required by the specialized research community becomes unmanageable by

general protein databases. Highly specialized information resources exist to fulfill that niche. If your favorite protein belongs to one of the following categories, you may want to try some of the resources that we list in Table 4-3.

Table 4-3	Some Specialized Protein Databases
Address	*Description*
imgt.cines.fr	IMGT, the International Immunogenetics database, specializes in immunoglobulins, t cell receptors, and major histocompatibility complex molecules of all vertebrate species.
rebase.neb.com	REBASE is a restriction/modification enzyme database.
afmb.cnrs-mrs.fr/CAZY	CAZy is an information resource on enzymes that degrade, modify, or create glycosidic bonds.
www.merops.co.uk	MEROPS is a database providing a wealth of information on proteases.
pkr.sdsc.edu/html/index.shtml	PKR, the Protein Kinase Resource, is a web-accessible compendium of information on the protein kinase family of enzymes.
nrr.georgetown.edu	NRR, the Nuclear Receptor Resource, is a collection of individual databases on members of the steroid and thyroid hormone receptor superfamily.
senselab.med.yale.edu/senselab	The Human Brain Database provides information on the proteins involved in neural processes, such as ions channels, membrane receptors of neurotransmitters and neuromodulators, and olfactory receptors (ORDB).
www.ncbi.nlm.nih.gov/COG	The COG (Clusters of Orthologous Groups) database regroups proteins shared by at least three major phylogenetic lineages and thus corresponds to ancient conserved domains.

Chapter 5

Working with a Single DNA Sequence

●●●

In This Chapter

▶ Identifying problems with your sequence

▶ Computing and displaying a restriction map

▶ Profiling your sequence for a variety of properties

▶ Finding ORFs, exons, and genes

▶ Assembling sequence fragments

●●●

Not everything that can be counted counts, and not everything that counts can be counted.

— Albert Einstein (1879-1955)

*P*roteins perform most biological functions, so biologists tend to consider DNA sequences kind of dull. They're mostly right. The truth of the matter is, 90 percent of the DNA you may use in your work is nothing more than an information storing device, such as a floppy disk or a DVD. Still, we probably shouldn't turn up our noses at information storing devices. DNA contains information that the cell knows how to put to effective use, just like your computer can turn a dull-looking DVD into the latest destroy-all-at-any-cost network game.

If you're interested in a protein, directly working out its amino-acid sequence is hell. Sequencing its gene and then deducing the protein sequence from the genetic code is much easier. This is why DNA sequences are so important. These days, working with DNA sequences is an obligatory intermediate step; you simply must know how to handle a nucleotide sequence in order to work with protein or RNA sequences. DNA sequences are the mothers of all sequences!

If you're reading this book, it's probably fair for us to assume that you're not a highly trained scientist working in one of those mega-super-duper genome sequencing centers churning out nucleotides by the millions everyday. This chapter is thus focused on the analysis of relatively small sequences — at most 100,000 nucleotides long. The skills necessary for dealing with genomic fragments of a reasonable size or with complementary DNA — cDNA, the DNA copy of a messenger RNA — is the name of the game in this chapter. This is the kind of sequence you can routinely determine in a well-equipped molecular biology lab or get done for a reasonable price by a sequencing company.

When you get an experimental sequence back, your first task is to make sure it's okay so that you don't waste a month trying to make sense of something completely botched and end up nowhere. This chapter begins with a section showing you how to check for the most common problems encountered in sequences fresh from the bench. But remember Murphy's Law: If something can go wrong, *and has not been discussed in this section,* it will happen to you!

Catching Errors Before It's Too Late

Sequencing a DNA fragment involves purifying it, cloning it into a vector (such as a plasmid), amplifying it into a biological host (most often a bacterium like E. coli), and finally submitting it to one of the various sequencing protocols, such as primer extension or dye-termination. During this process, accessory DNA segments are deliberately linked to your target DNA. In addition, many unintended events can occur, resulting in a sequence that doesn't faithfully correspond to the genetic information you intended to study.

The most common problem is that the sequence you get back has been contaminated with some of the sequence of the vector you used (or that somebody else used in the laboratory). Recognizing this problem and knowing how to deal with it is important — so important that we devote the next section to precisely this topic.

Removing vector sequences

The DNA (or cDNA) you send out for sequencing is inserted into a cloning vector — plasmid, phage, cosmid, BAC, PAC, or YAC — so that it can be manipulated. The sequences you get from such constructs inevitably include segments derived from these vectors, at least at one extremity or the other. You can detect and remove them from your sequence by simply running a search for similarity against the sequence of the vector you have been using. (As a responsible scientist, you're expected to have this information!)

However, your sequence might also be cross-contaminated by somebody else's vector — maybe from the bench next to yours. It's thus wiser to search your sequence not only for the vector you expect, but also for other possible vector contaminations, just in case. Here's how to do it using a nice interactive tool available at the NCBI (the National Center of Biotechnology Information):

1. **Point your browser to** www.ncbi.nlm.nih.gov/VecScreen/ VecScreen_docs.html.

 The entry page for VecScreen appears. It contains both a nice tutorial on sequence contamination (click the <u>contamination</u> link to read it) as well as an explanation of how VecScreen works. Basically, it performs a blastn search (see Chapter 7) of your sequence against a special database of all vectors and cloning adapter sequences known to man. This database is known as UniVec, and a simple click on the <u>UniVec</u> link on this page provides you with more details about this unique database.

 When you believe you're ready for the real thing, rather than just reading about the real thing, go grab your new sequence and proceed to Step 2.

2. **Point your browser to** www.ncbi.nlm.nih.gov/VecScreen/ VecScreen.html **or click the <u>VecScreen web site</u> link on the VecScreen entry page.**

 A typical sequence input window appears, ready for your input.

3. **Paste in your newly determined sequence in the input window, as shown in Figure 5-1.**

 You can use FASTA or raw format.

Paste in your new sequence.

Figure 5-1: The VecScreen input window.

4. Click the Run VecScreen button below the input window.

You immediately get back the BLAST formatting page, showing you that your VecScreen request was in fact a BLAST search in disguise. (For more on BLAST, see Chapter 7.)

It's better to use VecScreen from the start rather than trying to use blastn because the VecScreen server chooses parameter values that are best for vector detection.

5. Click the Format! button and wait.

Don't change any of the default parameters.

At this point — and depending on the workload of the BLAST server — you'll either get the standard waiting page (`This page will be automatically updated in` *xx* `seconds`) or the results of your search pop up immediately.

6. The two possible outcomes are

• **Non-significant similarity found**

This is good news! This message indicates that the sequence you submitted does not resemble any known vector/adapter. You can proceed to the next stage of its analysis, following the rest of this chapter.

• **An output listing matches of some kind**

This is bad news! Your query does contain some vector/query sequence.

A VecScreen contamination report output contains two parts. The top part displays the distribution of the contaminating sequences along the length of your query sequence, using a color-coded graph (Figure 5-2). The second part is a standard blastn output, providing you with the identity of the matched vector/adaptor as well as the detailed alignment.

The different colors on the graph correspond to the strong (red), moderate (purple), or weak (green) matches in the Univec database. A yellow code is used to delineate segments of suspect origin such as short segments in between two vector matches. The U87251 exhibits two regions of 100 percent identity with the popular PBR322 cloning vector. Most likely, it is a contaminated sequence that was submitted to GenBank before vector sequence filtering was systematically included in the submission process.

What to do next with contaminated sequences is fully explained in a complete tutorial that you can follow by clicking the <u>Interpretation of VecScreen Results</u> link provided at the top of any VecScreen positive output.

Your sequence displays two strong vector contaminations.

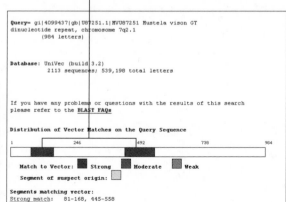

```
Query= gi|4099437|gb|U87251.1|MVU87251 Mustela vison GT
dinucleotide repeat, chromosome 7q2.1
       (984 letters)

Database: UniVec (build 3.2)
             2113 sequences; 539,198 total letters

If you have any problems or questions with the results of this search
please refer to the BLAST FAQs

Distribution of Vector Matches on the Query Sequence

1              246              492              738              984

Match to Vector:  ▮ Strong   ▮ Moderate   ▮ Weak
Segment of suspect origin: ▯

Segments matching vector:
Strong match:  81-168, 445-558
```

Figure 5-2:
VecScreen
output
display of
a conta-
minated
sequence.

In general, you have two options when finding contaminations. If the contaminated parts are at the extremity of your sequence and correspond to the vector you used, all is fine. Just remove these parts and proceed. If, on the other hand, contaminated parts are found elsewhere (like in Figure 5-2), or if the contaminating vector is not yours, you'd better consider this sequence to be the result of a chimeric clone or a DNA cross-contamination. In that case, throwing it away is the safest thing to do!

Cases when you shouldn't discard your sequence

In case of a VecScreen match, do not merely rely on the name of the UniVec database entry to decide that your experiment was contaminated by somebody else's vector. You may have used pUC19 as a cloning vector and see that VecScreen is reporting a strong match with pUR222 or lactose operon genes from E. coli. This is okay. Most commercial vectors are derived from the same initial natural plasmid and E. coli constructs. Their sequences must thus be 100 percent identical. The UniVec matches are simply reported in the order that they occur in the database. If your vector is not first, another entry (corresponding to a locally 100 percent identical sequence) will be reported instead.

Also, if you're working on genes that resemble the kinds of genes used to build vectors (such as antibiotic resistance genes, repressor/activator genes, or some biosynthetic genes), you should actually *expect* matches in the UniVec database. Don't throw away these sequences before interpreting the detailed alignment.

Why contamination is a big deal

Database organizations have the entire scientific community as clients and thus a worldwide responsibility about the quality of what gets included in their particular databases. If a contaminated sequence finds its way into GenBank, thousands of similarity searches will immediately start spreading erroneous results to unsuspecting biologists throughout the world. The next day, the numerous secondary databases taking their primary input from GenBank will, in turn, become polluted. An erroneous DNA sequence may create a false protein sequence that may induce a new erroneous protein family that may, in turn, wrongly cluster unrelated sequences, and so on down the line. These types of errors propagate very fast because most of the derived databases (of protein sequence, of domains, of polymorphisms) are automatically generated without anybody closely checking the results.

At one point, for instance, the SWISS-PROT database had to voluntarily incorporate conceptual translations (entries P39188 to P39195) of the quite common human genome Alu repeat as a way of detecting and combating further pollution by sequences from rearranged cDNA clones. BLAST searches of any Alu-containing cDNA sequence against the database now reports strong matches to these synthetic entries, thus alerting any unsuspecting author that his clone has potential problems.

Pollution of sequences (such as a bacterial gene fragment in a vertebrate genome) may lead to erroneous phylogenetic reconstruction (horizontal transfer!), wrong sequence contig assembly, or EST clustering — and more generally time wasted and effort used to sort out the problem that could have been better expended elsewhere. Fortunately, most contaminations arise from the intended use of vectors and adaptor oligonucleotides, the sequences of which are well known and easily identifiable. However, unintended events can introduce contamination from unknown viral, organelle, or microbial sources. Thus, be always suspicious of foul play each time a bacteria-like gene turns out in the middle of a rat chromosome, even if it isn't 100 percent identical to a good ol' E. coli plasmid sequence. The same remarks apply to sequences similar to Lambda-like bacteriophages.

In any case, beware of near 100 percent matches with any piece of the E.coli or yeast genome, even if these parts don't correspond to vectors. These are sure signs of cross-contamination. Modern molecular biology techniques are very good at amplifying pieces of DNA from the bench next to yours!

Computing/Verifying a Restriction Map

Here's another way of checking whether the sequence you get back from the sequencing lab/company actually corresponds to the piece of DNA you sent out: Compare the pattern of restriction enzyme digestion (also called a *restriction map*) that you experimentally observe with the pattern (*theoretical* restriction map) that you can compute from the sequence.

Computing a theoretical restriction map is easy. All you do is look for exact matches of a given restriction enzyme site within your sequence. All the known enzymes and sites are described in the REBASE database (`rebase.neb.com`). After locating the matches in the sequence, it's simply a matter of counting the number of resulting fragments (DNA segments between two successive sites) and computing their sizes. You can then compare the computer prediction with your lab experiment. The more restriction enzymes you test, the more certain you'll be that your sequence is correct.

You can use this approach to verify long sequences — such as bacterial genome sequences — that were assembled from many shorter ones, as happens with the Shotgun DNA sequencing protocol. If the sequence and its assembly are correct, the number and size of the predicted fragments should perfectly correspond to the experimental restriction map.

A couple of good Internet sites spare you all the work of having to collect the information about restriction enzyme sites and doing the matching yourself. These sites maintain their own current versions of the authoritative Restriction Enzyme Database, REBASE, so you don't have to worry about it. They scan your sequence for all the specific sites corresponding to an enzyme list that you can select according to various criteria.

One such Internet site we find especially useful offers a nice little interactive tool called Webcutter, developed by Max Heiman. Here's how to put it to work for you:

1. **Point your browser to** `www.firstmarket.com/cutter/cut2.html`.

 This gives you access to the most current version of Webcutter.

2. **Enter the sequence you'd like to check.**

 At the time of this writing, Webcutter offers you three ways to input your sequence:

 - Copy and paste it into the Sequence window displayed on this page.

 - Upload the sequence from your computer (Netscape users only).

 The sequence has to be in a text format, without its name or any FASTA-style header.

 - Directly fetch a GenBank entry by using its accession number or other identification methods. (The latter case applies if you're designing a cloning experiment for a gene already known.)

3. **Select the options that correspond to your problem.**

 Options include

 - Choosing whether to treat the input sequence as linear or circular

 - Selecting various output formats for the results (maps, table of enzymes)

- Filtering the results based on whether enzymes cut or do not cut your sequence — or if they cut your sequence a given number of times

- Specifying a subset of enzymes — based on various criteria — to be included in the analysis

4. **Click the Analyze Sequence button at the bottom of the form.**

 The result usually arrives very quickly and is — thankfully — self-explanatory.

The University of Massachusetts Medical School offers a similar service with a different interface at biotools.umassmed.edu/bioapps/rsites.html.

For a list of other restriction mapping servers, check out the REBASE Related Sites link on the REBASE home page (rebase.neb.com). All these servers are equivalent in terms of accuracy. Shop around until you find one you like best in terms of input/output format.

Designing PCR Primers

The polymerase chain reaction (PCR) is a wet lab strategy people use to amplify segments of DNA they find interesting. To make a PCR, you must

1. **Identify the DNA sequence you want to amplify.**

2. **Order primers from a DNA synthesis company.**

 Primers are small pieces of DNA (20-30 bases) that match the extremities of the complete sequence you're interested in. The most delicate step in a PCR is the design of good primers.

3. **Run your PCR experiment.**

These days, people use PCR for almost anything that has to do with wet lab sequence analysis: subcloning a gene in an expression vector, verifying its assembly and sequence, inducing or detecting mutations, or detecting its relatives in other organisms.

The trickiest step when you do a PCR is the design of the *primers* — the two small DNA fragments capable of firmly hybridizing on each side of your gene in a highly specific manner. Primer design programs are here to help you decide which portion of your large sequence makes the best primers, thus avoiding a series of potential problems too numerous to be listed here.

What is PCR?

Running a polymerase chain reaction (PCR) involves mixing a DNA template (containing the sequence that you mean to amplify), the primers, a cocktail of nucleotides and other biochemical compounds, and a heat resistant enzyme called DNA polymerase — all in a single little plastic tube. You then put that tube in a benchtop expensive machine called a *thermal cycler*. According to a preset program, this machine will make the tube go up and down in temperature. For each temperature cycle, the amount of copies of the DNA segment that you want to amplify is doubled. For instance, after 30 cycles, you have 2^{30} (one billion) more of it. This is why PCR is such a great technique in forensic science: Lick a stamp, and scientists will be able to sequence your DNA!

If you want to know more about PCR, visit `nature.umesci.maine.edu/forensics/p_intro.htm` for a great animated introductory course.

The Internet site of the University of Massachusetts Medical School (`biotools.umassmed.edu/`) provides a link to a very complete and easy-to-use tool for primer design. It is a Web implementation of the well-known program Primer3 developed by Steve Rosen and Helen Skaletsky. To put it to (hopefully good) use, do the following:

1. **Point your browser to** `biotools.umassmed.edu/`.

 The Bioinformatics Resources page of the University of Massachusetts Medical School appears, offering access to several sequence analysis programs that you may want to try by yourself later.

2. **Click the <u>Primer Selection Tool</u> link in the DNA Sequence Analysis section.**

 The Primer3 page dutifully appears. The form can look rather intimidating, as Figure 5-3 more than clearly shows. Don't let it scare you! In most cases, you can get by with the default settings.

3. **Paste your sequence in the Sequence window.**

4. **Click the Pick Primers button.**

 The results return very quickly. They include a map of the best oligonucleotide pair (left and right primers) according to the constraints listed in the form. In addition, four alternative solutions (by default) are proposed. The position, length, predicted melting temperature, and G+C percentage (the fraction of guanine and cytosine nucleotides) are given for each listed oligonucleotide.

Paste your sequence here.

Figure 5-3:
Primer3
input form.

Click here to get your primers.

If you're ready to modify the default setting, the form allows all reasonable experimental situations and constraints to be imposed upon the primer selection process, including

✔ Searching for only a left or right primer, or a single hybridization probe

✔ Proposing your own left or right primer

✔ Selecting sequence positions to be flanked or excluded

✔ Selecting a range of product sizes

✔ Imposing a range of oligonucleotide sizes, G+C percentages, or melting points

✔ And many more highly sophisticated options

After you've gotten your primers, your final responsibility is to verify that they will not hybridize anywhere except where you intend them to hybridize. For instance, you want to make sure that the selected oligonucleotide sequences aren't found outside of the gene you are interested in, or worse, resemble a frequent repeat in the DNA that you're going to amplify.

The techniques for avoiding these problems involves BLAST searches against the vector sequences, the relevant genomes as well as their most common repeats. You can use the NCBI BLAST server for this purpose. (For more on BLAST searches, see Chapter 7.)

Analyzing DNA Composition

After these analyses, with all the going back and forth from the computer to the bench they entail, you're now convinced that the sequence in your test tube is indeed the right stuff. The time has now come to analyze it! Whenever you have a new piece of a new genome, the first question people will ask you is: *What is the percentage of G+C?* Because the pairing between the guanosine (G) and cytosine (C) nucleotides is more stable than the pairing between adenosine (A) and thymidine (T) in the DNA ladder, this percentage determines how the DNA will behave in your experiments. The first level of analysis you can perform is thus simply to count the number of A (adenosine), G (guanosine), C (cytosine) and T (thymidine) in your sequence.

Establishing the G+C content of your sequence

Most often, people establish G+C content by using a program installed on their own computers. Many programs that to do this are commercial packages, available from an open source consortium (such as EMBOSS), or are written up by the computer science student you keep handy and pay with pizza. If you want to save the cost of the pizza, you can even do the coding yourself if you've learned the basics of one of the basic computer programming languages out there.

However, if you want to use an online service to establish the G+C content of your sequence for you, check out the Pasteur Institute's GEECEE interface at `bioweb.pasteur.fr/seqanal/interfaces/geecee.html`.

The EMBOSS package

EMBOSS (*European Molecular Biology Open Software Suite*) is a free open source software package specially developed for the needs of molecular biologists with little expertise in computer science. It consists of approximately 100 sequence analysis modules performing basic sequence handling tasks (such as word counting, and many more) that aren't always found on Web sites. The software automatically copes with data in a variety of formats. One caveat: You do have to install each module on a local computer running a Unix-related operating system (for example, Linux, Irix, Solaris, and so on). For more information on EMBOSS, including how to download its modules, go to: `www.hgmp.mrc.ac.uk/Software/EMBOSS/`.

Counting words in DNA sequences

After establishing the G+C content, the next step in complexity is to consider your DNA text as made up of overlapping "words." For instance, consider the following DNA sequence:

```
ATGGCTGACTGACTGACTGACTGAC......
```

You can read it as

```
A, T, G, G, C, T, G, A, C, T, G, A, C, T, G, A, ......
```

where counting the various components A, C, G, and T leads to the simple *nucleotide* statistics. But you can also read the sequence as

```
AT, TG, GG, GC, CT, TG, GA, AC, CT, TG, ....
```

and compute the frequency of the 16 different *dinucleotides* (2-letter words). Then again you can read the same sequence as made of triplets, such as

```
ATG, TGG, GGC, GCT, CTG, TGA, GAC, ACT, CTG, ...
```

and compute the frequency of the 64 different *trinucleotides* (3-letter words), and so on.

The Germany-based company Genomatix offers a nice free Web site where you can compute the nucleotide, dinucleotide, and trinucleotide frequencies of your DNA sequence. Here's how it's done:

1. **Prepare your DNA sequence in a file, using a text format.**

2. **Point your browser to** www.genomatix.de/cgi-bin/tools/tools.pl.

3. **Select the Create Statistics radio button under the Sequence Statistics heading.**

4. **Click the Start Task button at the bottom left of the page.**

 The Create Statistics page appears.

5. **Cut and paste your sequence into the Sequence window.**

 Use the format of your choice. (Here we use the *plain* — that is, raw — format.)

6. **Click the Load Sequence button.**

 A new Input form appears.

7. **Leave the Input form as is and click the Start This Task button.**

 The preset options compute the AT percent and GC percent, as well as the nucleotides, dinucleotides, and trinucleotides frequencies and presents the results, as shown in Figure 5-4.

Statistics for all input sequences (2800 basepairs in 1 sequences):

Base-Content:

The AT-content of this sequence is **65.71%**, the GC-content **34.29%**, others: **0.00%**

Mono-Nucleotides:

	in bps	in %
A	942	33.64%
C	457	16.32%
G	503	17.96%
T	898	32.07%

Di-Nucleotides:

		second nucleotide			
		A	C	G	T
first nucl.	A	363	120	174	285
	C	137	96	84	140
	G	164	116	97	126
	T	278	124	148	347

Tri-Nucleotides:

		second/third nucleotide															
		AA	AC	AG	AT	CA	CC	CG	CT	GA	GC	GG	GT	TA	TC	TG	TT
first nucl.	A	155	35	67	106	40	26	16	38	58	37	37	42	85	31	48	121
	C	47	26	32	32	19	24	26	27	18	26	22	18	38	25	32	45
	G	64	17	31	52	37	19	20	40	30	22	18	27	38	17	26	44
	T	97	42	44	95	41	26	22	35	58	31	20	39	117	51	42	137

Figure 5-4: Typical compositional analysis result.

Counting long words in DNA sequences

When analyzing DNA, it can be interesting to count words that are longer than three nucleotides, although it is rare to use words with sizes above six or eight. When their size is larger than 3, words are often referred to as k-tuples (6-letter words = hexamers = 6-tuple). There are 64 different 3-tuples, 256 4-tuples, 1,024 5-tuples, and 4,096 6-tuples. It makes no sense to compute the hexamer statistics on a short sequence where most hexamers occur, at most, only once. On the other hand, identifying hexamers (6-tuples) with unexpected high frequencies in a set of sequences (such as promoter regions) is often the starting point for discovering regulatory sequence motifs.

The Pasteur Institute, based in Paris, offers an online version of the EMBOSS program wordcount that allows you to compute the word frequency in your DNA sequence for any size (we tested it up to 20 letters). Here's how you get it to work for you:

1. **Point your browser to** `bioweb.pasteur.fr`.

 Click the English/American flag (top right) for the English version.

2. **Click the <u>DNA Sequence Analysis</u> link.**

3. **Click the <u>Codon usage, composition</u> link.**

4. **Click the <u>wordcount</u> link.**

 The wordcount page appears in all its glory.

5. **Enter your e-mail address and paste your sequence (raw format is okay) in the appropriate fields. Be sure to enter the word size you're interested in as well.**

6. **Click the Run Wordcount button at the bottom-left of the page.**

7. **Click the <u>wordcount.out</u> link to get your results as a simple list.**

Experimenting with other DNA composition analyses

While you're at the Pasteur Institute site, use the opportunity to further explore the DNA Sequence Analysis section. The site graciously allows you to interactively run other simple EMBOSS modules, such as *chips* (for codon usage statistics), or the CpG rich region finder *cpgreport*.

Finding internal repeats in your sequence

Another useful type of composition analysis involves locating segments that occur more than once within your sequence. Such segments are called *repeats*. There is no real difference between long words (6-tuple, 8-tuple, and larger) and repeats. The common-sense rule is that a repeat is a word long enough so that it's unlikely to occur very often by chance, given a random sequence. For instance, a GTC triplet found 4 times within a 500-nucleotide long sequence doesn't qualify as a repeat.

Sequence profiling and sliding windows

The programs highlighted in the first half of this chapter are computing word statistics over the entire sequence length. Be aware that you may sometimes encounter programs that do compute the same statistics but do so through a *window* — a succession of positions — sliding along the sequence. The unique statistics are thus replaced by a succession of values associated with each position of the window center.

If you plot these values against their corresponding sequence positions, you get a *profile* of whatever statistics you computed. Sometimes, these profiles are also called *landscapes*. Such profiling methods are used with DNA (and protein) sequences to identify local changes (peaks and valleys) in any properties: G+C percent, CpG frequency, DNA stability, bending propensity, and so on.

Another difference between word counting and repeat analysis is that repeats can be imperfect. Unlike words, similar repeats don't need to be identical. Finally, biologists like to distinguish *tandem repeats* (similar subsequences along the same DNA strand) from *inverted repeats* (similar sequences occurring on the direct and reverse strands). Biologists are interested in repeats because they are often involved in genome rearrangements or regulatory mechanisms of gene expression.

There are many different algorithms for finding repeats within a DNA (or protein) sequence. They all try to identify segments more similar to each other than would be expected by mere chance alone. The tricky part is in the scoring and ranking of the similar subsequence segments. Is the exact matching of five consecutive nucleotides good enough to be considered a repeat? And is it better than 9 out of 10, or 123 over 160? Which one do you want reported first? How far down the list of possible repeats do you want to go?

Finding repeats is a tricky business

Because there are no universal answers to the questions surrounding the precise nature of repeats, repeat-finding programs ask you to fix thresholds related to their scoring algorithms, repeat size, copy number, periodicity (distance between repeats), and other things you don't always understand. This makes them difficult to use. The default settings they provide may or may not work for your particular sequence.

In that respect, our quick survey of Web-based repeat finders — using a repeat-containing sequence we made just for this purpose — was amazingly unsuccessful on all the available servers. This doesn't mean that these servers are bad, but that the problem is truly complicated. Never believe a negative output from a repeat finder!

Fortunately, there is one simple universal (DNA/protein) approach that avoids all these difficulties. It is the *dot-plot* approach, which we show you in the following steps list. (For more on dot plots, see Chapter 8.)

For long nucleic acid sequences, you can use the dot-plot program provided by the Molecular Toolkit Web service at Colorado State University. Just do the following:

1. **Point your browser to** `arbl.cvmbs.colostate.edu/molkit/`.

 The Molecular Toolkit is a group of programs for analysis and manipulation of nucleic acid and protein sequence data. The programs are written in Java (1.0) and require that your browser support this language.

 This site is very useful if you need to perform some simple DNA sequence manipulation, such as reverse complementation (changing strands), protein translation, or if you want to display a restriction map. For now, continue through these steps.

2. **Click the <u>Dot Plots</u> link.**

 You obtain a straightforward input form, with two side-by-side boxes entitled DNA Number 1 and DNA Number 2, as shown in Figure 5-5.

Figure 5-5:
Input form
of the DNA
dot-plot
program.

3. **Copy your sequence from a .txt file or a Word document, by using Ctrl+C or the Copy button on the Word toolbar.**

4. **Paste your sequence in the DNA Number 1 window by using Ctrl+V or Word's Paste button.**

 The program is happy with the raw sequence format (nucleotide only).

5. **Copy your sequence into the DNA Number 2 window as well.**

 For this, you can use the special Copy DNA1 -> DNA2 button provided below the DNA Number 1 window.

6. **Click the Make Plot button, located below the DNA sequence boxes.**

 Your dot plot magically appears. That's all there is to it!

 Limit yourself to a sequence under 10,000 bp if you aren't prepared to wait for several minutes.

The dot plot is quickly performed in real time. It appears in the square below the input windows. For the example in Figure 5-6, we made up a sequence in which a large segment is imperfectly repeated three times. These repeats show up as three lines both in the upper and lower diagonal parts of the plot. We added three black lozenges on the main diagonal to indicate the location of the repeated segments. The long repeats stand out in comparison to non-significant shorter random repeats of length 9 to 15. Using your own sequence, you can experiment with the Window Size parameters (minimal size of the repeat) as well as the Mismatch Limit parameters (number of non-identical nucleotides) to see how they influence the graph appearance.

Triplet counting vs. codon frequency

We introduce the concept of counting trinucleotides (triplets or 3-tuples) in this chapter. These computations are always done by reading the DNA sequence on one strand, in an overlapping manner. This should not be confused with the computation of codon usage. (Remember that a *codon* is a triplet of nucleotides used in the context of the genetic code, when translating a DNA sequence into a protein; see Chapter 1.) The concept of codon usage thus only applies to DNA sequences corresponding to protein coding regions (for example, open reading frames or ORFs). It consists in computing the frequency of trinucleotides in a nonoverlapping manner, mimicking the protein translation process. Hence for a DNA sequence such as

 ATGTTTAGTGATGGACGCCCAGCATGC-
 GAC....

The codon frequency is computed using the following parsing:

 ATG, TTT, AGT, GAT, GGA, CGC,
 CCA, GCA, TGC, GAC, ...

The resulting table of codon usage tells you which codons are preferentially used to code for a given amino acid and, more important, which ones are rare. Codon usage varies among species. For instance, the codon usage in human genes is quite different from the one used by E. coli. For this reason, some human genes don't work well in these bacteria. Take a virtual trip to Japan at

 www.kazusa.or.jp/codon/
 countcodon.html

to test a simple codon usage program on your DNA sequence.

Figure 5-6:
Dot plot
showing a
segment
repeated
three times.

Using dot plots to identify inverted repeats

It's important to remember that you can also use a dot-plot program to search for inverted repeats — local similarities between the *direct* and *reverse* strand of a DNA sequence. To do this, you simply have to use the sequence along the X-axis (DNA Number 1), and its *reverse-complement* along the Y-axis (DNA Number 2). Use the Manipulate Sequences option at `arbl.cvmbs.colostate.edu/molkit/` to generate a sequence's *reverse-complement*. The only difference with the previous protocol is that the main diagonal disappears. Inverted repeats will again show up as off-diagonal segments.

Identifying genome-specific repeats in your sequence

Do not confuse the discovery of a new repeat in your sequence with the identification of a repeat from a preestablished list of repeats, like the Alu repeat family in the human genome. Discovering a new repeat has to do with the internal structure of your sequence, while identifying an established one is simply a matter of recognizing some local similarities between your sequence and a predefined reference database of repeats.

Repeated elements mostly occur in multicellular organisms. Vertebrate genomes are particularly rich in many families of repeated elements of various sizes. You can get a fairly complete picture of this topic by visiting the Repbase Web site of the Genetic Information Research Institute at `www.girinst.org`. This site gives you access to the CENSOR E-mail server. You can use this service by mailing your sequence to `censor@girinst.org`. The server compares your sequence against collections of human, rodent, plant, and invertebrate repeats.

A similar, albeit interactive, service is offered on the Repeat Masker server of the University of Washington at `repeatmasker.genome.washington.edu/`.

Finding Protein Coding Regions

Get ready — this is where the fun starts! We've checked our DNA sequence for contamination, verified its restriction maps, computed various composition statistics, and identified its repeats. It's time to move to the real exciting stuff! We can search if and where a protein is encoded within that obscure ATGCTACG gobbledygook.

A recipe to assess the significance of repeats

Let's say you found the sequence ATGC repeated 9 times in your 3000-bp long DNA sequence. Should you conclude that this motif has a special significance, or is there nothing particularly earthshaking about these 9 occurrences? Here's a simple recipe to quickly resolve this issue.

For the sake of simplicity, assume that your DNA consists of equal amounts of A, T, C, and G. This means that the random probability of an A at any given position is 1/4. Make the further assumption that all successive positions in the sequence behave independently. Starting at any position in the sequence, the random probability of observing ATGC is then $1/4*1/4*1/4*1/4 = 1/256$. In a random sequence, you thus expect to find ATGC every 265 bp, on average.

In a random 3000-bp long sequence, the expected number of ATGC occurring by chance is then $3000/256 = 11.7$. Thus, there's nothing special about observing 9 occurrences of ATGC within your sequence. Remember, this quick

computation is about *average* values, so don't conclude either that observing 9 ATGC while expecting 11.7 indicates that this motif is specifically *avoided*! Like always in statistical estimates, you must allow for fluctuations. The famous Poisson statistical distribution makes it possible to derive a simple common-sense rule: **Don't get too excited if what you observe is within a factor 2 of the expected value.** For instance, things start getting interesting if you observe 22 ATGC in your sequence — or only 5.

This quick recipe only applies to the assessment of perfect (identical) repeats. For imperfect repeats, you have to think in terms of a number of identical positions or more complex scoring systems. The calculation becomes way more complicated and requires real math training! Yet, you can still use this formula as lower and upper limits. (For instance, finding ATG[C or T] 13 times in a 3000-bp sequence is definitely not a big deal!)

To understand what is known as *protein discovery*, you must remember that protein coding genes have vastly different structures in microbes and multicellular organisms. In microbes, each protein is encoded by a simple DNA segment — from start to end — called an open reading frame (ORF). In animals and plant genes (vertebrates are the worst), proteins are encoded in several pieces called *exons*, separated by non-coding DNA segments called *introns*. That explains why the methods and programs used for finding proteins in microbes and higher eukaryotes are rather different.

This section starts with a simple strategy called *ORFing*, which shows you how to find protein coding regions in microbial DNA sequences or eukaryotic mRNA sequences.

ORFing your DNA sequence

To code for a protein, a DNA sequence must contain a translation Start codon (usually ATG) and not exhibit any of the Stop codons (usually TAA, TAG, TGA) in phase with the ATG for quite some length. This is the precise definition of an open reading frame. Given that proteins have an average length of about 350 residues and that very few of them are smaller than 100 residues, an additional minimal size criterion can be used, such as requesting that at least 300 nucleotides separate the Start from the Stop. This is what ORFing is all about. You can get some practice by using ORF Finder at NCBI (the National Center for Biotechnology Information), as we describe in the following steps list:

1. **Point your browser to** `www.ncbi.nlm.nih.gov/gorf/gorf.html`.

 The Input form is made up of a simple box, where you must paste your sequence (raw format), and a pull-down menu of genetic code choices. (We're ready to bet good money that you never knew there were 22 genetic code choices. So much for the universal code!) For now, let's stick with the standard genetic code (the default option).

2. **Copy your sequence from a .txt file or a Word document.**

 We used the first 5,000 bp of GenBank entry AE008569 (Rickettsia conorii genome). Alternatively, you can enter this accession number in the relevant input box, and indicate from 1 to 5,000 below the sequence input box. Then directly go to Step 4.

3. **Paste your sequence in the input box.**

 The program is happy with the raw sequence format (nucleotide only).

4. **Click the OrfFind button.**

 Figure 5-7 shows a typical output.

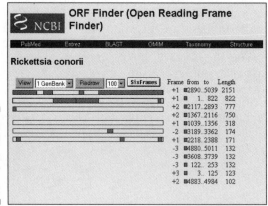

Figure 5-7:
Typical output of the ORF Finder program.

Your DNA sequence is displayed as six parallel horizontal bars, with each one corresponding to one of the six possible translation frames: +1, +2, +3 and on the reverse strand: –1, –2, and –3. The ORFs verifying the default size threshold (100 nucleotides) appear as small green rectangles on each bar. In addition to the graphical representation, the ORFs are also listed according to size. At this point, you're free to modify the size threshold (50, 100, 300).

Always click the Redraw button to refresh the output.

To examine each of the ORFs more closely, click the corresponding rectangle in the graphical display or in the list to the side. This expands the output to include the predicted amino-acid sequence and some supplementary buttons, as shown in Figure 5-8. These buttons let you screen some of the NCBI databases for sequences similar to this ORF, *right on the spot!* This is a great, simple tool for ORFing your sequence.

Before getting into more complicated programs, we want to remind you that this simple ORFing program is also good for finding higher organism protein-coding regions *if your sequence is a cDNA.* cDNAs (the image of mRNAs) don't include introns and have a microbe-like simple ORF structure. So you don't need to use another, more sophisticated program if you only want to delineate the protein-coding region within a human cDNA/mRNA sequence.

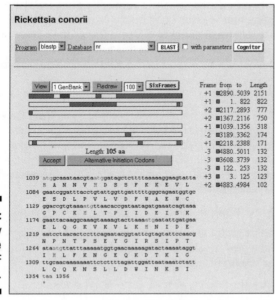

Figure 5-8: Secondary output of the NCBI ORF Finder.

Analyzing your DNA sequence with GeneMark

The simplest ORFing protocols can probably correctly identify 85 percent of the protein coding regions you may be interested in. However, there are a variety of situations that frequently occur where you may need to use a more sophisticated approach — the approach taken by GeneMark, for example. Such situations include

- Finding very short proteins
- Resolving ambiguous cases where overlapping ORFs are predicted in different reading frames — on the direct and reverse strand, for instance
- Wanting to pinpoint the exact Start codon (the most distal ATG isn't always the correct one)

GeneMark searches for coding regions using a criterion that's a bit more sophisticated than "being an uninterrupted reading frame longer than a certain length." This program also takes into account the statistical properties (very similar to word/codon usage) of your sequence and associates some sort of a probabilistic quality index to each candidate's ORFs. In the process, some small ORFs may get promoted, and the precise Start location may be redefined. The quality index measures the similarity between the candidate ORFs and an ideal gene model. This gene model is often implemented as a Markov Model, a mathematical concept beyond the scope of this book. The good news for you here is that these programs are easier to use than to understand. This is what we show you in the next steps list. Here goes:

1. **Point your browser to** `opal.biology.gatech.edu/GeneMark/`.

 This is the main GeneMark home page, coming to us from Georgia Tech in Atlanta. It offers different specialized versions of the program (each corresponds to a different gene model) for working on prokaryotic (microbes), eukaryotic (animals), cDNA (the DNA version of mRNA), or virus sequences. Let's say you want to analyze a sequence from a microbe that is little known.

2. **In the prokaryotic section, click the <u>Heuristic models</u> link.**

 This selects the program version corresponding to the analysis of sequences from a new organism. It brings you to a simple form with the usual sequence input box.

3. **Copy your sequence from a .txt file or a Word document.**

 We recommend using a microbial sequence about 5,000 bp in length.

4. **Paste the sequence in the input box.**

 The program is happy with the raw sequence format (nucleotide only) as well as many others.

5. **Click the Start GeneMark.hmm button to run the search.**

 After a while, the program outputs a list of predicted genes — with their locations — in a very simple format, as shown in Figure 5-9.

Figure 5-9:
Output of
GeneMark.

```
Gene Prediction Results

Information on input sequence

Sequence title: Wed Sep 25 13:56:18 EDT 2002
Length:        5040 bp
G+C percentage: 31.41 %

Parse predicted by GeneMark.hmm 2.0

GeneMark.hmm PROKARYOTIC (Version 2.1)
Sequence file name: sequence,   RBS: N
Model file name: heuristic_no_rbs.mat
Model organism: Heuristic_model
Wed Sep 25 13:56:28 2002

Predicted genes
  Gene   Strand   LeftEnd    RightEnd    Gene     Class
  #                                      Length
  1        +          1         822       822       1
  2        +       1039        1356       318       1
  3        +       1367        2116       750       1
  4        +       2117        2893       777       1
  5        +       2890       >5040      2151       1
```

If you're wondering what difference it makes to use a Markov model rather than a simple ORFing, the answer is in the differences between Figure 5-8 and Figure 5-9 — both of which relate to the same query sequence. GeneMark only retained the top five genes of the ORF Finder list!

Note that you can also run GeneMark (in the known model version) from its site at the European Bioinformatics Institute. The URL is www2.ebi.ac.uk/genemark/.

Finding internal exons in vertebrate genomic sequences

When it comes to predicting proteins from DNA, the most challenging problem around is analyzing a piece of human genomic DNA sequence. We already mentioned that vertebrate genes have a mosaic structure, where small bits of the protein are encoded by exons separated by non-coding introns. If you're looking at a human genomic sequence, your first question should be: Do I have a protein coding exon somewhere in there?

According to what molecular biologists have worked out, a protein coding exon is an ORF flanked by two specific signals known as *splice sites*. Several programs exist that are meant to recognize these exons. They all work on the same principle: On a first pass, they identify candidate ORFs with good compositional properties (word frequency, codon usage). This is basically what GeneMark does. On a second pass, they only retain those that are exactly flanked by good splice site sequences.

Vertebrate exons are small (150-bp long on average) and the sequences of their splice sites are variable. For this reason, finding exons is a more challenging problem than ORFing microbe DNA, which is why you should not expect an exon detection program to work nearly as well as something like ORF Finder or GeneMark, which have much larger targets to shoot for. Still, we'd like to show you what an exon detection program can do. Check out the following steps list, where we use Michael Zhang's program MZEF:

1. **Point your browser to** `argon.cshl.org/genefinder/`.

 This is the MZEF home page at Cold Spring Harbor Laboratory.

2. **In the central rectangle, click the <u>Human</u> link.**

 This selects the program version calibrated for human coding regions statistics. A simple input form is then displayed.

3. **Copy your sequence from a .txt file or a Word document.**

 If you don't have a sequence handy, you can fetch the sequence AF018429 from GenBank at NCBI (`www.ncbi.nlm.nih.gov`). This entry contains the exons 1 and 2 of the dUTPase gene.

 Make sure you get the sequence in FASTA format.

4. **Paste the sequence you copied into the input box.**

5. **Click the Submit button (below the sequence input box) to start your analysis.**

 The program quickly returns a minimally formatted output, as shown in Figure 5-10. MZEF correctly identified one exon between positions 1056-1172. The rest of the line consists of various quality values associated to the overall prediction, the various reading frames (FR1, FR2, FR3), the splice site, and the coding potential. The prediction is half correct because it missed one exon. Furthermore, the predicted exon actually starts at position 1018. Such a success rate (1/2, not entirely perfect) isn't so unusual for exon finding programs.

Exon predictors combining MZEF with other approaches are also available at the following sites:

- ✔ `corba.ebi.ac.uk/cgi-bin/sp/wrapper.cgi` (EBI)
- ✔ `genome.cs.mtu.edu/aat/aat.html` (Michigan Tech)

Figure 5-10:
MZEF output
on the
GenBank
AF018429
sequence
entry.

MZEF Results - human - gi|2443575|gb|AF018429.1|HSDUT1 Homo sapiens dUTPase gene, exons 1 2
Wed Sep 25 15:14:25 2002

Strand = 1
Overlap = 0
Prior Prob. = .02

Internal coding exons predicted by MZEF
Sequence_length: 1735 G+C_content: 0.610

Coordinates	P	Fr1	Fr2	Fr3	Orf	3ss	Cds	5ss
1056 - 1172	0.873	0.663	0.471	0.459	112	0.467	0.581	0.668

For more information about MZEF click here

Complete gene parsing for eukaryotic genomes

If you ever get into serious genome sequencing, you'll need the highest level of sophistication in computer-assisted annotation: genome-parsing software. These programs are designed to take large (100,000 to several million bp) pieces of a genomic sequence at once and predict the detailed exon/intron structure of entire genes. Like the MZEF program, these programs have a modular structure, where each module has been designed to recognize a given gene component: coding exons, first/last exons, promoter regions, poly-adenylation sites, and so on. The results of these independent modules are then combined into coherent gene structure predictions (limiting exon splicing to compatible reading frames, for instance) that are finally scored according to their similarity with an ideal gene model. Markov Models and dynamic programming optimization are the two key concepts at the basis of these programs. The most recent genome parsers, such as GenomeScan, also take into account protein sequence similarity information.

Despite their increasing algorithmic complexity, these programs remain easy to use; all you really have to do is paste your sequence into an input window, click a button, and voilà, you have parsed genes!

Analyzing your sequence with GenomeScan

To show you how simple it is, we're going to show you how to use the GenomeScan parsing software program. To get things ready, prepare your own vertebrate genome sequence in FASTA format. It has to be long enough (100,000 bp) to contain at least one complete gene. Alternatively, you can use the demonstration sequence available on the GenomeScan site.

Not all exons are detectable

It's important to remember that exon finding programs are mostly limited to the prediction of *internal coding exons*, that's to say exons that are entirely coding (ORF) and are flanked by two bona fide splice sites (at its 5' and 3' extremities).

Non-coding exons in the 5' and 3' untranslated part of transcripts, or partially coding exons (such as the first and last coding exons), are mostly invisible to these programs.

You also need protein sequences (also in FASTA format) that you know have some significant similarity with potential coding regions within your DNA sequence. You can find such proteins by doing a blastx comparison of your sequence to all known proteins. (For more on blastx, see Chapter 7.) Alternatively, you can use the demonstration set available on the GenomeScan site.

1. **Point your browser to** genes.mit.edu/genomescan.

 The GenomeScan home page at the Massachusetts Institute of Technology duly appears. GenomeScan is the successor to Chris Burge's GenScan. The main difference is the use of protein homology information.

2. **Click the GenomeScan Webserver link to call up the page with the input form.**

3. **Choose Vertebrate from the Organism pull-down menu.**

 This selects the program version calibrated for human coding regions statistics. A simple input form appears.

4. **Copy your DNA sequence from a .txt file or a Word document.**

 Don't forget the FASTA header.

5. **Paste the DNA sequence you copied into the DNA Sequence input box — the upper of the two larger input boxes.**

 Alternatively, you may click the DNA testfile link at the top of the page to use the demo sequence.

6. **Copy your homologous protein sequence set from a .txt file or a Word document.**

 Alternatively you may click the protein file link at the top of the page to use the demo protein sequences.

Each sequence must have a FASTA header. The set may contain multiple proteins so long as each is separated by a header on its own line. Files should contain less than one million bases.

7. **Paste your protein sequence into the Protein Sequence input box — the lower of the two larger input boxes.**

8. **Click the Run GenomeScan button at the bottom of the form to start the analysis.**

The output is a very long table listing all the components of the various predicted genes, with their coordinates and associated quality indices. This output is meant to be read by computer programs in the context of large scale sequencing projects and automated annotation. Fortunately, GenomeScan also provides a nice graphical output both as PDF and PostScript images.

Figure 5-11 shows an example of the graphical output generated from the demo DNA and protein sequences. Predicted genes (red arrows) and their predicted exons (red rectangles) are clearly indicated, together with their supportive evidence from blastx similarity (green rectangle below). A total of 5 genes are predicted along this 100,000-bp long vertebrate genomic sequence.

Figure 5-11:
Example of GenomeScan graphical output.

Assembling Sequence Fragments

Depending on their technology, current sequencing machines can only produce sequences at lengths of 500 to a 1000 nucleotides at a time. Thus, if your DNA sequence is longer than this, it has been obtained by assembling shorter overlapping fragments. Research dealing with the various assembly algorithms, as well as research devoted to the development of better assembly software, are important topics within the field of bioinformatics.

Managing large sequencing projects with public software

Taking what we would label the shotgun approach, a typical microbial genome sequence (4MB) is usually stitched together using over 50,000 individual fragments known as *reads*. The programs used for these genome sequencing or bigger projects (plant or animal genomes) take their input directly from the sequencing machine's *chromatograms* or *traces*. These traces are complex fluorescent profiles with peaks and valleys revealing the order and nature (A, T, G, C) of each nucleotide in the reads. Such programs include a base-calling system that associates a quality score to each nucleotide at each position of each read. They also include a data management system, as well as interactive editing and display tools. Because of their complexity, these complete genome sequencing software packages are not interactively available on Web sites; you need to install them locally on a dedicated powerful machine. Using them efficiently also requires some serious study. The most popular publicly available (for academics) genome-sequencing packages are

- The Staden Package: www.mrc-lmb.cam.ac.uk/pubseq/staden_home.html
- Phred/Phrap/Consed: www.phrap.org/
- The Acembly engine included in the popular AceDB suite: www.acedb.org/
- The Assembler from The Institute for Genome Research (TIGR): www.tigr.org/software/

Commercial genome sequencing software packages are also available from several companies:

 ↙ *Sequencher* from Gene Codes: `www.genecodes.com/`

 ↙ *Lasergene* from DNASTAR: `www.dnastar.com/`

 ↙ *GenomeAssembler* from Paracel: `www.paracel.com/`

We believe that showing you how to manage a large genome sequencing project is beyond the scope of this introductory book. However, you may still encounter the need to assemble a handful of sequences in your everyday life. For instance, you might be working on a cDNA that's just a few thousand bp in length, represented by a dozen PCR fragments. You may have generated a number of expression sequence tags (EST: one pass sequenced partial cDNAs) and want to see whether you can deduce a complete cDNA from them. Alternatively, you may want to assemble ESTs you extracted from a database. In the latter case, you may not have the corresponding traces and base-call quality scores.

What you're looking for is a simple program able to recognize significant overlaps between your fragments and assemble them into a single sequence, known as a *contig*. Of course, the fragment sequences might have been obtained from different DNA strands, making the visual recognition of overlapping segments quite difficult. The following section shows you how to do this with the nice EST assembler set up by the FIRC Institute of Molecular Oncology in Milano (IFOM).

Using the IFOM assembler

Prepare the set of sequences that you want to assemble in FASTA format as a .txt file or a Word document. Each individual sequence must be in FASTA format and have its own header.

1. **Point your browser to** `bio.ifom-firc.it/ASSEMBLY/assemble.html`.

 A straightforward input form appears, ready to do its duty. The maximum number of sequences you can enter in the sequence box is 10,000 (!), and their maximum length 30,000 bp (!). The time out for the interactive jobs occurs after five minutes. If the job takes longer, its output is sent to you by e-mail.

2. **Copy your fragment sets from a .txt file or a Word document.**

 Don't forget the FASTA headers

TIP

It doesn't matter whether or not you provide a mixture of sequences in different orientations. The program automatically tries them in both orientations.

3. **Paste the sets into the Sequence input box.**

The lower part of the form allows you to change the length of the minimal overlap (down to 20) as well as percent identity (down to 65). You can safely keep the default setting.

4. **Click the Submit Form button to run the assembly.**

This server implements the CAP3 assembly program from X. Huang.

If you provided a few short sequences, as shown in Figure 5-12, the program response is almost immediate.

The output begins with a summary of the resulting number of contigs and their constituents. In our example, all the fragments were assembled in a single contig, as shown in Figure 5-13.

The body of the output provides the details of the overlaps and contig structure. At the bottom of the output, the resulting contig consensus sequence is given. (See Figure 5-14.) This sequence can then be used for similarity searches against databases.

Paste the sequence fragment here.

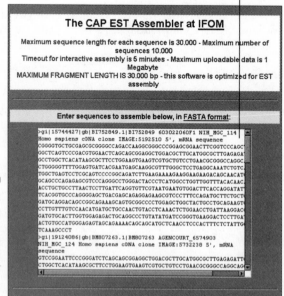

Figure 5-12:
Submitting
sequence
fragments to
the IFOM
Cap EST
assembler.

List of assembled fragments

The Cap Sequence Assembler at IFOM

Program Cap3 used

Assembly results:

Number of segment pairs = 30; number of pairwise comparisons = 9
'+' means given segment; '-' means reverse complement

Overlaps Containments No. of Constraints Supporting Overlap

****************** Contig 1 ***********************
GI|15744427|GB|BI752849.1|BI752849+
GI|19124086|GB|BM807263.1|BM807263+
GI|8147732|GB|AW958049.1|AW958049+
GI|8147734|GB|AW958051.1|AW958051+
GI|20393914|GB|BQ213074.1|BQ213074+
GI|13728017|GB|BG206330.1|BG206330+

DETAILED DISPLAY OF CONTIGS
****************** Contig 1 ********************

GI|15744427|GB|BI752+ CGGGGTGCTGCGAGCGCGGGGCCAGACCAAGGCGGGCCCGGAGCGGAACTTCGGTCCCAG

consensus CGGGGTGCTGCGAGCGCGGGGCCAGACCAAGGCGGGCCCGGAGCGGAACTTCGGTCCCAG

GI|15744427|GB|BI752+ CTCGGTCCCCGGCTCAGTCCCGACGTGGAACTCAGCAGCGGAGGCTGGACGCTTGCATGG
GI|19124086|GB|BM807+ GGATCTCAGCAGCGGAGGCTGGACGCTTGCATGG

consensus CTCGGTCCCCGGCTCAGTCCCGACGTGGAACTCAGCAGCGGAGGCTGGACGCTTGCATGG

GI|15744427|GB|BI752+ CGCTTGAGAGATTCCATCGTGCCTGGCTCACATAAGCGCTTCCTGGAAGTGAAGTCGTGC
GI|19124086|GB|BM807+ CGCTTGAGAGATTCCATCGTGCCTGGCTCACATAAGCGCTTCCTGGAAGTGAAGTCGTGC

consensus CGCTTGAGAGATTCCATCGTGCCTGGCTCACATAAGCGCTTCCTGGAAGTGAAGTCGTGC

Figure 5-13:
Top of the
Cap EST
assembler
output.

Overlap details

End of overlap details

consensus CAGCAAATATTTATTGGTCATCTTTTCTCCATAAGATAGTGTGATAAACACAGCATGAAT

GI|13728017|GB|BG206+ AAAGGTATTTTCCACACAGACAAGTGTTTTTTCACAAAATTATTNATTTTGNTGGGGCTG

consensus AAAGGTATTTTCCACACAGACAAGTGTTTTTTTCACAAAATTATTNATTTTGNTGGGGCTG

GI|13728017|GB|BG206+ TGGCGGCCGCTTCCTTTATGGGGGGGGAATTTAGAACCCGTTCCTGACGCGGGGGN

consensus TGGCGGCCGCTTCCTTTATGGGGGGGGAATTTAGAACCCGTTCCTGACGCGGGGGN

Singletons

Contigs

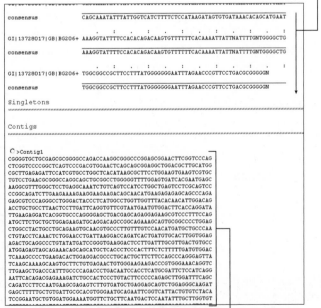

○ >Contig1
CGGGGTGCTGCGAGCGCGGGGCCAGACCAAGGCGGGCCCGGAGCGGAACTTCGGTCCCAG
CTCGGTCCCCGGCTCAGTCCCGACGTGGAACTCAGCAGCGGAGGCTGGACGCTTGCATGG
CGCTTGAGAGATTCCATCGTGCCTGGCTCACATAAGCGCTTCCTGGAAGTGAAGTCGTGC
TGTCCTGAACGCGGGGCCAGGCAGCTGCGGCCTGGGGGGTTTTGGAGTGATCACGAATGAGC
AAGGCGTTTGGGCTCCTGAGGCAAATCTGTCCAGTCCATCCTGGCTGAGTCCTCGCAGTCC
CCGGCAGATCTTGAAGAAAAGAAGGAAGAAGACAGCAACATGAAGAGAGAGCAGCCCAGA
GAGCGTCCCAGGGCCTGGGACTACCCTCATGGCCTGGTTGGTTTACACAACATTGGACAG
ACCTGCTGCCTTAACTCCTTGATTCAGGTGTTCGTAATGAATGTGGACTTCACCAGGATA
TTGAAGAGGATCACGGTGCCCAGGGGAGCTGACGAGCAGAGGGAGAAGCGTCCCTTTCCAG
ATGCTTCTGCTGCTGGAGAAGATGCAGGACAGCCGGCAGAATCAGTGCGGCCCCTGGAG
CTGGCCTACTGCCTGCAGAAGTGCAACGTGCCCTTGTTTGTCCAACATGATGCTGCCCAA
CTGTACCTCAAACTCTGGAACCTGATTAAGGACCAGATCACTGATGTGCACTTGGTGGAG
AGACTGCAGGCCCTGTATATGATCCGGGTGAAGGACTCCTTGATTTGCGTTGACTGTGCC
ATGGAGAGTAGCAGAAACAGCAGCATGCTCACCCTCCCACTTTCTCTTTTTGATGTGGAC
TCAAAGCCCCTGAAGACACTGGAGGACGCCCTGCACTGCTTCTTCCAGCCCAGGGAGTTA
TCAAGCAAAAGCAAGTGCTTCTGTGAGAACTGTGGGAAGAAGAACGACCCGTGGGAAACAGGTC
TTGAAGCTGACCCATTTGCCCCAGACCCTGACAATCCACCTCATGCGATTCTCCATCAGG
AATTCACAGACGGAAAAGATCTGCCACTCCCTGTACTTCCCCCAGAGCCTTGGATTTCAGC
CAGATCCTTCCAATGAAGCGAGAGTCTTGTGATGCTGAGGAGCAGTCTGGAGGGCAAGAT
GAGCTTTTTTGCTGTGATTGGCACGTGGGAATGCAGAATTCGGTCATTACTGTGTCTACA
TCCGGAATGCTGTGGATGGAAAATGGTTCTGCTTCAATGACTCCAATATTTGCTTTGGTGT

Figure 5-14:
Bottom of
the Cap EST
assembler
output: the
contig
consensus
sequence.

Assembled USP18 mRNA sequence

Beyond This Chapter

This chapter provides you with a good background on the various kinds of analyses people usually want to perform on their newly determined DNA sequences. There are, of course, many other things you can do and look for when working within a more specialized context. The most important area that isn't covered here concerns the search for biologically significant signals in the non-coding part of DNA sequences. This includes very hot topics, such as the prediction of promoter regions, regulatory elements, and/or protein binding sites.

Our main excuse for not presenting this area in more detail is that the success rate for these analyses is still quite low. Also, the best programs and relevant databases aren't always in the public domain or aren't available as Web servers. However, Table 5-1 provides you with some entry points that you can explore by yourself.

Table 5-1	**Web Sites for Searching Signals in DNA Sequences**	
Institution	*URL*	*Description*
German Research Center for Biotechnology	`transfac.gbf.de/ TRANSFAC/`	Search for transcriptional elements using TRANSFAC
Center for Information Technology, NIH, USA	`bimas.dcrt. nih.gov/molbio/ matrixs/`	Search for transcriptional elements using IMD
Center for Information Technology, NIH, USA	`bimas.dcrt. nih.gov/molbio/ proscan/`	Predict eukaryotic promoter regions
German National Research Center for Environment and Health	`www.gsf.de/biodv/ genomeinspector. html`	Detect distance correlations between sequence elements
Japan National Institute of Agrobiological Sciences	`www.dna.affrc. go.jp/htdocs/ PLACE/signalscan. html`	Detect regulatory signals in plant sequences
San Diego Supercomputer Center	`meme.sdsc.edu/ meme/website/`	Discover motifs in groups of related DNA or protein sequences
Université Libre de Bruxelles Belgium	`rsat.ulb.ac.be/ rsat/`	Analyze regulatory sequences

Chapter 6

Working with a Single Protein Sequence

Once you eliminate the impossible, whatever remains, no matter how improbable, must be the truth.

— Sherlock Holmes, as told to Sir Arthur Conan Doyle

*W*hen you're studying a protein, you turn yourself into an investigator. Basically, you want to find out everything you can before designing a new experiment in the lab. For instance, in order to make sure that the protein you have in your test tube really is what you think it is, you must carry out a simple physical analysis. Many online tools exist that help you predict the results of these analyses (and make sure you're on the right track). In this chapter, we show you where you can find these tools and how to use them. Among other things, we show you how to predict molecular weights, measure isoelectric points, count residues, or predict the outcome of a protease digestion.

Of course, there's more to a protein than its weight or its charge. While these properties do help you get a sense of what's what with your particular protein, uncovering a nice biological story takes more than that. An important

question you want to ask yourself is what your protein looks like when it's active. Does it need to be modified after being translated, does it contain coiled-coil elements, or is it a transmembrane protein? In this chapter, we show you how online tools can answer these questions.

The structure is also important, of course — so important that we've dedicated two chapters of this book to its analysis (Chapter 11 for protein structure analysis and Chapter 12 for RNA structure analysis). For this reason, we don't talk much about structures in this chapter.

Knowing what your protein looks like is one thing, but knowing what it actually does is another. For instance, you may want to know if your protein binds calcium or if it contains an enzymatic site. If it's an enzyme, there's no doubt that you need to know about its substrate (the kind of molecule it binds). To answer these types of questions, we show you three powerful methods to check whether your protein contains a domain with a known function.

The other powerful technique that can help you when it comes to guessing the function of a protein involves *similarity searches.* If you find in a database somewhere a well-characterized protein whose sequence is very similar to your protein, then you're allowed to say that most of what is true for this sequence is true for your sequence as well. Although we don't say anything about this type of analysis in this chapter, don't think that we're going to leave you high and dry. We cover everything that you need to know about similarity searches in Chapter 7.

Doing Biochemistry on a Computer

Two terrific places to go to if you want to do some biochemistry using a computer are

- The ExPASy (*Expert Protein Analysis System*) server at `www.expasy.ch`, with a specific page dedicated to protein analysis methods
- The Swiss EMBnet at `www.ch.embnet.org`

The people who created and maintain a big chunk of SWISS-PROT run these two sites. When it comes to looking closely at proteins, these folks know their job.

In this section, we assume that your heart is set on a protein you've never seen before. The only thing you know about this beauty is its sequence. Maybe this protein comes from a sequencing project or something similar. Before you open it up and check out the insides, you want to find out — and guess — as much about it as you can. That's the reason you're here reading this very informative chapter.

If you don't have your own protein yet, don't worry! For each example we use in this chapter, we provide you with a SWISS-PROT protein of your very own. This way, you can always compare the results of the program with the SWISS-PROT annotation.

In general, when using a program for the first time, it's advisable to run it on a well-characterized protein, simply to check if the program does what it says it does (and if it does it well). To find a suitable example, use one of the SRS servers as we describe in Chapter 2. This can give you a sense of how good the program you're using really is.

Predicting the main physico-chemical properties of a protein

ProtParam, a program that you can use online on the ExPASy server, is a convenient way to estimate every simple physico-chemical property (the *protein parameters*) that can be deducted from your protein sequence. It makes no complex and adventurous assumptions about your little protein: It simply counts for you.

Here's how it works:

1. **Point your favorite browser to** `www.expasy.ch/tools/#primary`.

 The Primary Structure Analysis section of the ExPASy Proteomics Tools page appears.

 At the end of this chapter, we give you an exhaustive list of all the available ExPASy mirror sites that you can use if the main ExPASy site is down.

2. **Click the <u>ProtParam</u> link near the top of the page.**

 The ProtParam Tool page appears.

3. **Enter your sequence in the search boxes provided.**

 Provide a sequence in one of two ways:

 - Enter the accession number, as shown in Figure 6-1. You can do this if your sequence exists in the SWISS-PROT sequence database. SWISS-PROT accession numbers usually start with a letter followed by five digits. In this example, we use a SWISS-PROT protein: the rat Syntaxin 1A (P32851).

 - Paste the sequence in raw format, as shown in Figure 6-2. If your sequence isn't a SWISS-PROT sequence, or if you don't know the sequence name, you can paste the sequence itself, using the other window. Note that the sequence must be entered in raw format and should not include anything but amino acids (white space and

numbers are tolerated and automatically removed). If your sequence is in FASTA format, DO NOT include the first line that starts with a greater than sign (>).

4. Press the Compute Parameters button.

If, in Step 2, you provided the accession number of the sequence, an intermediate page appears like the one shown in Figure 6-3.

5. If you have entered a SWISS-PROT accession number — and Figure 6-3 has dutifully appeared — enter the range of the analysis in the N-Terminal and C-Terminal fields. (The N-terminus is the left side of your protein sequence and the C-terminus is the right side.)

At the top of the page in Figure 6-3, you can find a list of the features contained in your sequence. This isn't a prediction but simply a display of the information contained in the SWISS-PROT entry. Below are two boxes that you can fill with the coordinates of the segment you're interested in. If you leave these two boxes blanks, ProtParam analyzes the complete sequence.

With coordinates like those in Figure 6-3, ProtParam analyzes the segment of your protein that starts on the 266th amino acid and ends on the 288th.

ProtParam tool

ProtParam (References) is a tool which allows the computation of various physical and chemical parameters for a given protein stored in SWISS-PROT or TrEMBL or for a user entered sequence. The computed parameters include the molecular weight, theoretical pI, amino acid composition, atomic composition, extinction coefficient, estimated half-life, instability index, aliphatic index and grand average of hydropathicity (GRAVY) (Disclaimer).

Please note that you may only fill out **one** of the following fields at a time.

Enter a SWISS-PROT/TrEMBL accession number (AC) (for example P05130) or a sequence identifier (ID) (for example KPC1_DROME):

P32851

Or you can paste your own sequence in the box below:

RESET Compute parameters

Figure 6-1:
Giving an accession number rather than a sequence to ProtParam.

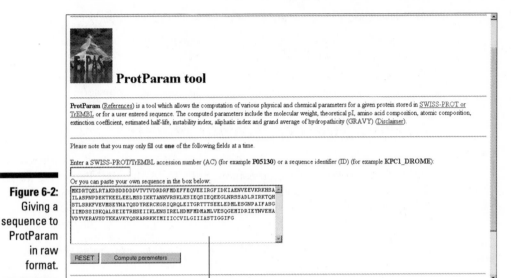

Figure 6-2:
Giving a
sequence to
ProtParam
in raw
format.

Paste your sequence here.

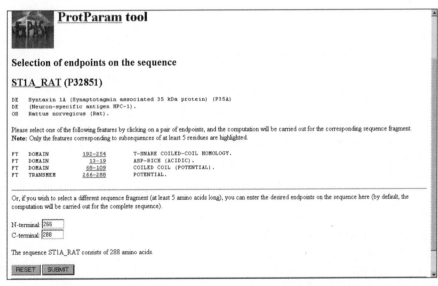

Figure 6-3:
Using
ranges with
ProtParam.

6. **Press the Submit button to proceed with the analysis.**

 The results, including the molecular weight, the extinction coefficient, and the estimated half-life appear on-screen.

7. **In your Web browser, choose File⇨Save As to save your results.**

Interpreting ProtParam results

If you've used Syntaxin as your sample protein in your trial run of ProtParam, you can now proudly pass on the knowledge that Syntaxin's theoretical molecular weight is 33067,4 Daltons, that its pI is 5.14, and that its aliphatic index is 83.96 — and this is only a sample of the large amount of information ProtParam can provide!

You can find detailed explanations on how ProtParam computes these parameters at `www.expasy.ch/tools/protpar-ref.html`. Among other issues, following are a few points you may want to pay attention to.

Molecular weight

The computer program is simply summing the weight of all the residues in the sequence. Remember that the program doesn't consider the following:

- ✔ It does NOT consider post-translational modifications like glycosylations or phosphorylations.
- ✔ It does NOT take into account complex maturations like the removal of a lead peptide or the excision of an intein.
- ✔ ProtParam does not know if your mature protein becomes a dimer (association of two proteins) or a trimer (association of three proteins) or any other kind of multimer.

Of course, these problems can eventually add up and lead to experimental results significantly different from the theory. In a way, that's where the fun begins . . . so keep your eyes open!

Extinction coefficients

An *extinction coefficient* tells you how much light your protein absorbs at a certain wavelength. Estimates of these values are useful if you need to follow your protein with a spectrophotometer when you purify it.

When you use this result, remember that the ProtParam value is only an indication. ProtParam predicts the extinction coefficient by summing the contribution of every amino acid as if each was independent. This calculation ignores the fact that the behavior of amino acids can be dramatically altered

depending on their immediate surroundings. This isn't something that can be predicted.

If you need the *exact* coefficient, you must measure it experimentally. On the other hand, for most proteins, the experimental coefficient is rarely very different from the theoretical one.

Instability

This parameter is only just a crude estimate of your protein stability in a test tube. When the index is below 40, the protein is usually stable. Above 40, it may not be stable.

Half-life

The *half-life* is a crude prediction of the time it takes for half of the amount of your protein present in a cell to disappear completely after its synthesis in the cell. This prediction is given for three types of organisms. You can safely extrapolate it to similar organisms.

The half-life value given here is meaningless if the degradation of your protein is part of a regulatory process. For instance, if the cell specifically directs a *protease* (a protein that digests other proteins) at your protein, your protein may disappear much faster than ProtParam tells you.

Digesting a protein in a computer

Protease digestions, where you use an enzyme to specifically cut your protein, can be useful if you're only interested in carrying out experiments on a portion of your protein. They can also be useful if you want to

- ✔ Separate the domains in your protein
- ✔ Identify potential post-translational modification by mass spectrometry
- ✔ Remove a tag protein when you express a fusion protein
- ✔ Make sure that the protein you're cloning isn't sensitive to some endogenous proteases

PeptideCutter is a very useful tool for this kind of purpose and is available from the ExPASy Web site at `www.expasy.ch/tools/#proteome`. Check it out; you'll find it easy to use. Paste in your sequence, choose an option or two, click a button, and your protein is sliced, diced, and chopped to your specifications.

Doing Primary Structure Analysis

The *primary structure* is the amino acid sequence of your protein. When you analyze the primary structure, you ignore the potential interactions between amino acids. Primary structure analysis doesn't tell you about the potential interactions between the amino acids in your sequence. These predictions are made by secondary or tertiary structure prediction. (See Chapter 11 for more on such predictions.)

You conduct a primary structure analysis to try to find segments in your protein that display a special composition. These segments can reveal some interesting properties of our protein, such as

- ✔ *Hydrophobic regions* that could be membrane spanning segments in proteins that anchor themselves into a membrane
- ✔ *Coiled-coil regions* that indicate potential protein-protein interaction
- ✔ *Hydrophilic stretches* that could be looping out at the surface of the protein

The methods we show you in this section, although less powerful than the state-of-the-art secondary structure prediction methods that we cover in Chapter 11, still have a lot going for them. For one, they're simple to use and simple to understand. When you use them, you can easily see what's going on and, therefore, easily avoid making mistakes.

Many of the methods we show you here rely on *sliding windows* technique. (For more on these guys, check out the sliding window side bar.) Predictions based on sliding windows aren't very sensitive or very precise, but they are very robust. If you see a strong signal when using a method based on sliding windows, chances are you're looking at something that's a genuine biological signal.

When the sliding operation is finished, the values associated with every amino acid are plotted against the sequence. Biologists name this display a property profile. (Do not confuse it with a domain profile, which is a formulation of a multiple sequence alignment.) If you're lucky, you may be able to identify transmembrane segments, loops, or coiled-coil regions by using sliding windows. Hydrophobicity is the most popular analysis because it's a good indicator of transmembrane segments or core regions within a protein.

Ironically, the main advantage of the methods that use sliding windows is also one of their main shortcomings: They don't interpret the results for you and provide only a raw signal.

Sliding windows

The sliding windows technique is the most ancient way of looking at sequences. The principle is very simple. What you need is a property and a list of values associated with each of the 20 amino acids. This property can be any measurable physico-chemical parameter, such as size, polarity, hydrophobicity, or even the propensity of amino acids to be in a specific structural state. The values in this table are the amino acid scale values. Many such tables exist that have been determined experimentally for almost any characteristic you may think of.

After you have this table, you choose a window size and slide it along your sequence. When the window is centered on an amino acid, the scale values associated with the amino acids it contains are summed up and averaged. The resulting value is associated with the central amino acid, and the window is shifted by one amino acid; this process goes on until the end of the sequence is reached. The following example illustrates how the window slides along the sequence. The asterisk (*) indicates the amino acid on which the window is centered.

```
Seq. AGVCFGTRESALPTFREDCYZK
      <---*---> Window 1
       <---*---> Window 2
        <---*---> Window 3
```

Because you have to do the interpretation yourself, two simple rules apply when interpreting the results of an analysis based on sliding windows:

- ✔ Be very strict and consider only strong signals.

- ✔ Check the robustness of your signal. Good signals aren't shy. They don't go away simply because you increase or decrease the window size by one amino acid or replace a property table with another similar table.

Another major weakness of the sliding window is the arbitrary window size: why 9 residues or 13 or 21? There is no magic answer to this question, but when it comes to choosing your window size, the following simple rule applies:

The window size must be in the same range as the size of the feature you're looking for.

For instance, if you're looking for transmembrane domains that are normally 21 amino acids long, you want to use a window size that is close to this value — in practice, 19 is the best window value for finding transmembrane regions. On the other hand, if you're interested in the structural features of globular proteins, a range between 7 and 11 would be more appropriate.

Looking for transmembrane segments

Predicting that your protein has transmembrane segments tells you much more about its function than almost any other simple prediction you can do.

When you know your protein is a transmembrane protein, you know that you can't work on it with the same techniques as if it were a globular protein. If you find one transmembrane segment at the N-terminus of your sequence, you can predict that the protein is secreted. On the other hand, many transmembrane domains in one protein may indicate a channel.

Because these predictions are so important, we show you two methods for testing for transmembrane segments: ProtScale (a very simple one) and TMHMM (one of the most complete). TMHMM is not part of the ExPASy server; it is a service offered to the community by the Technical University of Denmark.

✔ *ProtScale* uses a sliding window technique and one of many amino acid scale values. In this example, we use the hydrophobicity in order to identify groups of hydrophobic segments that are characteristic of transmembrane proteins. ProtScale doesn't predict anything for you; it returns a hydrophobicity profile and lets you do the interpretation.

✔ *TMHMM* is a state-of-the-art program that predicts transmembrane segments in your protein. TMHMM also tells you about the portions of your protein that are probably inside the cell and those that are probably outside.

Running Protscale

Let's start with Protscale:

1. **Point your browser to** www.expasy.ch/cgi-bin/protscale.pl.

 The ProtScale page duly appears.

2. **Enter your sequence accession number P78588 in the small search box.**

 You can paste the sequence in raw format or enter a SWISS-PROT Accession Number. (For an example of raw format, see Figure 6-2.)

 We've entered P78588, the accession number for FREL_CANAL protein, a protein that contains seven transmembrane segments.

3. **Scroll down the page and then select the radio button next to Hphob. / Kyte & Doolittle, as shown in Figure 6-4.**

 ProtScale gives a large choice of properties that you can test on your protein. The one we choose here is the most appropriate for predicting transmembrane helices.

4. **From the Window Size pull-down menu, choose 19.**

 This value is appropriate when looking for transmembrane domains.

5. **Press the Submit Button at the bottom of the page.**

6. **If you have entered a SWISS-PROT accession number, enter the range of the analysis.**

 If, in Step 2, you have provided the accession number of the sequence, an intermediate page appears like the one shown in Figure 6-3.

 At the top of the page in Figure 6-3, you can find a list of the features contained in your sequence. This isn't a prediction but simply a display of the information contained in the SWISS-PROT entry. Below are two boxes that you can fill with the coordinates of the segment you're interested in. If you leave these two boxes blanks, ProtParam analyzes the complete sequence.

7. **Click the Submit button to proceed with the analysis, and then wait for your browser to return the results.**

8. **At the bottom of the page, click Image in GIF format.**

9. **Choose File⇨Save As from your browser's main menu to save your results.**

Figure 6-4:
Choosing a
property
on the
ProtScale
server.

○ Polarity / Grantham	○ Refractivity
○ Recognition factors	○ Hphob. / Eisenberg et al.
○ Hphob. OMH / Sweet et al.	○ Hphob. / Hopp & Woods
⊙ Hphob. / Kyte & Doolittle	○ Hphob. / Manavalan et al.
○ Hphob. / Abraham & Leo	○ Hphob. / Black
○ Hphob. / Bull & Breese	○ Hphob. / Fauchere et al.
○ Hphob. / Guy	○ Hphob. / Janin
○ Hphob. / Miyazawa et al.	○ Hphob. / Rao & Argos
○ Hphob. / Roseman	○ Hphob. / Wolfenden et al.
○ Hphob. / Welling & al	○ Hphob. HPLC / Wilson & al
○ Hphob. HPLC / Parker & al	○ Hphob. HPLC pH3.4 / Cowan
○ Hphob. HPLC pH7.5 / Cowan	○ Hphob. / Rf mobility
○ HPLC / HFBA retention	○ HPLC / TFA retention
○ HPLC / retention pH 2.1	○ HPLC / retention pH 7.4
○ % buried residues	○ % accessible residues
○ Hphob. / Chothia	○ Hphob. / Rose & al
○ Ratio hetero end/side	○ Average area buried
○ Average flexibility	○ alpha-helix / Chou & Fasman
○ beta-sheet / Chou & Fasman	○ beta-turn / Chou & Fasman
○ alpha-helix / Deleage & Roux	○ beta-sheet / Deleage & Roux
○ beta-turn / Deleage & Roux	○ Coil / Deleage & Roux
○ alpha-helix / Levitt	○ beta-sheet / Levitt
○ beta-turn / Levitt	○ Total beta-strand
○ Antiparallel beta-strand	○ Parallel beta-strand
○ A.A. composition	○ A.A. comp. in SWISS-PROT
○ Relative mutability	

Window size: 19

5
7
9
11
13
15
17
19
21

Select window size here.

Interpreting ProtScale results

A good signal is a signal that is strong and not very sensitive to the parameters. For instance, Figure 6-5 shows the (strong) results you obtain on the SWISS-PROT protein P78588 when using the Kyte and Doolittle Hydrophobicity Scale.

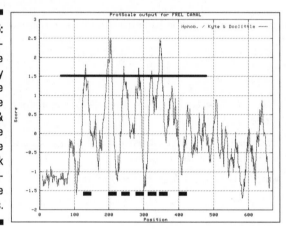

Figure 6-5: Hydrophobicity profile returned by ProtScale by using the Kyte & Doolittle Scale. The boxes mark known transmembrane segments.

In order to convince yourself that this result is meaningful, redo the comparison with another scale, like the one from Eisenberg, with the results you see in Figure 6-6. The details differ very little between these two graphs, but you can clearly see that the main features are well conserved and that the strongest peaks come roughly at the same position.

Figure 6-6: Hydrophobicity profile returned by ProtScale using the Eisenberg Scale.

In Figure 6-5, we've drawn a bold line that gives you a hint on how to interpret your results. The recommended threshold value when using Kyte and Doolittle is 1.6, but if, like us, you keep forgetting this magic value, here is a simple recipe:

1. **Place a piece of paper over your results.**

2. **Lower this piece of paper until the tips of the strongest peaks appear.**

3. **Keep lowering this threshold as long as you can see nice sharp peaks.**

With this simple method, you can identify unambiguously five out of seven transmembrane regions. On a sixth one, we could make an adventurous guess (It's easy to guess like this when you know the answer!), but the seventh transmembrane domain is impossible to find.

This isn't so surprising: Many proteins contain these seven transmembrane regions where six are easy to find and the seventh domain is notoriously difficult to predict.

Running TMHMM

TMHMM is maintained by the Center for Biological Analysis (CBS), the Technical University of Denmark. The CBS (www.cbs.dtu.dk/services/) offers many interesting services for sequence analysis that we encourage you to explore.

TMHMM is a state-of-the-art method that uses sophisticated mathematical models (Hidden Markov Models) to predict transmembrane regions. For a bit of sophistication, give it a try.

1. **Point your browser to** www.cbs.dtu.dk/services/TMHMM-2.0.

 The TMHMM page of the CBS site appears (Figure 6-7).

2. **Scroll down the page a bit to enter your sequence in the TMHMM search box.**

 TMHMM only recognizes the sequence in FASTA format. For the following example, we use the protein sequence FREL_CANAL. To obtain this sequence from SWISS-PROT:

 a. Open a new browser window.

 b. Open the page www.expasy.ch/cgi-bin/get-sprot-fasta? FREL_CANAL, which contains REL_CANAL in FASTA format.

 c. Copy the sequence onto the Clipboard.

 d. Paste the sequence in the TMHMM window (refer to Figure 6-7).

Figure 6-7:
The bottom
half of the
TMHMM
page.

3. **Select the Use Old Model check box.**

 We find this version to be slightly more accurate than the latest for this protein. This may change depending on the protein you're looking at.

4. **Keep the output format radio buttons to their default value (Extensive with graphic).**

5. **Press the Submit button.**

6. **When the Results page appears, choose File⇨ Save As from your browser main menu to save your results.**

 If you are using Internet Explorer, from the Save As Type pull-down menu, choose Web Page, Complete (*.htm, *.html). Your file is saved along with a directory that has the same name. Don't separate your file and this directory.

Interpreting results from TMHMM

Although results from TMHMM are slightly different from those you get from ProtScale, they are nevertheless in rough agreement. This is good news because the two methods use very different principles.

There are two major differences between ProtScale and TMHMM:

 ✔ TMHMM returns a precise prediction. This prediction is at the top of the output — just before the plotted graph — as shown in Figure 6-8.

 ✔ TMHMM predicts the segments that are inside the cell and the segments that are outside the cell.

```
# sp_P78588_FREL_CANAL  Exp number of AAs in TMHs: 117.5409
# sp_P78588_FREL_CANAL  Exp number, first 60 AAs:  0
# sp_P78588_FREL_CANAL  Total prob of N-in:        0.70138
sp_P78588_FREL_CANAL    TMHMM1.0    inside      1   116
sp_P78588_FREL_CANAL    TMHMM1.0    TMhelix   117   139
sp_P78588_FREL_CANAL    TMHMM1.0    outside   140   196
sp_P78588_FREL_CANAL    TMHMM1.0    TMhelix   197   215
sp_P78588_FREL_CANAL    TMHMM1.0    inside    216   277
sp_P78588_FREL_CANAL    TMHMM1.0    TMhelix   278   296
sp_P78588_FREL_CANAL    TMHMM1.0    outside   297   310
sp_P78588_FREL_CANAL    TMHMM1.0    TMhelix   311   333
sp_P78588_FREL_CANAL    TMHMM1.0    inside    334   339
sp_P78588_FREL_CANAL    TMHMM1.0    TMhelix   340   358
sp_P78588_FREL_CANAL    TMHMM1.0    outside   359   669
```

Figure 6-8:
TMHMM
results on
the SWISS-
PROT
Protein
FREL_
CANAL.

In this case, TMHMM fails to identify the segment that is in the middle (the one between amino acids 234 and 255), but it gives a fairly good estimation for five of the segments.

The graph in Figure 6-8 gives you the probability for each residue of your sequence to be in a transmembrane domain, inside, or outside of the cell. This probability is only a prediction; it's not necessarily correct.

If you really need an accurate prediction because you're about to design an experiment, we recommend running many predictions using different methods. Very different methods that give you the same results are usually a good indication that you're on the right track.

Looking for coiled-coil regions

Coiled-coil regions are portions of a protein that are formed by the intertwining of two or three alpha-helices. One of the reasons why it's interesting to find a coiled-coil regions is that these are often involved in protein-protein interactions. Another less glorious reason is that these coiled-coil regions can give false matches when you make a database search (see Chapter 7) and it can be a good thing to filter them out. If you want to predict these regions in your protein of interest, you can use the conveniently named COILS server at EMBnet. Check them out at `www.ch.embnet.org/software/COILS_form.html`

Predicting Post-Translational Modifications in Your Protein

Proteins often need to be modified before they become active in the cell. Biologists name the modifications *post-translational modifications* because they occur after the translation (synthesis) of the protein. These modifications may involve adding sugars, modifying amino acids, or removing pieces of the newly synthesized protein. If you're studying a new protein, you probably want to know about such matters. This may be very important if you want to clone and express a human protein in bacteria because, in order to be active, your protein may require some post-translational modifications that the bacteria itself cannot make.

One of the most useful tools for analyzing post-translational modifications is PROSITE. PROSITE, a database you can find on the ExPASy site, contains a list of short sequence motifs (also named patterns) that have been experimentally shown to be associated with some biological property. Many of these patterns are associated with post-translational modifications. On the ExPASy server (www.expasy.ch), you're able to compare your protein sequence with the collection of patterns contained in PROSITE. This way, you can find out which modifications your protein is likely to undergo.

When you do sequence analysis, there is something you must ALWAYS remember: Similar short sequences (for instance, less than 20 amino acids) don't ALWAYS have the same function. This means that if a small sequence has been shown to be an ATP binding in a protein, and if the protein you're analyzing contains a short segment with EXACTLY the same sequence, it doesn't NECESSARILY mean that your protein is also an ATP binding protein. It's only an indication that it *may* be an ATP binding protein. Of course, the longer the segment, the stronger the indication.

This warning on the meaning of short similarity regions also applies to PROSITE patterns — patterns that that are often quite short. We now show you how to check if your sequence contains any known PROSITE pattern.

Looking for PROSITE patterns

ScanProsite is a server that allows you to compare your protein with the list of patterns contained in the PROSITE database. Highly trained specialists designed each pattern in this database. If you find that your protein contains a PROSITE pattern, this can give you a pretty clear indication of its function.

The PROSITE patterns

When biologists first started having access to protein sequences, one of their first discoveries was that some small conserved sequences are often associated with important properties, such as cellular localization, ligand binding, and so on.

Amos Bairoch, the creator of SWISS-PROT, started exploiting this discovery while building his protein database. To help organize and annotate the proteins, he created a collection of small, well-conserved segments that he could use to classify and analyze new proteins. These special segments are known as *patterns* and they are still widely used today as a means of characterizing new proteins. PROSITE is the name Amos gave to this particular pattern database. These days, PROSITE no longer contains just patterns; it also includes a new type of sophisticated models named *profiles*. Profiles are more sophisticated than patterns because they describe every position of an entire protein family and not just a few highly conserved positions as patterns do (these domain profiles are very different from the property profiles we also describe in this chapter).

A pattern describes a small group of amino acids that constitute the signature of a function, a post-translational modification, a domain, or a protein family. Although these amino acids don't constitute the whole domain, they do provide you with a means of identification, just like when you recognize a Rolls Royce with a quick glance at the silver lady on its hood. Here is an example of a pattern:

$$[RK]-x-[ST]$$

A sequence that contains (gurus say *matches*) this pattern, must contain an arginine (R) or a lysine (K), followed by any residue (x), followed by a serine (S) or a threonine (T). This segment must be three amino acids long and can occur anywhere in the sequence. For instance, RGT, KCS, and KET are just some of the segments that match this pattern. The documentation that comes along with this pattern (the pattern documentation, or PDOC) tells us that it's characteristic of proteins that are phosphorylated with cAMP-dependent protein kinases.

Unfortunately, this pattern is so short that it occurs very often just by chance, even in proteins that are never phosphorylated. The consequence is that such a short pattern can only be helpful as a mean to strengthen other experimental data. It means nothing on its own.

Other patterns can be much more informative, especially when they are longer. For instance, the pattern for the prokaryotic C4 Zn-Finger

$$C-[DES]-x-C-x(3)-I-x(3)-R-$$
$$x(4)-P-x(4)-C-x(2)-C$$

This pattern is very sensitive: It allows the identification of ALL the proteins that contain this type of Zn-Finger. This pattern is also highly specific because it doesn't match any of the proteins that don't contain a Zn-Finger. It means that if this pattern matches your protein, you can be almost certain that it contains a Zn-Finger.

Today, patterns tend to be replaced with something more sophisticated: the profiles. Yet, patterns are so easy to formulate and to use that many people keep using them. The nice thing about patterns is that they enable you to test simple ideas without spending five years studying bioinformatics.

Here's how you can get ScanProsite working for you:

1. **Point your browser to** www.expasy.ch/tools/scanprosite/.

 The ScanProsite page of the ExPASy site appears. If you scroll down the page, you'll see two distinct sections:

 • The right section in Figure 6-9 is the place to go if you already have a pattern and you want to check whether other proteins in SWISS-PROT contain this pattern.

 • The left section in Figure 6-9 is up your alley if you have a protein sequence and you want to find out which PROSITE pattern it contains.

2. **Paste your sequence or enter its accession number into the search text box on the left (**in the Scan a Protein for PROSITE Matches **section).**

 We want to go from a protein sequence to a pattern, so we're staying with the search box on the left. You can use

 • Raw format (amino acids only, space and numbers are tolerated).

 • An accession number. For this example we use the SWISS-PROT protein P12259, which is the precursor of the Human Coagulation factor V.

3. **Put a check in the Patterns check box.**

 Other options here include profiles and rules. Rules are very short patterns that come along with guidelines. The current version of PROSITE only contains a few rules (less than 10).

Figure 6-9:
The
ScanProsite
Interface on
the ExPASy
Server.

For the purposes of this steps list, deselect the profile and the rule boxes. You can also use profiles, but when it comes to post translational modifications, patterns are usually more interesting than profiles.

4. **Press the Start the Scan button and wait.**

 An intermediate page appears. Keep waiting until your browser displays a Results page like the one in Figure 6-10.

 Wait for the results! Don't press any of the hyperlinks that appear on the intermediate page. The ScanProsite server can be very slow at some times of the day.

 If your results never arrive, you have the alternative of using any ExPASy site that is closer to you or running in a part of the world that is sleeping. (See the listing at the end of this chapter for alternative ExPASy addresses.)

5. **Choose File⇨Save As from your browser's main menu to save your results.**

Interpreting ScanProsite results

The main problem with most post-translational patterns is that they are short. The short sequences these patterns match may occur by chance and could have no real biological significance.

The rule is that similarities between short sequences must always be treated with suspicion (see the sidebar "The PROSITE patterns"). A good way to rule out many unlikely modifications is to read carefully the documentation associated with the pattern (the PDOC).

Understanding the ScanProsite output

The pattern documentation for each pattern (the PDOC) is available via the hyperlink displayed in the output. A portion of this output is shown in Figure 6-10. For each pattern found in PROSITE, ScanProsite reports:

- ✔ **PDOC#####:** The first hyperlink is the PDOC — short for pattern documentation — a high quality documentation that contains a lot of relevant information on your pattern and its potential biological function.

- ✔ **PS####:** A link to the PROSITE entry that corresponds to your pattern. This entry contains mostly technical information about the method used to derive the pattern.

- ✔ **The pattern name, in bold characters.**

- ✔ **A warning:** If your pattern is so short that it is likely to be irrelevant, a warning is issued here, similar to the warning you see for the AMIDATION site in Figure 6-10.

```
1810 -  1815   GMiySL
1867 -  1872   GSfkTL
1912 -  1917   GLstGI
1942 -  1947   GGsyNA
1943 -  1948   GSynAW
2020 -  2025   GNsdAS
2060 -  2065   GCevNG
2071 -  2076   GMenGK
2180 -  2185   GNtnTK
```

>PDOC00009 PS00009 **AMIDATION** Amidation site [pattern] [Warning: pattern with a high probability of occurrence].

```
 633 -   636   yGKR
```

>PDOC00076 PS00079 **MULTICOPPER_OXIDASE1** Multicopper oxidases signature 1 [pattern].

```
 304 -   324   GkWiIsSltpkhlqaGMqayI
1880 -  1900   GwWlLnTevgenqraGMqtpF
```

>PDOC00988 PS01285 **FA58C_1** Coagulation factors 5/8 type C domain (FA58C) signature 1 [pattern]. [PDB: 1czv 1czs 1czt]

```
1947 -  1980   AWsveklaaefaskp..WIqVDmqkeviItgIqTQG
2111 -  2140   AWqakannnkq......WLeIDllkikkItaIiTQG
```

>PDOC00988 PS01286 **FA58C_2** Coagulation factors 5/8 type C domain (FA58C) signature 2 [pattern]. [PDB: 1czv 1czs 1czt]

```
2045 -  2061   Ptraynrpt..LR1ELqGC
2205 -  2221   Pktwnqsit..LR1ELfGC
```

122 hits with 10 PROSITE entries

| ExPASy Home page | Site Map | Search ExPASy | Contact us | SWISS-PROT | PROSITE | Proteomics tools |

Figure 6-10:
The
ScanProsite
Output.

✔ **Structure:** Some patterns have been identified in proteins with a known 3-D structure. If this is the case with the pattern you're looking at, you can find here the PDB name of the structure(s) and a hyperlink to the 3-D representation of this structure. PDB stands for *the Protein Data Base*, a database that contains all the known protein 3-D structures. (See Chapter 11 for more information on the PDB). If you click one of these hyperlinks, a static GIF image of the corresponding structure appears in your browser. Figure 6-11 shows the image you can obtain by clicking the 1czv hyperlink (or use the address www.expasy.ch/cgi-bin/view-pdb?pdb=1czv&ps=PS01285). The image contains

- A 3-D structure of your protein using a ribbon representation

- The location of this particular pattern (marked in green) in the structure as a whole

✔ **The list:** A list of the segments that contain the patterns within your protein. The numbers indicate the position of the match within your sequence. Capital letters indicate residues that were specified by the pattern; lowercase letters indicate residues that weren't specified by the pattern.

Figure 6-11:
Localization
of a
PROSITE
pattern on
the 3-D
structure of
a protein.

Being careful with short patterns

The complete output obtained from P12259 gives us a good illustration of the irrelevance of most short patterns: This output indicates that 19 sites of myristillation exist in the Human Coagulation Factor V. Is this information you can trust or not?

If you read the corresponding PDOC (PDOC00008), you find that myristate is a fatty acid that is attached to the N-terminus (left end of the sequence) of a protein to anchor it into the membrane. None of the hits reported here are on the N-terminus. So you can safely conclude that none of these matches are genuine.

Using the species information

When you find a post-translational modification, make sure that this modification is consistent with the species the sequence comes from. The modifications that occur in prokaryotes are usually different from those that occur in eukaryotes. You must look for this information in the PDOC corresponding to your pattern. For instance, myristillation only happens in eukaryotes. If you find a myristillation site in a protein that comes from a prokaryote, you can safely ignore it.

Weak signals add up to give strong signals

Further along the output that ScanProsite returns, the pattern list contains two matches that are specific of coagulation factors: FA58C_1 and FA58C_2.

If you look at them individually, neither of these two patterns are really significant. On the other hand, the fact that you find two related patterns at so close a distance is very exciting: It's a good indication that both of these patterns are probably genuine.

This is a good example of the fact that two weak indications can make up for strong evidence when they are consistent (and independent!). It's a bit like two shortsighted witnesses telling you exactly the same thing about what they saw. For an investigator, this is even more interesting than the report of one eagle-eye witness, as long as the investigator can be sure that the two shortsighted witnesses never had a chance to talk with each other.

Eliminating weak patterns

When you have too many weak patterns, a good strategy is to build a multiple sequence alignment with related sequences. A pattern that corresponds to a genuine post-translational modification is normally well conserved. We provide you with a brief session on multiple alignments in Chapter 2 and most of what you need to know to build high quality multiple sequence alignments is in Chapter 9.

Everything is not in PROSITE

Also important to know is that the PROSITE pattern collection doesn't contain a description for every known post-translational modification. If you can't find what you need in PROSITE, the ExPASy server gives you links to other tools that are specialized in this type of analysis. (The links are located in the Post-Translational Modification Prediction section of PROSITE at `www.expasy.ch/tools/#ptm`.)

Finding Domains in Your Protein

Gurus define domains as independent globular folding units. For the rest of us, a *domain* is a portion of protein that can be active on its own if you remove it from the rest of the protein. A domain consists of at least 50 amino acids. Domains are like the various components of your kitchen: the oven, the microwave, the fridge, and so on. Altogether they constitute the complete kitchen, but they can also exist separately: You're allowed to remove the microwave from your kitchen if you just have to have that late-night popcorn in your lab.

Your average protein consists of two or three domains. Figure 6-12 is a snapshot of the TolB, a protein that is made up of two domains. Usually, each domain plays a specific role in the function of the protein. It may interact with other proteins, it may bind an ion like calcium or zinc, or it may contain an active site. It is common to have a catalytic domain associated with a binding domain and a regulatory domain. Imagine, if you will, a toaster, where you have the grill (catalytic), the toast holder (binding), and the switch (regulation).

Figure 6-12:
The TolB
protein is
made of two
domains.

A small fraction of all the existing domains have been experimentally characterized. We know what they do and how they do it. If you find such a domain within the protein you're interested in, this may give your research the boost it needs, even if, as a whole, your protein doesn't look like any other known protein.

Around the world are eight major collections of domains and three servers to search them. In principle and in practice, little difference lies between searching patterns and searching domains in your sequence.

Choosing the right collection of domains

Domains are so useful and important in biology that biologists have been trying to build nice collections of them for quite some time already. The problem is that a) it can be tricky to define a specific domain unambiguously, and b) gurus often find it hard to agree with each other. So now you know why various authors have described most of the important domains in slightly different manners.

The protein world: Kingdom or condominium?

The discovery of domains in the 1980s was a clear demonstration of something everybody had strongly suspected from the time Darwin started taking vacations in the Galapagos Island: **Nature is smart but lazy!** You can be sure that if there's a way to turn an old thingy into a brand new gizmo it needs, nature will find it!

Protein domains are a very nice illustration of this principle. Probably no more than a few thousand different types of domains exist. Nature builds new proteins with this collection like you used to play with your Lego bricks. Each type of domain is a special kind of brick characterized by a very stable shape. The main difference between a Lego brick and a domain is that evolution can easily modify an existing domain to accommodate a new function. These modifications don't change the overall fold of the domain, but they involve mutations that can partly change some of the domain properties — its binding affinity, for example.

For us, the consequences are that there's a large redundancy among the eight major collections of domains available today (Table 6-1). Life would be simpler if we could tell you that TastyDom (r) is the best and that you must not use PooPooDom (r)! Unfortunately, this isn't the way it works. Each collection has its pros and cons and eventually, to really understand your protein, you need to look at each collection! It takes time but, in the end, it can truly be worth the effort.

No strong methodological difference lies between all these collections. What makes each of them special is the know-how of their curators. Their diversity reflects the diversity of biology and the various interests of those who maintain them. Using only one of these collections would be like always eating in the same restaurant: even if the food is good, you miss something.

The InterPro project has made it possible to search simultaneously many domain databases. In Table 6-1, (IP) indicates the domain databases that can be accessed by InterPro. InterPro helps you make sure that you're not missing even the tiniest bit of information. You can also find out easily when some domain collections disagree on your sequence.

Unfortunately, InterPro is not exhaustive. The only way to make complete analyses of the domains contained in your sequence is to use the three major domain servers:

- ✔ InterProScan
- ✔ CD server
- ✔ Pfscan

It is highly probable that one (and only one) of these servers contains *the* bit of information that helps you make sense of your protein — meaning that you'll have to check out each server in turn. Throughout the rest of this chapter, we show you how to use each of these servers and how to get the best out of each.

Table 6-1	The Main Domain Collections		
Name	*Web Address*	*Size*	*Generation*
PROSITE-Profile (IP)	www.expasy.ch/prosite	1554	Manual
PfamA (IP)	www.sanger.ac.uk/Software/Pfam	3800	Manual
PfamB	www.sanger.ac.uk/Software/Pfam	88000	Automatic
PRINTs (IP)	www.bioinf.man.ac.uk/dbbrosers/PRINTS	1700	Manual
PRODOM (IP)	prodes.toulouse.inra.fr/prodom/doc	271051	Automatic
SMART (IP)	smart.embl-heidelberg.de	639	Manual
TIGRFAM (IP)	www.tigr.org/TIGRFAMs	1415	Manual
BLOCKs	www.blocks.fhcrc.org	2101	Automatic

Finding domains with InterProScan

You want to use the InterProScan server if you have a protein sequence and you want to know which domain this sequence may contain. InterProScan allows you to compare your sequence with InterPro, a domain database that includes most of the major domain collections available online. In a way, InterProScan is for protein sequences and domains what metasearch engines are for the WWW!

Here's how you can get InterProScan working for you:

1. **Point your browser to** www.ebi.ac.uk/interpro/scan.html.

 The InterProScan home page appears.

2. **Enter your sequence in the search text box.**

Most legal formats are recognized here (including the raw format), but the accession numbers don't work.

In the following example, we use the protein sequence FOSB_HUMAN. To obtain this sequence from SWISS-PROT:

a. Open a new Internet browser window.

b. Open the page `www.expasy.ch/cgi-bin/get-sprot-fasta? FOSB_HUMAN`, which contains FOSB_HUMAN in format FASTA.

c. Copy the sequence.

d. Paste your sequence in the InterProScan search text box, as shown in Figure 6-13.

3. **Set your Query mode by selecting the Interactive Run or the By Mail Server option.**

The query mode specifies how you want the server to return your results:

- *Interactive Run* means your results will be returned in your browser window.

- *By Mail Server* means your results will be returned at the address indicated in the e-mail address field

Whenever possible, choose the Interactive Run option. Use the By Mail Server option only if InterProScan fails to return a result. Before taking this step, however, make sure that you provide a sequence with the right format.

Figure 6-13:
The
InterProScan
home page.

4. Enter your e-mail address in the Email Address field.

The server requires an e-mail address, even if you're using the interactive mode.

5. Click the Submit Job button.

An intermediate page pops up that informs you of how long your search has been running.

Do not press any button while you wait; the results are automatically displayed using SRS (the Sequence Retrieval System), when ready.

6. Save your results.

DO NOT use the Save Button that appears on the left part of the screen within the result window (Figure 6-14).

DO NOT press the Printer Friendly button if you want to save your results into a file.

To save your page:

If you are using Netscape, use the Select File⇨Save As option in your browser.

If you are using Internet Explorer:

a. Select File⇨Save As.

b. From the Save As Type pull-down menu, choose Web Page, Complete (*.htm, *.html).

Figure 6-14: Output of the InterProScan Server.

Internet Explorer produces two objects:

- A file named Query Result.html
- A directory name Query Result

The directory contains all the images needed to display the .html file. Never separate the directory and the .html file.

If you have a PDF printer installed, you can print the result page to a PDF file. Remember to select the landscape format on the Paper Size tab in the Page Setup dialog box (File⇨Page Setup) before you print by choosing File⇨Print.

Interpreting InterProScan results

Because InterProScan returns a large amount of information, it's something of an art form to be able to interpret its results correctly. Before clicking any hyperlink, make sure that you understand exactly what the result page contains.

Understanding the InterProScan output

Figure 6-14 shows the results you obtain when scanning the protein FOSB_HUMAN with the InterProScan server. It illustrates well the variety of results one may expect when comparing a sequence with domain databases.

The left part of the output (where the Link, Save, and View buttons are) has to do with SRS. SRS (Sequence Retrieval System), which we explain in Chapter 2, is the program used by the InterProScan server to store its information. You don't need to know much about SRS to use InterProScan.

In the output, each line represents a match with one domain from one of the domain databases. The information reported includes the following:

✔ A large (yellow) box that indicates the type of diagnosis provided: Family or Domain. Some domains or signatures are specific of a complete protein family while others are simply specific of a domain. When several domains from different domain databases describe the same thing, the InterPro database groups them in the same box, like the IPR 004827 domain.

✔ The IPR#### hyperlink points to the InterPro documentation. The InterPro documentation is an attempt to summarize the information spread in the documentations of various domain databases. It's a good idea to always read the InterPro documentation AND the individual documentations.

✔ In front of each line is a hyperlink that can take you to the domain entry in its database. For instance, if you click PS00138, your browser takes you to the entry 138 of the PROSITE database, where you can find the individual PROSITE documentation.

✔ On every line, little colored boxes show you where the match occurred on your sequence. Their size is proportional to the size of the domain. If you pass the mouse pointer over these rectangles, the coordinates of the match (in your sequence) appear on-screen.

✔ The name at the end of each line is the name of the entry in the corresponding database.

✔ The name in the last column is the first line of the IPR documentation. It gives you some indication on the biological meaning of the match.

Making sense of the inconsistencies in the InterProScan output

You can see in Figure 6-14 that all the domain databases agree on the presence of a series of Leucine zippers that are roughly in the middle of the sequence.

On the other hand, none of the domain databases agrees *exactly* on the boundaries of these matches. Deciding who is right and who is wrong is always a bit of an art, but in this case a majority vote could be a good indication. Now you know why using all the domain collections — and not just one — is so important.

Another reason why seven databases are better than one is that there's always a good chance that one of the databases may contain a domain that isn't in the others. This is especially true with exotic domains that may not have yet been included in all the domain collections.

Of course, the fact that a domain is or is not reported isn't 100 percent proof of anything (see the sidebar "A common mistake when scanning domains"). To make sure that a reported hit is genuine, we recommend taking a close look at the alignment between the profile and your sequence. Unfortunately, InterProScan doesn't let you do this, which is why you also need to use the other servers like CD search and Pfscan that we describe in the next section.

A brief history of domain collections

It all started at the end of the 1980s with Amos Bairoch and his PROSITE collection of protein patterns. Gradually, scientists realized that patterns were only the most striking features of protein domains. It is true that patterns are simple and powerful, but they don't provide a very good trade-off between *specificity* (matching only the real domains) and *sensitivity* (matching the real domain everywhere it occurs) when you use them to search for domains in a protein.

This is the reason why biologists decided to substitute patterns with a more general formulation of a domain, something that lets you know which amino acid is to be expected at every position of the domain and with which probability. This way, you can say that position 10 is more likely to be a serine than an asparagine but that it's very unlikely to ever be an alanine. On the other hand, position 40 may be just as likely to be a serine or an alanine. When you assign values to these probabilities, you have a profile like the one Gribskov described in 1987 (don't confuse these profiles with the property profiles that ProtScale produces).

A profile is the mathematical formulation of a domain (or a protein family). Gurus say that profiles are *position specific scoring schemes* and comparing them with sequences is possible. Biologists use multiple sequence alignments to build profiles. This is the reason why a profile knows the positions that are very conserved and those that are not. All the profile collections are based on a multiple sequence alignment collection.

BLOCKs, the early 1990s brainchild of Steve and Georgia Henikoff, was one of the first collections of multiple alignments to be assembled in the U.S.A. BLOCKs contains many short multiple alignments (blocks) that can be turned into profiles and used to recognize protein families. BLOCKs was soon followed by PRINTs, another similar database developed in the UK: It was also in the early 1990s that some scientists in Cambridge (UK) and in Toulouse (France), started working on Prodom, a collection of automatically generated blocks. Prodom is meant to be an exhaustive collection of domains that includes everything one can find through sequence comparison on NR (the complete nonredundant protein database). In Prodom, it doesn't matter whether a domain is associated with a known function or not. Prodom is very useful to guess the domain segmentation of your protein.

While the people in Toulouse kept doing Prodom, those in Cambridge initiated a new database: Pfam (pronounced pea-fam). Pfam is a collection of high quality profiles manually annotated (PfamA) as well as a collection of automatically generated profiles (PfamB). Today Pfam is an Anglo-Swedish-American project that provides the community with the largest collection of manually generated domains. The Pfam people also took advantage of a new type of very powerful profiles that appeared in the mid1990s: the Hidden Markov Model (HMM). HMMs aren't only used in biology, but their impact on domain analysis was so important that HMMs have become synonymous for profiles in bioinformatics.

This was also the time when the PROSITE people decided to include a profile section in their database. Today, most of the profile collections are made of HMMs. The most recent additions are SMART and TIGRFAM.

Finding domains with the CD server

The CD (Conserved Domain) server of the NCBI follows the same principle as the InterProScan. The main advantage of the CD server is that reported hits come along with a score that helps you discriminate the good from the spurious matches. On the downside, the CD server doesn't integrate as many databases as InterProScan, although it contains quite a few domains that have been contributed by the NCBI that you can't find anywhere else.

If you're interested in giving the CD server a try, here's how you'd go about it:

1. **Point your browser to** `www.ncbi.nlm.nih.gov/Structure/cdd/wrpsb.cgi`.

 You can also access this server from the main NCBI BLAST page. Point your browser to `www.ncbi.nlm.nih.gov/BLAST/` and click the <u>Search the Conserved Domain Database using RPS-BLAST</u> link. This hyperlink is roughly in the middle of the page.

2. **Paste your sequence or its identifier in the large box near the top.**

 For this example, we chose the protein FOSB_HUMAN (P53539). If you want to use this sequence: Type **P53539** in the Sequence box, as shown in Figure 6-15.

3. **Deselect the Low Complexity check box, located under the Advanced Options for the BLAST Server heading.**

 In many interesting domains, an amino acid is over represented. For instance, there are more leucines than expected by chance in the leucine zippers, or more glycines than expected by chance in the glycine-rich domains, and so on, for many domains.

 Because repeated residues make the sequence simpler, sequences that contain them are said to be *low-complexity* sequences. If you keep the low-complexity filtering, you may loose these domains.

4. **Click the Submit Query button.**

 The Submit Query button isn't at the bottom of the page, but near the top of the page.

5. **Save your results.**

 To save your page, if you are using Netscape, select the File⇨Save As option. If you are using Internet Explorer:

 a. Select File⇨Save As.

 b. From the Save As Type pull-down menu, choose Web Page, Complete (*.htm, *.html).

Figure 6-15:
The CD
search
server at
NCBI.

Interpreting and understanding CD server results

The principle of the output from the NCBI CD server is very similar to the output of the BLAST program, which we cover more fully in Chapter 7.

The CD server output can be divided into 3 sections:

✔ **The Graphic:** The Graphic display, as shown in Figure 6-16, shows the regions of your protein that match a domain:

• Red domains are from SMART.

• Blue domains are from Pfam.

• Ragged ends indicate partial matches.

If you pass the mouse pointer over a domain, its complete name appears in the little window along with its score. The score is an *E-value*: It tells you how many times you can expect a hit that good by sheer chance only, given your sequence and the database you scanned. You can interpret your score by using the following rules:

- The lower the E-value, the better the score and the more likely your hit is to be genuine.

- E-values need to be below 0.01 to mean something.

- The E-value may need to be much lower than 0.01 if you remove the low-complexity filtering (Step 3).

- Treat ragged end matches with suspicion, especially if they occur in low-complexity regions. In our example, none of the ragged end matches is significant.

✔ **The Hit List:** The hit list, as shown in Figure 6-17, reports the domains that match your sequence, sorted by E-value. The best hits come first on top. If you click on the hyperlink, a page pops up that contains the documentation associated with this entry.

✔ **The Alignments:** The alignments come at the bottom of the CD server output. The top sequence is your sequence (query) and the bottom sequence is the consensus sequence of the domain (subject).

Figure 6-16:
Graphic output from the NCBI CD server.

```
gnl|CDD|6326 pfam00992, Troponin, Troponin. Troponin (Tn) contains three su...   41.4  4e-05
gnl|CDD|6326 pfam00992, Troponin, Troponin. Troponin (Tn) contains three su...   34.5  0.006
gnl|CDD|4908 pfam03148, Tektin, Tektin family. Tektins are cytoskeletal pro...   37.6  7e-04
gnl|CDD|8526 pfam03131, bZIP_Maf, bZIP Maf transcription factor. Maf transc...   37.0  0.001
gnl|CDD|8066 pfam01576, Myosin_tail, Myosin tail. The myosin molecule is a ...   34.9  0.004
gnl|CDD|8066 pfam01576, Myosin_tail, Myosin tail. The myosin molecule is a ...   34.5  0.006
gnl|CDD|4540 pfam01496, V_ATPase_sub_a, V-type ATPase 116kDa subunit family...   34.1  0.008

•gnl|CDD|311, smart00338, BRLZ, basic region leucin zipper;

          CD-Length = 65 residues,  96.9% aligned
          Score = 77.6 bits (191), Expect = 5e-16

Query:  153  EEEEKRRVRRERNKLAAAKCRNRRRELTDRLQAETDQLEEEKAELESEIAELQKEKERLE  212
Sbjct:    1  EKDEKRRRRRERNREAARRSRERKKAYIEELERKVEQLEEENERLKKEIEQLRRELEKLK  60

Query:  213  FVL  215
Sbjct:   61  SEL  63

•gnl|CDD|5934, pfam00170, bZIP, bZIP transcription factor. The Pfam entry includes the basic region and the leucine
 zipper region.
```

Figure 6-17:
The Hit
List and
Alignments
output from
the NCBI CD
server.

A common mistake when scanning domains

Don't forget that when a domain server reports some hits between your sequence and a domain, this information always results from an alignment between this domain (profile) and your sequence. If the program believes the alignment is good enough, it will report the match. Otherwise, it will not.

Unfortunately, neither the profile nor the programs are perfect and mistakes occur. The server can tell you that your protein contains a specific domain when this isn't true, or it may fail to report a domain that is nonetheless present in your protein. If you trust the results blindly, there's always a chance you may get it wrong.

The InterProScan server isn't very helpful when it comes to avoiding that type of mistake: It

doesn't report the score of the hits and does not even display the alignments. The other servers we show you in this chapter (CD and Pfscan) give you some score along with the hits. This score has a statistical meaning and it informs you how likely it is that your match may have occurred by chance only and is devoid of a biological meaning.

Conservative interpretations (only believing very high scores) are almost always correct. On the other hand, if the score isn't so good, or if only a portion of the domain matches with your sequence, you need additional evidence to make sure that what you see is real.

Finding domains with Pfscan

The people who developed Pfscan are those who maintain PROSITE. The Pfscan server provides you with the most powerful interface available for PROSITE. Pfscan also includes some domains that have not yet been released *officially* via InterPro.

If you found nothing on the InterProScan server or on the CD server, then you've come to the right place. Pfscan is your last chance! Here's how to take advantage of it:

1. **Point your browser to** hits.isb-sib.ch/cgi-bin/PFSCAN.

 The Pfscan home page appears.

2. **Paste your sequence or its ID number into the Protein Sequence Input text box, as shown in Figure 6-18.**

 You have the choice among the following formats:

 - Sequence ID

 - Raw format (Residues only, space and numbers tolerated)

 - FASTA

 Pfscan automatically recognizes the chosen format.

 For this example, you can use the FOSB_HUMAN. To do so, type **P53539** in the Protein Sequence Input box.

Figure 6-18:
Pfscan
server at the
Swiss
Institute of
Bioinfor-
matics.

3. In the Database of Motifs section, select the domain collection you want.

PROSITE is much smaller than Pfam. Selecting PROSITE gives a much quicker answer than if you also select Pfam.

4. In the Display Matches in SEView section, select Yes.

SEView is a graphical representation of the output. It is similar to the one returned by the InterProScan server. SEView isn't selected by default, so don't forget to select the box if you want to see it!

Your browser must be able to run Java Applets in order to use SEView.

5. Click the Scan Button.

6. Print your results.

Unfortunately, the pages returned by Pfscan can be difficult to save because they contain JAVA elements that all Internet browsers do not handle equally well. The best solution here is to print your results to a PDF file.

Interpreting and understanding the Pfscan results

Pfscan provides one of the richest outputs available for domain analysis. This comes at the cost of an interface that is slightly more difficult to interpret than that provided by CD or InterProScan.

Here are some things to keep in mind when interpreting Pfscan output:

- ✔ **SEView:** SEView is only here if you select Yes in Step 4.

 SEView is a JAVA applet that makes it easy to see where the matches occur in your sequences. In Figure 6-19, each rectangle represents the matched domains. If you click one of them, the name and the score of the match are displayed at the top of the applet.

- ✔ **The Hit List:** Pfscan uses the normalized score:
 - High score means good match.
 - Only scores above 7 are considered good.

 Pfscan doesn't sort the hits according to their scores, meaning that the best matches aren't necessarily on the top. Relevant hits start with a red exclamation mark "!".

- ✔ **Match Location:** The Match location is a text representation of the SEView applet. The profile names are active hyperlinks. If you click any of them, the links take you to the profile documentation.

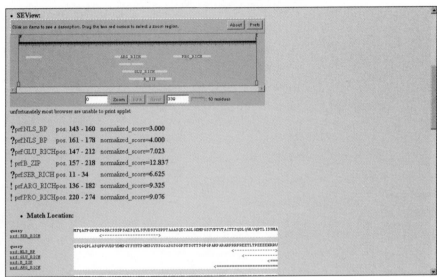

Figure 6-19:
Output of
Pfscan (top).

✔ **PROSITE Profiles representation:** For each reported match, Pfscan provides a *match detail.* This is a very original feature of this server. The match detail is a graph on which you can see immediately whether your sequence has the right residues at the right place.

To see how this works, take a look at the entry that corresponds to the B_ZIP domain, as shown in Figure 6-20.

The bars that are *above* the sequence (green) indicate the most common amino acid for the domain at this position. The higher the bar the more conserved this amino acid. When the bar is plain, your sequence contains the right amino acid at this position.

The bars that are *below* the sequence indicate positions where your sequence does NOT contain the right amino acid. The length of the bar indicates how unlikely your residue is to be at this position. A long bar means that the amino acid found in your sequence should not be here.

This representation gives you an instant idea whether key residues (such as active sites) are conserved or not in your domain.

Drawing conclusions from the Pfscan output

In Pfscan, the graphic representation is a very powerful means of making sure you're drawing the right conclusions. For instance, with the graph on Figure 6-20, we can safely conclude that our sequence contains all the right residues for having a Basic-Leucine Zipper.

Figure 6-20:
Output of
Pfscan
(bottom).

If a very conserved residue (high top bar) is replaced by a very unlikely residue in your sequence (high bottom bar), be cautious.

Note that Pfscan also returns two domains that aren't detected by the CD server or InterProScan: the Proline-Rich Domain and the Arginine-Rich Domain . A simple glance at the corresponding sequences is enough to convince us that these are indeed Proline and Arginine rich domains (see Figure 6-20). This is a good example of the advantage of using more than one domain analysis server: Now we know two more things about our protein.

More Protein Analysis for Free over the Internet

An extremely large number of resources exist over the Internet to do sequence analysis online and for free; we've listed a few in Table 6-2. The following links are only a sample of the most stable sites available to you.

In general, if your work depends on one of these sites, we suggest that you choose a very stable resource. As a rule, avoid sites that run from a personal home page (`www.something.somewhere/~somebody`) as they happen to be the less reliable.

Table 6-2	Protein Sequence Analysis over the Internet	
Name	*Site*	*Description*
ExPASy	www.expasy.ch/tools	Proteins
Pbil	npsa-pbil.ibcp.fr	Proteins
PIR	pir.georgetown.edu	Proteins
CBS	www.cbs.dtu.dk/services	Proteins
Hits	hits.isb-sib.ch/	Proteins
InterPro	www.ebi.ac.uk/interpro/scan.html	Domains
CD search	www.ncbi.nlm.nih.gov/Structure/cdd/wrpsb.cgi	Domains

ExPASy is such an interesting resource that it can sometimes become over-loaded. You may always use one of its mirrors (us.expasy.org/ for the United States, ca.expasy.org/ for Canada, cn.expasy.org for China, kr.expasy.org/ for Korea, au.expasy.org/ for Australia) when this happens.

Part III

Becoming a Pro in Sequence Analysis

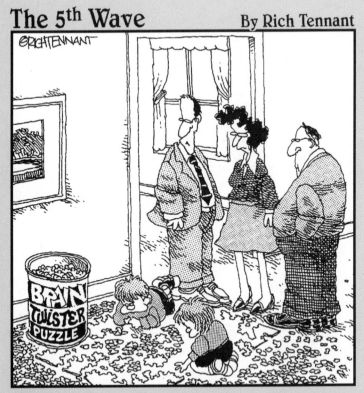

The 5th Wave By Rich Tennant

"That reminds me. I have to do some multiple sequence alignments later on."

In this part . . .

The most powerful tools used in bioinformatics rely on sequence comparison methods. In this part, we show you how you can search a database by sequence comparison using BLAST or how you can use pairwise comparison techniques, such as Dotlet, to compare two sequences. We also show you multiple alignment programs — such as ClustalW or Tcoffee — that compare many sequences at the same time. If all these things are new to you, beware: They'll forever change the way you do biology!

Chapter 7

Similarity Searches on Sequence Databases

*I*f you already have a protein or DNA sequence and you want to find other sequences that look like it, then you've come to the right place. In this chapter, we show you how to use BLAST to compare your sequence with every sequence contained in a sequence database and keep the best results. You can do this in two clicks of a mouse. We also show you situations where you can use PSI-BLAST, a more powerful version of BLAST, to ask biological questions.

On the other hand, if you only know the name of your sequence or if you can only describe this sequence with words (Mouse serine and protease for instance) rather than with its actual amino-acid sequence, then you first need to go and look for this sequence in a database. We explain how to use Entrez or SRS (the Sequence Retrieval System) to look for a sequence in Chapter 2.

Understanding the Importance of Similarity

Similar sequences often derive from the same ancestral sequence. This means that if your sequences are similar, they probably have the same ancestor, share the same structure, and have a similar biological function. This principle even

works when the sequences come from very different organisms. For you, this means that you can extrapolate something you know about a DNA or a protein sequence to every other DNA and protein sequence that is similar.

For example, imagine that your favorite sequence looks very much like another one that somebody has studied in another lab. Because these two sequences are similar, you can say, "If something is true for that sequence, it is probably true for mine as well!" Just imagine how much time you can save: Studying a gene in the lab takes years; searching a database for similarity takes seconds. And you're not even cheating!

When two proteins or gene sequences are very similar, biologists call them *homologues*, which is a fancy word for two proteins or gene sequences that have the same ancestor, similar function, and similar structure. The snag comes in deciding how similar is very similar. If your sequences are more than 100 amino acids long (or 100 nucleotides long), the rule says that you can label proteins as homologous if 25 percent of the amino acids are identical and DNA as similar if 70 percent of the nucleotides are identical. If your values are below these stated values, your guess is as good as any. Fans of Rod Serling take note: For proteins, the range of identity below 25 percent is named the *twilight zone*.

In the twilight zone, nothing is sure about the meaning of observed similarities. For instance, some proteins with less than 15 percent identical amino acids have exactly the same 3-D structure while some proteins with 20 percent identical residues have different structures. Homology or non-homology is never granted in the twilight zone.

The 25 percent figure of the twilight zone is mostly a common-sense indicator. In reality, things are slightly more complicated. In most cases, to make sure two sequences are true homologues, you also need to use some other information reported by the search. These include

- ✔ The *Expectation value* (E-value), which tells you how likely it is that the similarity between your sequence and a database sequence is due to chance
- ✔ The length of the segments similar between the two sequences
- ✔ The patterns of amino acid conservation
- ✔ The number of insertions/deletions

In the rest of this chapter, we show you how to interpret this information when conducting a database search.

The Most Popular Data-Mining Tool Ever: BLAST

Thirty years ago, biologists would search a database by printing its whole content on paper, taping it onto their office wall, writing down their own query sequence on a piece of paper, and spending a few hours manually scanning the wall and drinking coffee. With millions of sequences available, those days are gone. Fortunately, you're now able to use computers to run the most successful bioinformatics tool ever: BLAST, the *Basic Local Alignment and Search Tool*, which is expressly designed to do the paper-on-the-wall scanning for you.

In this section, we show you two simple ways to do a BLAST search (by using a protein or a DNA sequence), and, most importantly, we show you how to interpret your results. At the end of the section, we also give you plenty of ideas on how to use BLAST for doing most of the things you could need (except making coffee).

BLASTing protein sequences

BLASTing protein sequences is what you want to do if you already have a protein sequence and you want to find other similar protein sequences in a sequence database. You should be happy to know that BLAST is at its best with proteins. Two flavors of BLAST that exist and deal with proteins are

- **blastp:** Compares a protein sequence with a protein database.
- **tblastn:** Compares a protein sequence with a nucleotide database.

If you aren't sure what to do with your protein sequence, use blastp. In doubt, always use blastp! Nevertheless, if you suspect that blastp may not be the best choice, determine what you're trying to discover and see Table 7-1.

Table 7-1	Choosing the Right BLAST Flavor for Proteins
What You Want	*The Right BLAST Flavor*
I want to find out something about the function of my protein.	Use **blastp** to compare your protein with other proteins contained in the databases.
I want to discover new genes encoding simple proteins.	Use **tblastn** to compare your protein with DNA sequences translated into their six possible reading frames (three on each strand).

tblastn and the coding DNA

There are six ways to translate a DNA sequence into a protein: You have the three reading frames on the (+) strand — also named the *direct strand* — and there are the three other reading frames on the opposite strand, the *reverse strand*. (See Chapter 1 for more on the notion of DNA polarity.) These six possible translations each yield a different protein.

Now you may ask: What happens when **tblastn** compares a protein sequence with a DNA sequence? The answer is simple: **tblastn** does

EVERYTHING and automatically turns any DNA sequence into six proteins.

This way, you never need to worry about the orientation or the frame of the DNA sequences in the databases because tblastn takes care of it for you. The only thing you need to worry about is your query sequence: It must be in the 5'-3' orientation if it is a DNA sequence, or N-terminus to C-terminus if it is a protein sequence.

If your sequence comes from a database, it is always in the right orientation.

In this section, we show you how to use two of the most popular blastp online services:

- ✔ The blastp server from its home, the National Center for Biotechnology Information (NCBI) in the USA
- ✔ The blastp server from the Swiss EMBnet server

These two servers have slightly different interfaces. If you know how to use them both, you may feel at ease using any of the BLAST servers that we list at the end of this chapter.

Two good reasons for knowing how to use many BLAST servers are

- ✔ *The databases*: The BLAST servers don't give you access to the same databases. If you don't find the database you need on one server, you can always look for it on another server.
- ✔ *The speed*: The most popular servers are often overcrowded. When this happens, searching somewhere else is always an option.

It is a rarity for two different BLAST servers to return the same answer when you do the same query. The main reason is that the database versions often differ a little, although differences in the BLAST program version can have an effect as well. Results rarely change in a dramatic way, but they change a little. It's just one of those discrepancies that you must learn to live with.

Running the NCBI blastp

In this example, you are asking the following question: "Are there any proteins similar to the hamster nucleolin in the protein database SWISS-PROT?" Good question. Here's how you find the answer:

1. **Point your browser to the NCBI BLAST server at**
 www.ncbi.nlm.nih.gov/BLAST.

 The BLAST page at the NCBI appears, as shown in Figure 7-1.

Figure 7-1: The BLAST home page at NCBI.

2. **Under the Protein BLAST heading, click the <u>Standard protein-protein BLAST [blastp]</u> link.**

 A search page similar to the one in Figure 7-2 appears.

3. **Paste your sequence in the search window.**

 You have two ways to pass on a sequence to the blastp server:

 - If your sequence is already in a database, you can give its ID number to blastp. For our example, we use the hamster *nucleolin*, a protein from SWISS-PROT whose accession number is P09405. To use the nucleolin, enter **P09405** in the sequence box, as shown in Figure 7-2.

 - If your sequence is not in a database, you must provide it in the FASTA format. Figure 7-3 shows you what the hamster nucleoline sequence looks like in FASTA format.

 The sequence you give to blastp is the *query sequence*.

 Sequences similar to the query that blastp returns are the *hits* or *matches*.

 The database you search is the *target database*.

Figure 7-2:
The blastp
server at
NCBI.

4. **Choose SWISS-PROT from the Choose Database pull-down menu.**

5. **Deselect the Do CD-Search box.**

 CD search is a feature for searching Conserved Domains. We recommend you deselect this box if this is one of your first BLAST searches because the output may be confusing. We explain CD search in detail in Chapter 6.

6. **Click the BLAST! button (and wait).**

 An intermediate page similar to Figure 7-4 appears.

 Sometimes the NCBI BLAST server gets very busy. Fortunately, NCBI is not the only BLAST server around; you can always try your luck on another server. (See Table 7-6, at the end of this chapter, for a listing of BLAST servers around the world.)

Enter your sequence here.

Figure 7-3:
Providing
the NCBI
server with
a sequence
in FASTA
format.

Figure 7-4:
The BLAST
intermediate
result page.

If the server nearest you is too slow, you can take advantage of the world's various time zones. The following table indicates the part of the world that's sleeping while you do your work in the morning or in the afternoon.

Your Location	*Morning*	*Afternoon*
USA	Japan	Europe
Europe	USA	Japan
Japan	Europe	USA

7. Click the Format! button on the intermediate page and wait for the results.

When you click the Format! button, a new browser window opens. As soon as the search is complete, BLAST displays your results in this new window. Unfortunately, the search may take more time than the form indicates.

A typical search against NR or SWISS-PROT takes a few minutes. If BLAST tells you that the completion of your search may take more than 20 minutes, you're probably better off trying your luck some other time or on another BLAST server.

DO NOT press any button while you wait!

If you get no reply, DO NOT resubmit the same query several times in a row — it will only make things worse for everybody (including you)!

8. Save your results.

Do not try to print the results; the alignment section may be HUGE. If you want to keep a trace of this search, save it in a file by using the File⇨Save As option of your browser.

Unfortunately, saving a complete NCBI BLAST page isn't easy. Saved pages are normally unable to redisplay the active graphic display when you reload them in your browser. Here are two (still rather unsatisfactory) solutions:

- **Save the graphic display:** Pass the mouse over the graphic display, right-click, and then choose Save Picture As from the context menu that appears, as shown in Figure 7-5. The saved image is in GIF format. Although you can then include the display in electronic documents, be aware that all active hyperlinks are now lost.

- **Print your output to a PDF file:** If you have a PDF printer installed, you can print your output to a PDF file. Depending on the PDF printer you use, hyperlinks may be lost.

Figure 7-5:
Saving the
NCBI
graphic
display.

Running the EMBnet blastp

The BLAST server running on the EMBnet-ExPASy server is similar to the one running at NCBI, but it uses a different interface. This interface is very similar to the one used by many BLAST servers around the world. If you know how to use the EMBnet blastp, you know how to use almost any blastp server in the world.

The main difference between the NCBI blastp interface and the EMBnet interface shown in Figure 7-6 is that the EMBnet blastp interface gives you many more choices. The downside is that you are the one who has to make sure that the various selected parameters are compatible. In the following steps, we show you how to avoid any confusion.

1. **Point your browser to** `www.ch.embnet.org/software/bBLAST.html`.

 The EMBnet Basic BLAST page appears.

2. **Choose blastp from the Select the Program pull-down menu.**

 The pull-down menu lets you select the flavor of BLAST that you want to use.

3. **After having made sure that you've selected the Protein Databases radio button, select the protein database that you want to work with.**

Figure 7-6:
The EMBnet
BLAST
interface.

[Figure showing the EMBnet BLAST interface form with the following visible fields:]

Please select the program: blastp — *Program*

Please select the database:
- ○ DNA databases: Please select
- ⦿ Protein databases: SwissProt
- ☑ Gapped alignment on/off: blosum62 — *Select matrix*
- ☑ BLAST filter on/off: SwissProt ID or AC — *Select format*
- ☑ Graphic output on/off: Nucleolin — *Query title (option)*
- Paste your sequence here: (or ID or accession number) P09405
- E-mail address ☐

The database you select must be compatible with the BLAST program that you choose in Step 2. The EMBnet server does not check this for you. The kind of database you can search with a protein sequence as a query depends on the flavor of BLAST that you selected in Step 2:

BLAST flavor	Database type
blastp	Protein
tblastn	DNA

Selecting a genome database can make it easier to analyze your results. Genome databases are those with the word genome in their names, such as C.Elegans Genome. This can be especially useful if your query belongs to a very large family. If you know which specie you're interested in, choose it there. If you can't find it, use the advanced version of this server:

```
www.ch.embnet.org/software/aBLAST.html
```

If you can't find the database you need there either, then look for it on other BLAST servers like the one from the European Bioinformatics Institute.

4. **Choose SWISS-PROT ID or AC from the Select Format pull-down menu.**

You have the choice between several formats: Plain text is a FASTA format without the first line (the one that starts with >). All the other formats refer to accession numbers, not sequences.

5. **Enter the ID number (or the sequence itself) in the sequence box.**

 We're using the P09405 as our ID number. If you have a sequence, paste it from the Clipboard into the same window.

6. **Click the Run BLAST button.**

7. **Save your results.**

 BLAST outputs can be HUGE: Never print them (unless you plan to rid your whole department of its stock of paper).

 When saving BLAST results, preserving the graphic display is notoriously difficult. If you want to keep this display, save it separately: Pass the mouse pointer over the image, right-click, and then choose Save Picture As from the context menu that appears.

 If you have access to a PDF printer program, print your result into a PDF file; this is the best way to keep a printable electronic record.

Understanding your BLAST output

All the flavors of BLAST return a similar output. This output is rather complicated and can be confusing even for experienced users. Figure 7-7 shows you an overview of a complete BLAST output with its four distinctive sections. If you understand everything in this output, we think you can call yourself a bioinformaticist and let others suspect you have magic powers!

An overview of the BLAST output

You'll find four sections in the output of most BLAST servers. These sections always appear in the same order, shown in Figure 7-7, and include

1. *A graphic display:* Shows you where your query is similar to other sequences. Depending on the server you use, this display can change a lot. It may also be absent on some servers.

2. *A hit list:* The name of sequences similar to your query, ranked by similarity.

3. *The alignments:* Every alignment between your query and the reported hits.

4. *The parameters:* A list of the various parameters used for the search.

Each of these elements contains a lot of information. Knowing what matters in each of these displays can be very useful to your research.

Figure 7-7:
Schematic
represent-
ation of
the main
components
of a BLAST
output
(does not
include the
parameters
section at
the bottom).

```
                                        Color Key for Alignment Scores
                                <40      40-50     50-80     80-200    >=200
              1_16890
                        0      100      200      300      400      500      600      700

                                                                         Score      E
Sequences producing significant alignments:                             (bits)  Value

gi|12084110|pdb|1FJE|B  Chain B, Solution Structure Of Nucle...          300    9e-82
gi|11514580|pdb|1FJ7|A  Chain A, Solution Structure Of Nucle...          162    4e-40
gi|11514582|pdb|1FJC|A  Chain A, Solution Structure Of Nucle...          148    5e-36

                                   Alignments

>gi|12084110|pdb|1FJE|B   Chain B, Solution Structure Of Nucleolin Rbd12 In Complex With Snre
              Rna
              Length = 175

 Score =  300 bits (768), Expect = 9e-82
 Identities = 167/171 (97%), Positives = 168/171 (97%)

Query: 300 VEGSEPTTPFNLFIGNLNPNKSVNELKFAISELFAKNDLAVVDVRTGTNRKFGYVDFESA 359
           VEGSE TTPFNLFIGNLNPNKSV ELK AISELFAKNDLAVVDVRTGTNRKFGYVDFESA
Sbjct: 5   VEGSESTTPFNLFIGNLNPNKSVAELKVAISELFAKNDLAVVDVRTGTNRKFGYVDFESA 64
```

The graphic display

The display in Figure 7-8 helps you visualize your results. Your query sequence is on the top. Each bar represents the portion of another sequence similar to your query sequence and the region of your sequence where this similarity occurs. Red bars indicate the most similar sequences, pink bars indicate matches that are a bit less good, and green bars indicate matches that are not impressive at all. The blue and the black bars indicate the matches that have the worst scores.

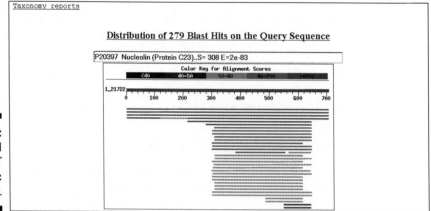

Figure 7-8:
The NCBI
BLAST
graphic
display.

Taxonomy reports

Distribution of 279 Blast Hits on the Query Sequence

P20397 Nucleolin (Protein C23)..S= 308 E=2e-83

Red, pink, and green matches are usually the good ones. Black hits are the bad hits: proteins that have so little in common with the query that their alignment probably means nothing from a biological point of view (they come from the twilight zone).

The nice thing about this display is that it can help you see that some matches do not extend over the complete length of your sequence. It is a useful tool to discover domains. For instance, here in Figure 7-8, the top hits are proteins homologous to the nucleolin. Near the bottom, the shorter hits correspond to the domain in the nucleolin that binds RNA. These hits indicate proteins that also contain this RNA binding domain but are otherwise unrelated to the nucleolin.

This display from the NCBI is active: If you pass the mouse over a bar, the name of the corresponding sequence appears in the little window at the top; if you click on this bar, the browser takes you to the corresponding alignment within this same page. (Please note that the graphical display you get from EMBnet is not active.)

The hit list

A hit list like the one shown in Figure 7-9 is the favorite part of most BLAST users. It immediately tells you whether your sequence looks like something already in the database and whether or not you can trust it as a good hit. Each line contains four important features:

```
                                                      Score    E
Sequences producing significant alignments:          (bits)   Value

gi|128843|sp|P09405|NUCL_MOUSE   Nucleolin (Protein C23)    676   0.0
gi|128844|sp|P13383|NUCL_RAT     Nucleolin (Protein C23)    627   e-179
gi|128842|sp|P08199|NUCL_MESAU   Nucleolin (Protein C23)    598   e-171
gi|128841|sp|P19338|NUCL_HUMAN   Nucleolin (Protein C23)    509   e-144
gi|128840|sp|P15771|NUCL_CHICK   Nucleolin (Protein C23)    326   1e-88
gi|464252|sp|P20397|NUCL_XENLA   Nucleolin (Protein C23)    308   2e-83
gi|12229875|sp|Q13310|PAB4_HUMAN  Polyadenylate-binding prot...  101  6e-21
gi|3183544|sp|P11940|PAB1_HUMAN   Polyadenylate-binding prote... 96   2e-19
gi|417441|sp|P04147|PABP_YEAST    Polyadenylate-binding protei... 96   3e-19
gi|129535|sp|P29341|PAB1_MOUSE    Polyadenylate-binding protei... 96   3e-19
gi|12229876|sp|Q15097|PAB2_HUMAN  Polyadenylate-binding prot...  93   1e-18
gi|1171978|sp|P42731|PAB2_ARATH   Polyadenylate-binding prote... 92   3e-18
gi|1352709|sp|P20965|PABP_XENLA   Polyadenylate-binding prote... 89   2e-17
gi|3123239|sp|P31209|PABP_SCHPO   Polyadenylate-binding prote... 87   1e-16
gi|128576|sp|P27476|NSR1_YEAST    Nuclear localization sequenc... 81   6e-15
gi|1171979|sp|Q05196|PAB5_ARATH   Polyadenylate-binding prote... 79   4e-14
gi|585638|sp|P21187|PABP_DROME    Polyadenylate-binding protei... 78   7e-14
gi|12229883|sp|Q9ZQA8|PABX_ARATH  Probable polyadenylate-bin...  74   8e-13
gi|133249|sp|P19684|ROC5_NICSY    33 kDa ribonucleoprotein, ch... 74   1e-12
gi|12643628|sp|O64380|PAB3_ARATH  Polyadenylate-binding prot...  74   1e-12
gi|417556|sp|P32588|PUB1_YEAST    Nuclear and cytoplasmic poly... 73   1e-12
gi|133247|sp|P28644|ROC1_SPIOL    28 kDa ribonucleoprotein, ch... 66   3e-10
gi|13124200|sp|Q12926|ELV2_HUMAN  ELAV-like protein 2 (Hu-an... 64   8e-10
gi|123734|sp|P26378|ELV4_HUMAN    ELAV-like protein 4 (Paraneo... 64   8e-10
gi|13124206|sp|Q60899|ELV2_MOUSE  ELAV-like protein 2 (Hu-an... 64   8e-10
gi|2500580|sp|O61701|ELV4_MOUSE   ELAV-like protein 4 (Parane... 64   9e-10
```

Figure 7-9: The NCBI BLAST hit list.

✔ **The sequence accession number and the name:** This hyperlink takes you to the database entry that contains this sequence. In this entry, you may find very important annotated information describing the sequence. A SWISS-PROT link (|sp| in Figure 7-9) may tempt you with potentially useful information.

✔ **Description:** A description that comes from the annotation lets you know at a glance whether this finding could be interesting for you. Of course, you never have a guarantee that this annotation is correct; we recommend that you carefully check the complete annotation before getting overly excited.

✔ **The bit score:** A measure of the statistical significance of the alignment. The higher the bit score, the more similar the two sequences. Matches below 50 bits are very unreliable.

✔ **The E-value (the expectation value):** By estimating the number of times you could have expected such a good match only by chance (given the database), the E-value provides you with the most important measure of statistical significance. (See the nearby sidebar, "E-values, similarity, and homology.") The lower the E-value, the more similar the sequences and the more confidence you can have that this hit is really homologous to your query. For instance, sequences identical to your query have an E-value very close to 0. Matches above 0.001 are often close to the twilight zone.

The alignments

No matter what we say about E-values and statistical significance, a number is only a number. When it comes to telling the real story, biologists only trust alignments. Biologists are convinced that alignments cannot lie, and this is mostly true — if you know how to look at them.

BLAST displays the alignments for a BLAST search just below the hit list, as shown in Figure 7-10. In every BLAST alignment, you can find the following features:

✔ **The percent identity:** This gives you a concrete substitute for the E-value. Remember the magic number: More than 25 percent identity is good news. The *Positives* field gives a measure of the fraction of residues that are either identical or similar — represented with a + on the actual alignment. The *Gaps* field shows residues that were not aligned.

✔ **Length:** This is the length of the alignment, which indicates how long are the two segments of your sequences that BLAST has aligned. It is the best way to see whether a hit means something. Sometimes, very short alignments can come up with high E-values and not be very meaningful.

E-values, similarity, and homology

A high level of similarity between two sequences often indicates that the two have evolved from a common ancestor and have the same overall 3-D structure. Biologists call these sequences *homologues*. In theory, homologue sequences often have similar biochemical functions.

In practice, if you are studying a protein, the most desirable object in the universe is a very well-characterized protein sequence that's clearly homologous to your protein. People search databases in hopes of finding this special sequence.

That's all very fine, but how do you show that your two sequences are homologous? Think of homologous sequences as relatives in a family. We all know that relatives tend to look alike, but we also know that two persons with the same eye color aren't necessarily siblings. On the other hand, if they have the same type of hair, the same facial features, and so on, we can be tempted to conclude that they are true relatives. It works the same way with sequences.

How similar must sequences be in order to be considered homologous? The answer is clear: More than *25 percent* of the amino acids present for proteins and more than *70 percent* of the nucleotides present for DNA. Above this limit, you can be almost sure that two proteins have the same structure and the same common ancestor. Below this limit lies the *twilight zone* — nobody can really be sure whether the observed similarity means anything or not in the twilight zone.

Warning: Be careful! The *25 percent* and *70 percent* limits only work for sequences that contain more than 100 amino acids or nucleotides. You might frequently get near-perfect identity between short segments (10 residues, for instance) of totally unrelated proteins or DNA sequences.

Everybody loves percent identities because they are so easy to see by eye. Unfortunately, they are

not always a good indicator. For instance, they tell you nothing when many similar — but not identical — amino acids are aligned together. Moreover, how do you tell the difference between 60 matched residues spread over a 100-residue segment, and 120 matches spread over a 200-residue segment? The longest is probably more meaningful, but the percent identity says nothing about this.

Gurus invented E-values so that we'd have a criterion more objective than percent similarity. E-values (short for *expectation values*) are a powerful tool for comparing pairwise alignments with different similarities and different lengths. They also help you decide how much you can trust your conclusion on homology. With E-values, you know when you can get excited or when you should wait and see.

The E-value has a very concrete meaning: It is the number of times your database match may have occurred just by chance. We consider a match that's very unlikely to occur just by chance to be a very good match. This is why results associated with the *lowest* E-values are the best. We say they're the most *significant* because we know we can trust them enough to infer homology.

In theory, alignments associated with E-values lower than one should all be trusted. In practice, this is not true because BLAST uses an approximate formula for computing the E-values and strongly underestimates them. In the sequence world, a similarity with an E-value above 10^{-4} (0.0001) is not necessarily interesting. If you want to be certain of the homology, your E-value must be lower than 10^{-4}.

BLAST isn't the only program that uses E-values. You may come across them almost any time that you compare two sequences and even with programs that compare sequences and domains. The principle is always the same.

```
>gi|464252|sp|P20397|NUCL_XENLA   Nucleolin (Protein C23)
         Length = 651

Score =  308 bits (789), Expect = 2e-83
Identities = 203/381 (53%), Positives = 267/381 (69%), Gaps = 22/381 (5%)

Query: 280  KRKKEMTXXXXXXXXXXXVEGSEPTTPFNLFIGNLNPNKSVNELKFAISELFAKNDLAV 339
            KRKKEM       +    ++FIGNLN  K  +ELK A+ E F+K +L +
Sbjct: 209  KRKKEM----PKTIPEAKKTKTDTASEGLSIFIGNLNSTKEFDELKDALREFFSKKNLTI 264

Query: 340  VDVRTGTNRKFGYVDFESAEDLEKALELTGLKVFGNEIKLEKP------KGRDSKKVRAA 393
            D+R G ++KFGYVDF S E++EKAL+LTG K+ G E+K+EK       K  ++KK R +
Sbjct: 265  QDIRIGNSKKFGYVDFSSEEEVEKALKLTGKKILGTEVKIEKAMAFDKNKTAENKKERDS 324

Query: 394  RTLLAKNLSFNITEDELKEVFEDAMEIRL-VSQDGKSKGIAYIEFKSEADAEKNLEEKQG 452
            RTL  KN+ ++ T +EL+E+FE+A +IR+    +DG +KGIAY+EF +E +A K LEEKQG
Sbjct: 325  RTLFVKNIPYSTTVEELQEIFENAKDIRIPTGKDGSNKGIAYVEFSNEDEANKALEEKQG 384

Query: 453  AEIDGRSVSLYYTGEKGQRQERTGKTSTWSGESKTLVLSNLSYSATKETLEEVFEKATFI 512
            AEI+GRS+ + +TGEK Q    +G    G+SK LV++NLSYSAT+++L EVFEKAT I
Sbjct: 385  AEIEGRSIFVDFTGEKSQ---NSGNKKGPEGDSKVLVVNNLSYSATEDSLREVFEKATSI 441

Query: 513  KVPQNPHGKPKGYAFIEFASFEDAKEALNSCNKMEIEGRTIRLELQ------GSNSRSQ 565
            ++PQN G+ KG+AFIEF+S EDAK+A++SCN  EIEGR+IRLE      S Q
Sbjct: 442  RIPQN-QGRAKGFAFIEFSSAEDAKDAMDSCNNTEIEGRSIRLEFSQGGGPQGGGRGGSA 500

Query: 566  PSKTLFVKGLSEDTTEETLKESFEGSVRARIVTDRETGSSKGFGFVDFNSXXXXXXXXXX 625
            SKTLFV+GLSEDTTEETLKE+F+GSV  ARIVTDR+TG++KGFGFVDF++
```

Figure 7-10: Pairwise alignments reported by BLAST.

- **The top sequence:** Your query.

- **The bottom sequence:** The hit or the subject.

- **The line in between the sequences:** Between the two sequences is a line that contains a (+) sign for similar amino acids, a letter for identical residues, or a space for mismatches.

- **The XXXXX regions:** In your query, BLAST masks automatically the regions with many identical residues with Xs. Gurus name these regions the *low-complexity segments*. They can cause troubles in the search. The Xs tell BLAST to ignore the corresponding segments. The masking occurs only in the query sequence.

- **The numbers:** Numbers to the side of the sequences indicate the coordinates of the match on your query sequence and on the hit sequence. BLAST makes local alignments that may contain only a portion of the query and the hit.

A good alignment should not contain too many gaps and should have a few patches of high similarity, rather than isolated identical residues spread here and there. This criterion is especially useful if the alignment is close to the twilight zone.

The parameters

You don't need to worry much about the meaning of the information that comes at the bottom of the result page. If you've changed the default parameters of BLAST, this part keeps track of it for you, making sure that you can reproduce this search if you need it.

Nobody can guarantee that you can exactly reproduce a result, even if you use the same BLAST server. The upgrade of any of the components in the server can modify the results of the search. Components that may be upgraded include

- ✔ The databases
- ✔ The BLAST program itself
- ✔ The default parameters of the server

In real life, you have no control over these parameters. They always end up changing with time. At least the information returned at the bottom of the search may give a clue about what could be going wrong.

BLASTing DNA sequences

If you think that BLASTing a DNA sequence is like BLASTing a protein sequence, you're right and you're not right at the same time. While it's true that the principle is the same and that in practice BLASTing DNA requires operations similar to BLASTing proteins, BLASTing DNA does not always work so well. It is faster and more accurate to BLAST proteins (blastp) rather than nucleotides.

Everybody's favorite mistake!

BLAST does local alignments. If you forget this (and we all do sometimes), you may run into big troubles. Here's the reason why: Imagine you have three proteins. The first one has two domains (A and B), the second protein has only one A domain, and the third protein has only a B domain.

To simplify, we can write the three sequences like this:

```
Sequence 1: AAAAAABBBBBB
Sequence 2: AAAAAA
Sequence 3: BBBBBB
```

Now, if you run a BLAST with Sequence 1 as a query, you will find a good hit with Sequence 2 (because of domain A) and a good hit with Sequence 3 (because of domain B). The E-values of both of these hits may be very high if domain A and domain B are long and well conserved.

So far so good . . . If you say that Sequence 1 is homologous to Sequence 2 and 3, there's nothing wrong with that (as long as you remember that sequence 1 is not homologous over its entire length to Sequence 2 or 3). If this is an exciting result, you may get carried away, forget to look at the alignments, and conclude that Sequence 2 and Sequence 3 are similar to one another because they are both similar to Sequence 1. This would be wrong of course!

In fact, this is one of those classic situations when the friends of my friends are not necessarily my friends. Gurus prefer saying that *"BLAST hits are not transitive, unless the alignments are overlapping."* Because most proteins contain more than one domain, you're wise to be careful when looking at a BLAST output — don't blindly think that all the reported hits belong to the same big family.

If you know the reading frame in your sequence, you're better off translating the sequence yourself and BLASTing with a protein sequence (see the corresponding section, "BLASTing protein sequences," earlier in this chapter). If this is not the case, then you must choose one of the following BLAST programs (see Table 7-2) that deal with nucleotides.

Table 7-2	Different BLAST Programs Available for DNA Sequences		
Program	**Query**	**Database**	**Usage**
blastn	DNA	DNA	Very similar DNA sequences
tblastx	TDNA	TDNA	Protein discovery and ESTs
blastx	TDNA	Protein	Analysis of the query DNA sequence

T is for translated. It means that you hand over a DNA sequence to BLAST and BLAST translates this sequence into its six frames (three on one strand and three on the other one). When there is a Stop codon, BLAST replaces it with an X. The six-frames translation means that you needn't worry about the DNA strand you give to BLAST. The program tries each possible frame for you anyway.

Choosing the right flavor of BLAST for your DNA sequence

Although the list of programs in Table 7-2 with all the weird acronyms can look complicated, don't let it intimidate you and remember that nothing ever breaks in the cyberlab! Just ask yourself the questions in Table 7-3 in the order listed to determine which is the best program to use.

Table 7-3	Choosing the Right Flavor of BLAST for DNA
Question	**Answer**
Am I interested in non-coding DNA?	Yes: Use **blastn**. Never forget that blastn is only for closely related DNA sequences (more than 70 percent identical)
Do I want to discover new proteins?	Yes: Use **tblastx**.
Do I want to discover proteins encoded in my query DNA sequence?	Yes: Use **blastx**
Am I unsure of the quality of my DNA?	Yes: Use **blastx** if you suspect your DNA sequence is the coding for a protein but that it may contain sequencing errors.

Blastx can correct sequencing errors for you. If you aren't sure of your protein sequence, blastx may reveal its similarity to a better-sequenced piece of DNA. If you find that your sequence contains an unexpected frameshift, don't get immediately excited. Sequencing errors are the cause of most of the observed frameshifts. You need to double-check in the lab to be sure.

Making the right choices for BLASTing DNA

From a server point of view, there's no real difference between BLASTing a protein and BLASTing a DNA sequence. The outline is exactly the same and you can follow blindly the steps in the "BLASTing protein sequences" section, earlier in the chapter. Simply remember the following points:

- **Pick the right database:** Choose a database that's compatible with the BLAST program you want to use. Unless you use blastx, it must be a DNA sequence database. Most BLAST servers do not check to make sure that you did this properly and, consequently, let you wait forever for results that never come.

- **Restrict your search:** Database searches on DNA are slower. When possible, restrict your search to the subset of the database that you're interested in. For instance, if you're interested in the Drosophila (fruit fly), search only the Drosophila genome.

- **Shop around:** Don't hesitate to shop around in order to find a BLAST server containing the database that you're interested in.

- **Use filtering:** Genomic sequences are full of repetitions. Make sure that you're using some filtering.

The BLAST way of doing things

People who do bioinformatics analysis know that *there is almost nothing you can't do with BLAST!* Whatever biological question you're asking, chances are that BLAST can give you some ideas before you start using more complicated programs or spend time designing costly experiments. In Table 7-4, we give you four examples. For each of these applications, there is an exhaustive and complicated *good* solution and then there's the BLAST way: Quick and efficient, it can be enough in most cases. Of course, if you have too much time on your hands, you can always go for the hard way.

Table 7-4	Asking Biological Questions with BLAST	
What You Need to Do	**The BLAST Way**	**The Complicated Alternative**
Finding genes in a genome	Cut your genome sequence in little (2-5kb) overlapping sequences. Use blastx to BLAST each piece of genome against NR (the Non Redundant protein database). This works better if you have no introns (bacteria).	Run gene prediction software.
Predicting a protein function	Use blastp to BLAST your protein sequence against SWISS-PROT. If you get a good hit (more than 25 percent identity) over the complete length of the protein, you've solved your problem and you know that your protein has the same function as the SWISS-PROT protein.	Conduct domain analysis or wet-lab experiments.
Predicting a protein 3-D structure	Use blastp to BLAST your protein against PDB (the database of protein structure). If you get a good hit (more than 25 percent identity), you know that your protein and this good hit have a similar 3-D structure.	Conduct homology modeling, X-ray, or NMR analysis of your protein.
Finding protein family members	Use blastp (or its more powerful cousin PSI-BLAST) and run it against NR (the nonredundant protein family). After you have all the members of the family, you can make a multiple sequence alignment (see Chapter 9) and draw a phylogenetic tree.	Clone new family members using PCR techniques.

Controlling BLAST: Choosing the Right Parameters

As they say, *power is nothing without control*. This saying summarizes fairly well how you can get the best out of BLAST: by controlling it. In this section, we show you the main parameters in BLAST, what they do, and how you can change them to suit your needs.

When using BLAST, the #1 rule to follow is to define as clearly as possible what you want to do. Treat BLAST like an ancient oracle. To get the answer you need, your question must be crystal clear. Tables 7-3 and 7-4 may help you fine-tune your questions so that they're as clear as possible.

The default parameters that BLAST uses are quite optimal and well tested. If you don't get anything meaningful with them, don't expect any miracles by changing them. Table 7-5 lists the main reasons why you may want to change the default parameters, along with the parameters you may change. These reasons are valid for DNA and protein sequences alike.

Table 7-5 Some Reasons to Change BLAST Default Parameters

Reason	*Parameters to Change*
The sequence you're interested in contains many identical residues; it has a biased composition.	Sequence filter (automatic masking)
BLAST doesn't report any results.	Change the substitution matrix or the gap penalties.
Your match has a borderline E-value.	Change the substitution matrix or the gap penalties to check the match robustness.
BLAST reports too many matches.	Change the database you're searching OR filter the reported entries by keyword OR increase the number of reported matches OR increase Expect, the E-value threshold.

Controlling the sequence masking

When BLAST searches databases, it makes an important assumption: BLAST assumes that all your sequences are *average* sequences. For instance, if you're searching protein sequences, BLAST supposes that the average protein composition of any protein is the same as the average composition of the whole database. In practice, this assumption is too far from reality to work all the time.

Masking protein sequences

Many proteins contain patches known as low-complexity (or low-*entropy*) regions. For instance, these regions can be segments that contain many prolines or many glutamic acid residues. If BLAST aligns two proline-rich domains,

this alignment gets a very good E-value because of the high number of identical amino acids it contains. Unfortunately, there is a good chance that these two proline-rich domains are not related at all. In fact, these one-amino-acid-rich domains are notorious for fooling BLAST.

In order to avoid this problem, BLAST filters out low-complexity regions when analyzing proteins. To do that, it replaces the regions with Xs in the sequence. If you're specifically interested in the low-complexity regions and don't want these regions filtered out from your search, you must deselect the corresponding Low Complexity check box next to Choose Filter in the Options for Advanced Blasting section of the blastp search page, as shown in Figure 7-11.

Imagine that you have just cloned and sequenced a protein. In the sequence of this protein, you find a stretch of 10 Prolines: **PPPPPPPPPP**. At this point, you're probably wondering whether this is a sequencing error or a mistake of some kind. You would feel much more confident if you had a valid protein sequence that contains the same stretch of amino acids.

To find this protein, write your own sequence **PPPPPPPPPPPPPPPPPP** and give it to blastp as a query. Do not forget to deselect the Low Complexity check box.

Some domains are very common in protein sequences, such as Zn Fingers in tandems or Fibronectines domains. If your protein contains this type of domain, BLAST reports many matches with proteins that contain the same domains but are otherwise unrelated. To make your search more interesting, filtering out these domains is a good idea. Here's how:

1. **Use CD search, InterProScan, or Pfscan to find domains in your protein.**

 See Chapter 6 for more information about using these searches.

2. **Read the domain documentation to find how widespread the domains you found are.**

3. **Replace the sequence of the less informative domains with Xs or rewrite it in lowercase and select the Mask Lower Case check box next to Choose Filter in the BLAST form. (See Figure 7-11.)**

4. **Run a standard blastp as shown at the beginning of this chapter.**

 When your results are returned, interpret them as shown in the "Understanding your BLAST output" section, earlier in this chapter.

Figure 7-11:
The advanced parameters of blastp at NCBI.

Masking DNA sequences

If you think keeping protein sequence results significant can be a problem, things are even worse when you're dealing with DNA sequences because DNA is full of sequences that are repeated many times, and yet such similarities don't actually mean very much when you come right down to it. Most genomes contain many such repeats, especially the human genome — 60 percent of it are repeats! If you want to avoid the interference of that many repeats, secect the Human Repeats check box that appears in the blastn page, as shown in Figure 7-12. Selecting this box will cause the human repeats to be filtered out from your sequences.

Figure 7-12:
Filtering possibilities for a DNA sequence while using blastn.

Large-scale genome sequencing is not always as smooth as gurus want us to believe. For instance, in many cases, the final genome ends up containing bits and pieces of the organism used to clone it. This is why, when looking at a human genome, you may find bits of yeast and E. coli. This simply means that before you get very excited about a gene you've just found in the human genome (or in any genome), you may want to make sure that this gene is not a contamination brought in by the organism used for the cloning. Imagine the embarrassment if you discover that the cancer susceptibility gene you've been studying for three years is in fact an E. coli protein!

Changing the BLAST alignment parameters

Figure 7-11 shows you all the parameters that you can change on the NCBI BLAST server. These include the Expect value, the Word Size, and the means of filtering. Among these parameters, two important ones have to do with the way BLAST makes the alignments; these are the *gap penalties* (named *gap cost* on the NCBI BLAST server page) and the *substitution matrix* (named simply *matrix* on the NCBI BLAST server page). If you change either of these parameters, BLAST may report different hits. The hits that are the most likely to change from one run to the next are the hits that were borderline (E-value higher than 0.0001).

The best reason to play with these parameters is to check the robustness of a hit that's borderline. If this match does not go away when you change the substitution matrix or the gap penalties, then it has better chances of being a match that's biologically meaningful.

Controlling the BLAST output

If your query belongs to a large protein family, the BLAST output may give you troubles because the databases contain too many sequences nearly identical to yours.

Sometimes, this wealth of homologous sequences can prevent you from seeing a homologous sequence less closely related but associated with experimental information. If this is something that's happening to you, refer to Table 7-5, where the last entry gives you four solutions for correcting this problem. In the following sections, we show you how to implement these solutions.

Choosing the right database

Choose a database that is suited to your needs. If BLAST reports too many hits, you may find a way around this quandary by searching SWISS-PROT rather than NR (SWISS-PROT is 100 times smaller than NR), or by searching only one genome.

Limit by Entrez query

Use the Limit Results by Entrez Query box (see Figure 7-13) or its associated pull-down menu if you get too many hits when you're interested in only one type of organism.

You can use this box to combine keywords with the two Boolean operators AND and NOT. For instance, if you want BLAST to report proteases only and to ignore proteases from the HIV virus, type **"protease NOT hiv1[Organism]"**. If you want to find more about this, see the online documentation on Entrez queries at www.ncbi.nlm.nih.gov/entrez/query/static/help/helpdoc. html#Writing_Advanced_Search_Statements

Expect

This is the cutoff for reporting hits. BLAST doesn't report hits with an E-value higher than what you indicate in the corresponding box. This parameter forces BLAST to report only good hits with a low cutoff.

Be aware that if your search yields more than 100 hits, you also need to increase the number of reported Descriptions in the Format section, as shown in Figure 7-13.

Figure 7-13: Reformatting options of BLAST.

Making Iterative BLAST with PSI-BLAST

Sometimes BLAST just isn't enough. Imagine, for instance, that you want to catch all the members of a very large protein family, starting with the one sequence that you have. When running BLAST, you catch only the most closely related sequences. How about finding the other distant cousins, the ones that your query sequence can't recognize?

PSI-BLAST does just that; it first looks for sequences that are closely related to yours and then, gradually and carefully, it extends the circle of friends to include sequences that didn't look so interesting at first but that happen to be related in the end.

Each time PSI-BLAST makes a new attempt to recruit new members of the family, it does so by using the most common properties among the members already recruited. Each new round is an *iteration*.

How PSI-BLAST really works

In standard BLAST pairwise sequence alignments, the program computes scores by reference to an amino acid substitution matrix. This matrix allocates a maximal reward for identity matches, positive weights for matching similar amino acids (leucine and isoleucine, for instance), and negative penalties for matching very different amino acids (glycine and tryptophane, for instance). More than 100 such matrices exist. The most popular are the PAM and the BLOSUM series.

The first iteration of PSI-BLAST is a simple blastp by using the BLOSUM62 substitution matrix. When this is done, PSI-BLAST analyzes the alignments position by position and deduces a new substitution matrix for each position of the query. For instance, PSI-BLAST may find that the cysteine on position 5 (in the query) is always conserved while another cysteine on

position 25 can be aligned with a serine in some of the hits. The new position-specific substitution matrix computed this way penalizes strongly substitutions on position 5 but tolerates them on position 25. For PSI-BLAST, these two cysteines are now different. In PSI-BLAST, PSI stands for *Position Specific Iterated* BLAST.

When it does the next iteration, PSI-BLAST uses the new scoring matrix that is specific of each position in the query. This way the program knows about the important positions, and it can include distant family members on the grounds that they are conserved where it really matters. After each iteration, the position specific matrix is updated. PSI-BLAST is thus simultaneously identifying increasingly distant relative sequences, and building a position-specific substitution pattern that is specific of the corresponding protein family.

PSI-BLASTing protein sequences

In the following example, we use PSI-BLAST to ask whether the leghemoglobin, a protein that sequesters oxygen in plants, is related to the human hemoglobin.

1. **Point your browser to the main NCBI BLAST page at** www.ncbi.nlm. nih.gov/BLAST.

2. **Click the <u>PSI- and PHI-BLAST</u> link (the second link under protein BLAST).**

 The PSI-BLAST page appears, as shown in Figure 7-14.

3. **Enter your sequence accession number or paste the complete sequence in the window.**

 The procedure to give a sequence to PSI-BLAST is exactly the same as the procedure blastp uses. (Refer to Figures 7-2 and 7-3.)

 - If your sequence is already in a database, you can give its ID. For this example, we use the human hemoglobin, a protein from SWISS-PROT whose accession number is P01922.

 - If your sequence isn't in a database, you must provide it in the FASTA format.

Figure 7-14:
Giving a
sequence to
PSI-BLAST.

4. **Choose SWISS-PROT from the Choose Database pull-down menu, (refer to Figure 7-14).**

 The aim of the search is to hit a SWISS-PROT sequence annotated as a leghemoglobin.

 If the first PSI-BLAST run finds no homologue to your sequence, you can use NR (the complete nonredundant protein sequence database) instead of SWISS-PROT. NR may contain intermediates between your sequence and another SWISS-PROT sequence that PSI-BLAST may catch after a few iterations.

5. **Scroll down to the bottom of the PSI-BLAST page (the format section) and choose the maximum number of hits, as shown in Figure 7-15.**

 Most databases contain a large number of hemoglobin proteins; this used to be one of the most studied proteins. You want to make sure they do not hide the leghemoglobin.

6. **Click the BLAST! Button.**

 The time it takes can be longer than what it says on-screen. Be patient!

 An intermediate page (entitled Reformatting BLAST) appears, containing a Format! button.

7. **Click this Format! button.**

 A new page appears in a new window entitled Results of BLAST. This is where your results are displayed when ready.

 Be sure to keep the Reformatting BLAST window: You're going to need it.

Figure 7-15:
Changing
the
maximum
number of
hits
reported by
PSI-BLAST.

Format

Show ☑ Graphical Overview ☑ Linkout ☑ NCBI-gi Alignment ▾ in HTML ▾ format

Number of: Descriptions 100 ▾ Alignments 250 ▾

0
10
50
100
250
500
1000

Alignment view: Pairwise

Format for PSI-BLAST ☑ with inclu... shold 0.005

Limit results by entrez query: _____ or select from: (none) ▾

Expect value range: _____

Layout: Two Windows ▾ Formatting options on page with results: None ▾

Autoformat: Semi-auto ▾

Send results by e-mail _____

8. **Inspect the results**.

There are many very similar sequences and only a few distantly related. Getting excited here is difficult; we have many hemoglobin proteins and a few myoglobins, as we might have expected.

Note in Figure 7-16 that in PSI-BLAST, the hit list is different from BLAST. It contains three new features associated with each sequence:

A check box: If this box is selected, PSI-BLAST is going to use the corresponding sequence to derive the position-specific matrix for the next iteration of PSI-BLAST.

The New symbol: It shows you sequences that PSI-BLAST reports as hits for the first time.

The green pill: It shows you sequences that PSI-BLAST has already used to obtain the current result.

9. **Click the Run PSI-BLAST Iteration 2 Button (near the top of the page).**

The Reformatting BLAST window pops up.

10. **Click the Format! button on the Reformatting BLAST window.**

The results appear in the Result of BLAST window.

The time it takes can be much longer than what it says on-screen. Be patient!

11. **Continue repeating Steps 9-10.**

The leghemoglobin appears after the third iteration, but it doesn't have a very good E-value. However, after the fifth iteration, its E-value becomes much better, suggesting that the leghemoglobin is related to the hemoglobin family.

Figure 7-16:
The hit
list in
PSI-BLAST.

Avoiding mistakes when running PSI-BLAST

In the leghemoglobin example (in the preceding section), the reason that keeps us going from one iteration to the next is the fact that the annotated sequences added by every iteration are obviously hemoglobin related. This is a sign that we are going in the right direction.

On the other hand, if after the second iteration, every new hit is an alcohol dehydrogenase, it means that it's time to give up! Unfortunately, no simple rule governs when to stop or when to continue. The only source of help is the annotation. Anything strange happening means that you must stop or that you have found something interesting.

Choosing the right sequences

The main difficulty in PSI-BLAST is deciding which sequences you can keep from one iteration to the next. It is a decision PSI-BLAST controls automatically with the E-value threshold (Inclusion Threshold in the Format section), but the check box associated with each match gives you the possibility to manually alter the automatic procedure.

This Inclusion Threshold is a mixed blessing for the following reasons:

✔ If you set this threshold too low, PSI-BLAST doesn't catch any sequence.

✔ If you set this threshold too high, PSI-BLAST catches unrelated sequences.

Catching unrelated sequences is really like catching fleas: It gets worse and worse. The first bad sequence you catch attracts other sequences unrelated to your query, and things rapidly get out of control.

The only good reason to lower the threshold is if the first iteration did not give any result whatsoever. Otherwise, a good strategy is to keep the default settings and inspect the hits that are below the threshold. After each iteration, you can also add the hits that seem to be relevant (according to their annotation) and remove those that are selected but seem to be irrelevant. It takes time and it is a bit of an art, but sometimes it pays off.

Changing the query is also a means to check your results. If you use one of the matches as an initial query, you can reasonably expect that PSI-BLAST may find your original query after a few iterations.

Avoiding the confusion of multi-domain proteins

Most proteins are multi-domains. A domain is something like a small autonomous protein that can fold on its own. The same domain may occur in very different proteins.

If your protein is multi-domain, PSI-BLAST can run into troubles because each domain competes with the others, and they may all recruit different sequences from unrelated proteins. The result you get from this is usually very difficult to interpret.

When this happens, the best strategy is to cut your protein according to its domain structure and to treat each sequence independently. You can find where the domains are by using the NCBI CD server or the EMBnet pfscan. (See Chapter 6 for more details.)

As a rule, if you didn't find any domain and if your protein is larger than 200 amino acids, cut it into smaller pieces 200 amino acids long and analyze them one by one. The snag of this strategy is that it leaves you with the boring task of putting together all the pieces of the puzzle. But sometimes, this is where the fun begins.

Similarity Searches for Free over the Internet

If you've tried to run the examples in this chapter on the NCBI Web server, you know that sometimes there's just too much traffic to do useful work. Fortunately, BLAST servers keep popping up all around the world. Those we show in Table 7-6 constitute only a sample of the most robust and up-to-date servers.

Table 7-6	BLAST and PSI-BLAST Servers around the World	
Country/ Continent	**Program**	**URL**
USA	BLAST/PSI-BLAST	www.ncbi.nlm.nih.org/BLAST
USA	BLAST	genome.wustl.edu/gsc/Blast
EUROPE	BLAST	www.ch.embnet.org/software/ bBLAST.html
EUROPE	BLAST	www.ebi.ac.uk/blast2
JAPAN	BLAST/PSI-BLAST	www.ddbj.nig.ac.jp/E-mail/ homology.html

Although it's by far the most popular, BLAST isn't the only method available for searching databases. Three main alternatives to BLAST are

- ✔ Smith and Waterman (ssearch): It's slower and arguably more accurate than BLAST.

- ✔ FASTA: It's a bit slower than BLAST but more accurate when making DNA comparisons.

- ✔ BLAT: Use this for locating cDNA rapidly in a genome or finding close (mammalian versus mammalian) proteins in a genome.

In practice, the interface for using these methods is very similar to those used for BLAST, and you should have no problem adapting the content of this chapter to any similarity-based database search over the Internet. In Table 7-7, we give you a sample of the most stable and up-to-date servers that run these services. Be aware that none of these programs has any BLAST-type facility for filtering low-complexity regions.

Table 7-7	**Alternative Methods for Homology Searches**	
Country/ Continent	*Program*	*Address*
USA	FASTA	`fasta.bioch.Virginia.edu/fasta`
EUROPE	FASTA	`www.ebi.ac.uk/fasta33`
EUROPE	SSEARCH	`www.ch.embnet.org/software/ GMFDF_form.html`
JAPAN	SSEARCH/ FASTA	`www.ddbj.nig.ac.jp/E-mail/ homology.html`
USA	BLAT	`genome.ucsc.edu`

Chapter 8

Comparing Two Sequences

● ●

In This Chapter

▶ Making a dot plot over the Internet

▶ Identifying repeats in a protein sequence

▶ Finding out about local and global alignments

▶ Finding the best local alignments between your two sequences

▶ Finding out about the gap penalty and the substitution matrix

● ●

Your true value depends entirely on what you are compared with.

— Bob Wells

*I*n this chapter, we show you some of the methods available for comparing two protein sequences — or two DNA sequences — with each other. In their own peculiar jargon, bioinformatics gurus call these analyses *pairwise comparisons* because they relate to a pair of sequences. They are extremely useful and can help you

✔ Convince yourself that two sequences are in fact homologous

✔ Find out that your sequences share a domain

✔ Identify the exact location of common features, such as disulfide bridges or catalytic active sites

✔ Compare a gene and its product

Typically, biologists use pairwise comparisons to refine database search results and carry out detailed analysis. It is common practice to use pairwise techniques after a database search in order to check whether some of the reported matches are really interesting or are just a flash in the pan.

In this chapter, we show you which method is best suited to your needs and also show you how you can use the Internet to run a method on the sequences that you're interested in.

Making Sure You Have the Right Sequences and the Right Methods

The most delicate decision you need to make when doing a pairwise analysis is the choice of the two sequences you want to compare. Choosing two sequences is a bit like arranging a boxing game between two opponents: The idea is to get the most exciting fight.

You don't use pairwise comparison methods to *discover* a sequence that would be homologous to a sequence you already have. This would take too much time. If you want to compare your sequence with every other sequence in a database, use a database search program such as BLAST. (See Chapter 7 for more on BLAST.)

If you've made your way through Chapter 7, you're now in a position to argue that database search programs merely do pairwise comparisons between a query sequence and all the sequences within a database, meaning that there is no real need to make extra pairwise comparisons, is there? Although it's true that programs like BLAST search databases through pairwise comparisons, these programs are *optimized for speed, not for alignment accuracy.* The programs we describe here are just the opposite: They are optimized for giving the most accurate possible result, which is why you want to apply them to carefully selected sequences.

Choosing the right sequences

A good reason for making a pairwise comparison between two sequences is when you have a strong suspicion that these sequences are homologous. Two sequences are homologous when they share a common ancestor. (See Chapter 7 for a complete description.) Such sequences often have similar 3-D structures and related functions.

The best way to find a sequence homologous to a sequence you already have is to make a database search with the BLAST program. Select your sequences by using the following (conservative) criteria:

- **DNA sequence:** At least 70 percent identity over more than 100 bases between the hit and the query or an E-value lower than 10^{-4}. (For more on E-values, see Chapter 7.)

- **Protein sequence:** More than 25 percent identity over more than 100 amino acids between the hit and the query or an E-value lower than 10^{-4}.

Unfortunately, these criteria are only an indication of a match, not a guarantee. If your hit has a score very close to the threshold, it may not be homologous to the query, or it may be so distantly related that aligning it correctly is difficult. This is where pairwise comparison methods come into the picture: They help you decide how meaningful a database hit really is.

Making a database search is like scanning the classified ads for a new apartment. You may find a few things interesting, but you can't be sure that you've found the perfect (or perhaps just adequate) living space until you visit. The "penthouse with air conditioning and an astonishing view" may in fact be a run-down shed with gaping holes in the wall on top of a factory building.

Choosing a sequence from a database search output is all about making the right compromise between new information and certainty: You want a sequence that's similar enough to your query so that you can be sure it's homologous. On the other hand, you want this sequence to be interesting from a biological point of view so that the comparisons reveal something new to you.

In short, the object of your desire is an extensively annotated SWISS-PROT sequence that matches your query with an E-value much lower than 10^{-4}. In real life, we often have to compromise.

No written rule says you must select your sequences from a database search. Doing so is simply very convenient. An alternative is to select your two sequences by using an experimental criterion: because they have the same function or other similar property. For this purpose, you may use the sequence retrieval system (*SRS*) that enables you to identify sequences by using keywords. (See Chapter 2 for more on SRS.)

If you have only one sequence, it is possible (and sometimes advisable) to make a pairwise comparison on this sequence alone. All you need to do is pretend that the second sequence is identical to the first one. This may seem a bit silly, but it's a good way to discover interesting features in your sequence, such as

✔ Repeated domains

✔ Regions with a small motif repeated many times (low complexity)

✔ Palindromes (portions of DNA that are repeated in different orientations) and potential secondary structures in RNA

Internal repeats can help elucidate the function of your protein. For instance, the discovery of a repeated region in the breast cancer susceptibility gene of type 1 (Brca1) made it possible to build a domain profile for this domain and to identify the involvement of Brca1 in DNA repair mechanisms. On its own, this story makes a good case for not overlooking self-comparisons!

Choosing the right method

Comparing sequences is one of the most difficult tasks biologists ever came across. In fact, the problem is not entirely solved yet, and this is why there is more than one solution. Table 8-1 lists three methods and indicates the situations for which each is best suited.

Table 8-1	Different Types of Pairwise Comparisons
Method Name	*Situation*
Dot plot	**General exploration of your sequence:** Discovering repeats Finding long insertion/deletions Extracting portions of sequences to make a multiple alignment
Local alignments	**Comparing sequences with partial homology:** Making high quality alignments Making residue-per-residue analysis
Global alignments	**Comparing two sequences over their entire length:** Identifying long insertion/deletions Checking the quality of your data Identifying every mutation in your sequences

It's a good strategy to always start with a *dot plot*. (We get into the dot-plot details in the following section.) Dot plots are very powerful because they give you an instant general picture. With a bit of experience, a simple glance at a dot plot immediately tells you what is really going on. It's an ideal tool for deciding on the next step of your analysis.

Yet, a dot plot is not enough for a fine-grained examination. Dot plots do not really produce alignments; they simply give generic indications. They don't unambiguously tell how amino acids or nucleotides correspond to one another between your two sequences.

In order to ferret out this important information, you need to make an alignment. There are two kinds of alignments: the global alignments (where the two sequences are aligned along their entire lengths) and the local alignment (where the program only aligns the most similar portions of the two sequences).

Global alignments are usually easier to interpret than the local ones. However, it only makes sense to build a global alignment if your sequences are related over their entire lengths and if they don't contain very long insertions or deletions. If you don't know how to choose between global and local alignments, always go for the local methods.

 When you're comparing two sequences, start making a dot plot of each sequence against itself. This makes it easier to identify potential repeats within each sequence. With this information in hand, interpreting the dot plot of the two sequences is much simpler.

The main difficulty when using the methods we list in Table 8-1 is the proper choice of the parameters. Unfortunately, there is no absolute rule; producing the exact comparison you need is mostly a trial and error process. In the remainder of this chapter, we show you how to play with these parameters until they produce a result you can be satisfied with.

Making a Dot Plot

Dot plots are the simplest means of comparing two sequences. In fact, the dot plot — the kind you make using pencil and paper — is the only type of sequence comparison that you could do in the days when no one had true computer access. (See the sidebar "A DIY guide to dot plots.") Dot-plot techniques are closely related to sliding window methods — where a *window* is a portion of your sequence that you compare with other sequence portions of similar size. (For more on sliding windows, see Chapter 6.) Dot plots have two major advantages: They are simple and don't require any biological hypothesis. This way, it is hard to make a mistake when making a dot plot. If you aren't sure what a dot plot really is, a simple glance at the figure shown in the sidebar "A DIY guide to dot plots" may give you an idea.

The interpretation of a dot plot relies on the most sophisticated detection device available to you: your eyes. When it comes to identifying patterns like diagonals, nothing beats the human eye, even when the signal is noisy. This is so true that anytime you see a good signal on a dot plot, you can bet your life something interesting is going on, even if the statistics are too smart to see anything!

Choosing the right dot-plot flavor

In this section, we introduce a very powerful tool named Dotlet. Dotlet is very convenient because it's easy to use, free of charge, doesn't need any installation, and runs on (almost) any computer that has access to the Internet.

Dotlet is ideal for sequences that are less than 10,000 amino acids or nucleotides long. This means that it can be helpful for most proteins but is restricted to small DNA sequences.

A DIY guide to dot plots

Making a dot plot is very much like having your sequences playing tick-tack-toe against each other. For instance, imagine that you want to compare these two sequences:

 Sequence 1: THEFATCAT
 Sequence 2: THEFASTCAT

It isn't so obvious to tell that they are almost identical, except for the extra S in Sequence 2. Of course, it gets much worse with real sequences.

To make a dot plot, take a piece of paper, draw a grid and write out Sequence 1 along the top with one letter above each column. Next, write out Sequence 2 on the left side with one letter next to each line, like in the accompanying figure.

This is where the game begins. Consider each cell one by one, and cross it if the top sequence and the side sequence contain the same symbol. When this is done, you get something very much like what's shown in the figure.

Of course, as in any tick-tack-toe game, the longest diagonals win! They indicate segments of similarity between your two sequences. The figure below also makes it very easy to see that the two sequences are homologous over their entire length, with the exception of an extra residue in the vertical sequence.

	T	H	E	F	A	T	C	A	T
T	X					X			X
H		X							
E			X						
F				X					
A					X			X	
S									
T	X					X		X	
C							X		
A								X	
T	X					X		X	

If your sequences are longer than 10,000 characters, no adequate tool is available online and you need to install a more powerful version of Dotlet on your computer. At least two such packages are available for free: Dotter and Dottup. You can find their respective specifications (and URL addresses) in Table 8-2.

- ✔ **Dotter:** The ancestor of Dotlet, Dotter is very similar in spirit to Dotlet and adapted to protein and medium-size DNA sequences.

- ✔ **Dottup:** A bit less user friendly, Dottup enables you to compare entire genomes.

Table 8-2	Various Flavors of Dot-Plot Programs			
Name	*Used For*	*Range*	*URL*	*Platforms*
Dotlet	Proteins, DNA	10,000	www.ch.embnet.org	All
Dnadot	Proteins, DNA	100,000	arbl.cvmbs.colostate.edu/molkit/dnadot/	All
Dotter	Proteins, DNA	100,000	www.cgr.ki.se/cgr/groups/sonnhammer/Dotter.html	Unix, Linux, Windows
Dottup	Complete genomes, DNA	>100,000	www.emboss.org	Unix, Linux

Using Dotlet over the Internet

Dotlet is one of the most user-friendly dot plot programs available over the Internet. It is a Java applet that you can access on the EMBnet server. An applet is a little program that loads automatically in your Internet browser and then runs on your own machine. The nice thing about it is that you don't need to install anything, as long as you have Java running (which most Internet browsers do by default these days).

Dotlet isn't a complicated program, but it contains many features. If you don't use these features properly, you may pass very close to the result you need but miss it. The general rule when using Dotlet is that you want to find a clear signal and that you should keep playing with the parameters until you get one.

In the upcoming steps list, we show you some examples of what we mean by the term *clear signal* and give some tips on how best to obtain one.

Downloading Dotlet

You can download Dotlet in one click of a mouse, simply by pointing your browser to www.isrec.isb-sib.ch/java/dotlet/Dotlet.html. When you open this page, your browser also downloads the Dotlet program. This may take a few minutes if you're on a slow connection.

When the download is finished, your browser displays a page like the one shown in Figure 8-1.

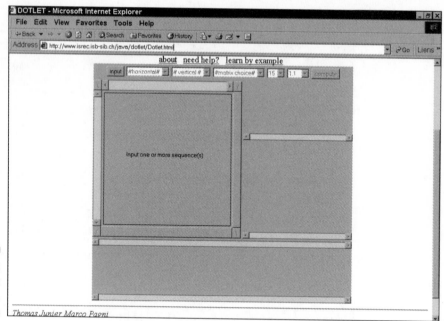

Figure 8-1:
The Dotlet
page.

If your browser does not display this page properly, it means that it can't run Java; you may need to update to the latest version of Internet Explorer or Netscape in order to get Java running on your computer.

Dotlet is not a server but a Java applet. Everything Dotlet does it does on your own computer, not on the EMBnet server. For you, this has three consequences:

✔ You can use Dotlet offline after you download it once. This is useful if you work on a laptop that isn't always connected.

✔ The speed of the program depends on your own computer. The faster your computer, the faster Dotlet runs. Differences become apparent with sequences longer than 1,000 residues.

✔ When you use Dotlet, *you are not sending your sequences anywhere!* Each sequence you compare with Dotlet stays on your computer and isn't sent to the EMBnet server. If you work in a company that's at all interested in protecting proprietary information, you know this is a VERY GOOD thing!

To make sure Dotlet isn't communicating with any server when you use it, choose File➪Work Offline from your browser menu after you've downloaded Dotlet.

Unfortunately, Mac users are out of luck when it comes to using Dotlet. If you can, try using Dotlet from a PC.

Entering your sequence in Dotlet

In the following steps list, we compare two sequences that contain a kinase domain. We're going to assume that you've already got Dotlet up and running in your browser (if not, the address is `www.isrec.isb-sib.ch/java/dotlet/Dotlet.html`):

1. **Open a new browser window by pressing Ctrl+N on your keyboard.**

 You need access to two browser windows, which is why we start with this step.

2. **Type the address** `www.expasy.ch/cgi-bin/get-sprot-fasta?P05049` **into the new window you opened in Step 1 and press the Return key.**

 Type in the exact URL, do not omit the '?'. This URL gives you a direct access to SWISS-PROT if you know the ID number of your sequence. For our example, the ID number is P05049.

 After a bit of a wait, a new window appears displaying the sequence associated with the ID number you just entered.

3. **Copy the sequence displayed in your browser using Ctrl-C.**

4. **Return to your first browser window (the Dotlet window shown in Figure 8-1) and click the Input button at the top left.**

 A pop-up window appears like the one in Figure 8-2.

5. **Paste your sequence in the pop-up window using Ctrl-V and enter a title.**

 You must enter your sequence in raw format (only the sequence) as shown in Figure 8-2.

Figure 8-2:
The sequence window in Dotlet.

Name is the title of your sequence on the dot plot.

6. **Press the OK button in the Sequence Window.**

 When you press the OK button, the window disappears; this means your sequence is now loaded in Dotlet.

 If you want to compare a sequence only with itself, go directly to Step 13.

7. **Type the address** `www.expasy.ch/cgi-bin/get-sprot-fasta?P08246` **into the new window you opened in Step 1 and press the Return key.**

This is the second sequence you want to load in Dotlet.

8. **Copy the sequence displayed in your browser.**

9. **In the Dotlet window (Figure 8-2), click the Input button.**

10. **Paste your sequence using Ctrl-V in the pop-up window that appears and enter a title.**

11. **Click the OK button in the pop-up Window.**

When you click the OK button, the window disappears; this only means that your sequence is now loaded in Dotlet — it hasn't disappeared for good!

12. **Back in your original browser window — the Dotlet window — choose the two sequences you want to compare from the two drop-down windows nearest to the Input button, as shown in Figure 8-3.**

Figure 8-3 shows you how to use the pull-down menus in order to choose which sequences you want to compare. The names in the menus are those you provided in the Title box in Steps 5 and 10.

The sequence you choose in the first pull-down menu is the horizontal sequence (top of the dot plot, left to right).

The sequence you choose in the second menu is the vertical sequence (left side of the dot plot, top to bottom).

Figure 8-3:
The sequence pull-down menus in Dotlet.

13. **In the Dotlet window, click the Compute button at the far right.**

Dotlet never takes action automatically (except when you change the threshold). Any time you change one of its parameters, you must inform Dotlet by clicking the Compute button.

The output looks like what's shown in Figure 8-4. This figure shows Dotlet's three windows:

- **The dot-plot window:** The large window on the left contains your dot plot.

- **The threshold window:** This small window on the right is the one you must use to tune the sensitivity of your dot plot.

- **The alignment window:** This is the long narrow window below that shows which pair of residues from the two sequences corresponds to each dot in the dot-plot window.

Figure 8-4:
Default
output of
Dotlet.

The default output of Dotlet is usually difficult to interpret. To make it talk some sense, you need to refine it. We show you how to do this in the next section.

Fine-tuning your Dotlet

The sequence of instructions we show here isn't necessarily optimal, but it follows a top-down approach. We start with settings that you can roughly adjust and finish with settings that you must finely tune to yield an informative dot plot.

1. Set the zoom factor.

By default, Dotlet has a zoom factor of 1:1. This means that 1 pixel corresponds to 1 residue.

If your sequences are long, you may not see the entire dot plot. This is exactly what happens in Figure 8-4. You can tell this is so because the scroll bar down the left side is *active* — you'd need to use the rectangular box in the scroll bar to scroll down to see the rest of the display. To see the entire dot plot, do the following:

a. **In the Zoom selector — the pull-down menu to the left of the Compute button — choose 1:2.**

b. **Click the Compute button.**

Dotlet now displays a complete dot plot of your two sequences, as shown in Figure 8-5.

Seeing the overall picture is useful, even if it looks a bit small. After you get a feeling for the regions that are interesting in your dot plot, you can always zoom back.

Figure 8-5:
Choosing the right zoom factor in Dotlet.

2. **Choose the window size.**

When Dotlet compares two amino acids (or nucleotides) (see the sidebar "The inside of Dotlet"), it also compares the amino acids (or nucleotides) in the immediate vicinity. The window size controls the size of this neighborhood. To use our example, you need to change the default value:

a. **Choose 51 from the Window pull-down menu (the menu to the left of the Zoom selector).**

b. **Click the Compute button.**

Long windows make clean dot plots. They make sure that two similar amino acids (or nucleotides) only yield a dot if the amino acids (or nucleotides) around them are also similar. Gurus say that long windows make the dot plot more *stringent*.

The size of a window should be within the same range as the size of the elements you're looking for. For instance, if you're looking for conserved domains in proteins, a size of 50 amino acids or higher is appropriate.

Shorter windows are more sensitive but bring some noise with them. They may help if you're looking at very distantly related protein domains. It is convenient to start with a large window and narrow it a little until the signal you're looking for appears.

After changing the window size, you MUST also tune the threshold (see Step 3). There is little point in changing the window size if you don't adjust the threshold accordingly.

3. **Adjust the threshold.**

 The *threshold* is a value that defines the color of each dot in the dot-plot window. Dots with a score above the threshold are white while those with a score below are black. The threshold is by far the most powerful and the most delicate parameter when making a dot plot. The very nice thing about Dotlet is that it lets you adjust the threshold dynamically and see the effect in real time. (You don't even have to click the Compute button!)

 On the top of the threshold window — the window on the right side for the Dotlet display in Figure 8-4 — there is a little horizontal scroll bar with a cursor. This is the *top cursor*. On the bottom of this same window, there is another small horizontal scroll bar with another small cursor. This is the *bottom cursor*.

 The horizontal axis in the threshold window represents the score. Low scores are to the left, high scores to the right. (See the sidebar "The inside of Dotlet" for more details.) The big peak on the blue curve shows you the most common score for two residues chosen randomly.

 You can imagine the threshold as two vertical lines going through the top and the bottom cursor. Its position controls the appearance of the dot-plot window (the large window on the left in Figure 8-4). In this dot-plot window:

 - Dots with a score below the bottom cursor threshold are black.
 - Dots with a score above the top cursor threshold are white.
 - Dots with a score between the two cursors thresholds are gray.

 The game is all about setting the threshold so that interesting residues appear as white dots on a black background in the dot-plot window on the left. You can use the following procedure to achieve this effect:

 a. **Drag the bottom cursor in the threshold window entirely to the right.**

 This makes the dot plot on the left entirely black.

b. **Click on the right side of the threshold window, by the middle.**

c. **Hold down your mouse button as you slowly drag the threshold to the left.**

While you do this, keep an eye on the dot plot to see a conserved segment appear, as shown in Figure 8-6.

Figure 8-6:
Dotlet with
a proper
parameter
setting.

When the top cursor and the bottom cursor are on the top of each other, dots in the dot-plot windoware black or white. If you shift one of the cursors to the left or to the right, the dots whose scores are between these two thresholds appear in gray. This can help smooth the plot and make it look nicer for publication.

If you want to see black dots on a white background, shift the top cursor slightly to the right of the bottom cursor. This inverts the color chart.

4. **Save your dot plot.**

Saving your results is the most delicate part in Dotlet. All browsers are not able to print these results, and we didn't find one able to save the display as a file. As far as we know, there are only two ways to keep a record of your display:

Print the browser content into a PDF file. To do so, you must have a PDF printer installed on your computer.

Use the ol' screen-capture technique. You can use it with any graphic display that gives you trouble when trying to save. Follow this procedure to do a screen capture:

a. **Press the PrntScrn button on your keyboard (usually top right).**

b. **In Windows, choose Start⇨Programs⇨Accessories⇨Paint to open the Windows Paint program.**

The Start button is at the bottom-left of your screen.

You can use any graphic editor you like, but Paint comes by default on any PC running Windows.

c. **Press Ctrl+V.**

d. **Click Yes when the program asks you about resizing the bitmap.**

e. **Choose File⇨Save As to open the Save As dialog box.**

f. **Type a name for your file in the File Name field.**

g. **Choose JPEG from the Save As Type pull-down menu.**

h. **Click Save.**

You can now print your dot plot or include it in any type of document.

Interpreting your results

P05049 and P08246, the two proteins we chose for the previous steps list (in the section "Entering your sequence in Dotlet"), are very distantly related. If you use BLAST to find one with the other, you get a match with an E-value of 10^{-4}. If you have been through Chapter 7 — the chapter where we introduce you to the intricacies of BLAST — you already know that getting really excited over such a match isn't a good idea. It could even be a false positive.

Knowing that P05049 is a serine protease, it would be interesting to check the protease activity of P08246 in the wet lab. The problem is that when you launch a costly experiment on the basis of such a weak BLAST match, you're definitely taking a chance. If you fail, the probabilities are high that your boss will ask the lethal question, "What in the world where you thinking of!"

This is the reason why — before doing this experiment — it would be reassuring to make sure that P05049 and P08246 are genuinely homologous *in the serine protease region*.

Making a dot plot is a good way to answer this question. A dot plot like the one we produced in the previous steps list is very much to the point. Figure 8-6 makes a good case for the existence of a conserved domain between the two proteins. You can make sure of this by looking at their SWISS-PROT annotation (accessible at www.expasy.ch). SWISS-PROT says that P05049 and P08246 both contain a serine protease domain in the region that the dot plot indicates.

TECHNICAL STUFF

The inside of Dotlet

The big questions you want answered when comparing two sequences are a) whether the two amino acids or nucleotides you're looking at are similar or not, and b) if they are similar, does this similarity actually mean something from a biological point of view?

For instance, if your two sequences contain 10 cysteines each, it's pretty obvious that each cysteine of each sequence is — at most — homologous to one cysteine in the other sequence. But who is related to whom? This is a tricky question. You could try to solve it by comparing not only the cysteines but also the residues that are around them. You would then say that two cysteines are homologous when their neighborhood is very similar. The neighborhood size has a name: It is the famous *window*. But having a window isn't enough; you also need some criteria to decide when two windows are so similar that you can consider their central amino acids (or nucleotides) to be homologous. This is exactly what the *threshold* does for you: It draws a line between the pairs of amino acids (or nucleotides) whose windows are so similar that they must appear on the dot plot, and the pairs that are not similar enough to appear on the dot plot. The nice thing in Dotlet is that the line is not drawn forever: You can shift it at will.

Using windows is quite smart, but Dotlet is even smarter than that because it enables you to guess where you should put your threshold. Dotlet gives you this information with the blue and purple curves that appear on the threshold window. The blue curve shows you the distribution of the scores of all the possible pairs of windows in your sequences. Because the majority of these window pairs are unrelated, the most common score corresponds to two unrelated windows. This is the main blue peak, which indicates the score you may expect given two unrelated amino acids and their surrounding window. Clearly, you want your threshold to be above this value (to the right of this peak).

The purple curve is a logarithmic representation of the blue curve: It magnifies small peaks and shrinks the large ones. This purple line shows you whether there is a high-scoring peak that corresponds to an unusual concentration of high-scoring window pairs in your two sequences. Such a peak is a sure sign that your sequences contain a few homologous domains. This minor purple peak only appears if your sequences contain a strongly conserved domain. If you see it, you want to set the threshold somewhere between the big blue peak and the purple one on its right side. If you don't see it, it's because this peak is often too small to be seen.

Of course, in real life you may not have a SWISS-PROT annotation for the two proteins. In fact, in the real world, you may not have any annotation at all! That should not deter you from making a dot plot. Even if you don't know its function, the identification of a conserved domain between two protein sequences is a good way to locate the specific region responsible for a function.

Doing biological analysis with a dot plot

The preceding example shows you how to spot a distantly related domain shared by two proteins. Dot plots give very characteristic representations of biological relationships. Using them is mostly a matter of recognizing a few graphic patterns and knowing which situations they correspond to.

In the following example, we show you some of these situations. For each one, we assume that you have entered your sequences in Dotlet by using a procedure like the one we describe in the section "Using Dotlet over the Internet."

Identifying tandem repeats

Proteins commonly contain a small domain repeated many times over. It seems that internal duplication is often used by evolution to create new proteins or increase the efficiency with which they perform their function. Domains that you may often find duplicated include the Fibronectin domain (we like to refer to it as molecular Velcro), EF-hands that bind calcium, or Zn fingers involved in DNA binding.

Dot plots constitute by far the best way to spot repeated domains. In Figure 8-7, we show you the YE73 human protein sequence, potential human transcription factor (involved in RNA transcription) that contains 13 very conserved Zn finger domains in tandem.

To analyze this sequence, follow these steps:

1. **Point your browser to** www.expasy.ch/cgi-bin/get-sprot-fasta? Q9P255.

 Q9P255 is the ID number for the YE73 human protein sequence.

2. **Enter the sequence into Dotlet following the procedure that we outline in the section "Using Dotlet over the Internet."**

3. **Set the threshold and window setting so that they match those in Figure 8-7.**

 You want a window size of 21, a zoom of 1:2, and a gray scale on the top and bottom cursors of the threshold window of 39%.

 The pattern on Figure 8-7 is typical of tandem repeats. Notice the following features:

 - The main diagonal represents the sequence against itself.

 - Repeats appear as long continuous diagonals above and below the main diagonal.

- The diagonals are evenly spaced.

- The collection of diagonals is bound by the shape of a square.

Figure 8-7:
Looking at
tandem
repeats
with Dotlet.

Anytime you see this type of graphic pattern, you know you're looking at tandem repeats. If this is the case, you can use this collection of diagonals to draw three interesting deductions:

✔ The number of repeats is equal to the number of diagonals above (or below) the main diagonal (including the main diagonal).

✔ The distance between two adjacent diagonals represents the size of the repeat if no spacing falls between the repeats.

✔ The shortest diagonal gives you the coordinates of a single repeat unit.

YE73 is an extreme example because it contains highly conserved repeats. Life isn't always so simple, and, in most cases, the repeated domains you come across aren't so similar because of accumulated mutations. Proteins that contain distantly related repeats produce more complicated dot-plot patterns. In these proteins, each repeat unit doesn't necessarily recognize every other unit.

Figure 8-8 shows you such an example. Tf3a is a transcription factor that also contains Zn finger domains in tandem. You can find the sequence at www.expasy.ch/cgi-bin/get-sprot-fasta?P03001.

Tf3a tells the story of Zn finger domains that have been duplicated and have diverged since then. These Zn fingers are less conserved and give a pattern much less regular than what we obtained with YE73.

Figure 8-8:
Looking at
the poorly
conserved
tandem
repeats of
Tf3a with
Dotlet.

Tandem domains are not the only form of repeated domains in proteins or DNA. It is safe to assume that any kind of repetition you can think of probably occurs in proteins or in nucleic acids. If it has not yet been found, you can bet that it is about to be discovered!

The signature of a repetition is always the same: It is a diagonal that lies off the main diagonal when you compare the sequence with itself.

Repeated domains can help you elucidate the function of your protein. If you find a repeated domain with no known function — and no similarity with any characterized protein — you can do the following:

1. **Extract each repeat unit.**

2. **Make a multiple alignment of the domains. (See Chapter 9 for more on how to do this.)**

3. **Identify conserved positions in the domain.**

4. **Turn your domain into a PROSITE pattern or a profile. (See Chapter 6 for more on PROSITE patterns.)**

5. **Scan SWISS-PROT in order to check whether this pattern is associated with a function.**

Finding low-complexity regions in proteins

Low-complexity regions are segments that contain a certain amino acid more often than you would expect in *normal* proteins. For instance, a segment of 100 amino acids that contains 45 prolines is definitely a low-complexity segment.

Low-complexity regions often have biological functions like protein-protein interaction (Leucine Zipper) or nonspecific DNA/RNA binding (ARG-rich domains).

The good news about low-complexity regions is that seeing them on a dot plot is very easy. When you compare a sequence with itself, low-complexity regions pop up as squares. You can see one of these in Figure 8-9.

To reproduce the results you see on Figure 8-9, follow these steps:

1. **Point your browser to** `www.expasy.ch/cgi-bin/get-sprot-fasta?` `P21997.`

 P21997 is the ID number for the sulfated surface glycoprotein 185

2. **Enter the sequence into Dotlet following the procedure that we outline in the section "Using Dotlet over the Internet."**

3. **Set the settings so that they match those in Figure 8-9.**

 You need a window setting of 7, a zoom setting of 1:2, and a gray scale on the top and bottom cursors of the threshold window of 46%.

Figure 8-9:
Identifying
low-
complexity
segments
with Dotlet.

Analyzing nucleic acids with Dotlet

By nature, dot plots are well suited for mapping genes onto their products. Unfortunately, Dotlet can be a bit slow for such applications, which routinely require the comparison of sequences longer than 10,000 residues.

Dotlet contains two useful features for analyzing nucleotide sequences:

✔ **If you compare a protein sequence and a nucleotide sequence:**
Dotlet automatically translates the nucleotide sequences into its three
possible protein reading frames. This way you can pinpoint exons/
introns boundaries.

✔ **If you compare a nucleotide sequence with itself:** Dotlet automatically
replaces one of the sequences with its complementary sequence. This
way, if your sequence contains two complementary strands (like a hairpin
stem in an RNA structure), they appear like diagonals going from the top-
right corner toward the bottom left (perpendicular to the main diagonal).

If you want to use Dotlet for more refined sequence analysis, we recommend
the excellent online tutorial produced by Dotlet authors, available at
www.ch.embnet.org/java/dotlet/dotlet_examples.html.

Making Local Alignments over the Internet

If you've gone through the dot-plot section of this chapter, you know that
making a dot plot is an ideal way to get an overview of the relationship
between your sequences. Nonetheless, dot plots aren't predictive methods:
They show a signal but don't tell you what it means. To make this interpreta-
tion, you need a method that produces an alignment.

There are two kinds of alignments: global alignments, where the two
sequences are aligned over their entire lengths, and local alignments, where
the program only aligns the most similar portions of your two sequences and
ignores the rest.

If you have a dot plot showing that your two sequences are related over their
entire lengths, you may not need to make a local alignment, and you can go
directly to the global alignment section of this chapter. If you are not sure,
stay here!

There are two major reasons for analyzing your sequences with local
alignments:

✔ To compare two distantly related sequences that share only a few
non-contiguous domains

✔ To analyze repeated elements within a single sequence

Local alignment methods do exactly what their name implies: You give them
two sequences, and they output an alignment of the two most similar portions
of these sequences.

The nice thing about local alignment methods is that they automatically get rid of the amino acids or nucleotides that they can't align.

In theory, a local alignment should correspond to one of the diagonals that appear on the dot plot. In practice, however, things aren't so simple; you may see a signal on the dot plot that doesn't correspond to any local alignment, and sometimes local alignment programs report alignments that don't appear on the dot plot. Only when the dot plot and the alignment agree perfectly can you be sure.

Another rule with sequence alignments is that you should trust them only when they are *clearly correct*. In this section, we show you what we mean by clearly correct. The only situation where you can use a suspicious alignment is when structural or experimental information exists that supports this alignment.

Choosing the right local alignment flavor

Two types of methods exist for making local pairwise alignments: a fast, heuristic method named BLAST and a slower, more accurate one named Lalign. Table 8-3 lists the pros and cons of these two methods so that you can select the one that suits you best.

The version of BLAST that compares two sequences is the same one that you can use for searching a database. It's adapted so that you can restrict it to two sequences only, but it doesn't generate alignments different from those BLAST reports when searching complete databases. In short, if you want to check a hit, BLASTing two sequences can't tell you anything more than the database search. You can access this special flavor of BLAST at www.ncbi.nlm.nih.gov/blast/bl2seq/bl2.html.

Table 8-3	Using BLAST or Lalign for a Pairwise Comparison	
Property	*BLAST*	*Lalign*
Speed	Very fast	Slower
Size of sequences you can use	Very long sequences	Shorter sequences
Score	E-value	Matrix score and E-value
Alignments	Reports the best	Reports the ten best (or more)
Sequence type	Best with DNA	Best with proteins

Using Lalign to find the ten best local alignments

The main problem when using BLAST is that this program only returns one alignment: the best scoring one. This is very useful for searching databases, but it may not be so interesting if the region of the sequences you are interested in belong to a *good* local alignment but not to the *best* local alignment. It's a bit like a music store that would only sell you one CD — the number one bestseller of the week.

Lalign is the ideal complement to BLAST. It is slower but more accurate, and it returns as many local alignments as you want (the best scoring one, the second best scoring, and so on, up to a number that you specify). Don't worry about the speed issue. On sequences shorter than a thousand symbols, you may not notice any difference between the bl2seq server and Lalign.

In general, Lalign is great when it comes to analyzing complicated proteins full of repeats. In the following steps list, we use Lalign to extract local alignments from two distantly related sequences that both contain a serine protease domain:

1. **Point your browser to** www.ch.embnet.org/software/ LALIGN_form.html.

 The Lalign page of ch.EMBnet.org appears.

2. **Choose Local from the alignment method options, as shown in Figure 8-10.**

 You can use the Lalign interface to do local and global alignments.

This is William Pearson's *lalign* program. A manual page for this program is available here. The lalign program implements the algorithm of Huang and Miller, published in Adv. Appl. Math. (1991) 12:337-357.
This program is part of the FASTA package of sequence analysis program. The complete package is available by anonymous ftp from ftp.virginia.edu.

Usage: Paste your two sequences in one of the supported formats into the sequence fields below
and press the "Run lalign" button.
Make sure that both format buttons (next to the sequence fields)
shows the correct formats

Choose the alignment method : ⊙ local (default) ○ global ○ global without end-gap penalty

Number of reported sub-alignments : 3

Scoring matrix : default

Opening gap penalty : -14 (default -14)

Extending gap penalty : -4 (default -4)

Figure 8-10: The Lalign home page.

3. **Choose the number of reported subalignments.**

 You should choose a number that corresponds roughly to the number of diagonals you observed on the dot plot. If you want Lalign to report 10 local alignments, select 10.

4. **Select the default substitution matrix.**

 The substitution matrix controls the cost of the mutations when Lalign aligns the sequences. (For more on this topic, see the sidebar Substitution matrices and the cost of mutations.)

 The default matrix is an identity matrix for DNA and a BLOSUM45 for proteins. In most cases, this default is entirely appropriate.

 If you decide to change the matrix, use the following guidelines:

 - BLOSUM is generally better than PAM for local alignments.

 - *High* BLOSUM indexes make Lalign more stringent and are adapted to similar sequences. When you use BLOSUM80 instead of BLOSUM45, local alignments tend to be shorter.

 - *Low* PAM indexes have the same effect as *high* BLOSUM indexes. The BLOSUM index indicates the percent identity between the sequences you're comparing while the PAM index is proportional to the number of mutations.

5. **Set the gap-opening penalty.**

 The *gap-opening penalty d*efines the cost for opening a gap in one of the sequences.

 If you set the gap-opening penalty higher than its default value, local alignments that contain gaps may be split into several shorter alignments.

 The default values have been tuned on the default matrix. If you change the matrix, Lalign does not automatically readjust the values. For instance, if you use a BLOSUM with an index lower than 45, you must decrease the gap-opening value and set it lower than 14.

 Unfortunately, there is no rule for predicting the penalty that best suits your sequences. Optimal penalties vary from sequence to sequence, and finding the most adequate value is a matter of empirical trial and error.

6. **Set the gap-extension penalty.**

 The *gap-extension penalty* is an extra penalty proportional to the length of the gap. This penalty is added to the gap-opening penalty to yield the complete gap cost.

 The gap-extension penalty must always be lower (about ten times lower) than the gap-opening penalty.

When you compare distantly related sequences, a high gap-opening penalty and a very low gap-extension penalty often give the best results. They indicate that gaps should be penalized more on their existence than on their length.

When you compare closely related proteins, you want gaps to be penalized on their length (gap extension) *and* on their occurrence (gap opening). Yet bear in mind that changing gap penalties only have a noticeable effect when comparing distantly related sequences.

7. **Choose SWISS-PROT ID or AC from the Input Sequence Format pull-down menu.**

 Use the menu to indicate the format that you want to use. If you want to use your own sequence, select the Plain Text, which is the same as raw format.

8. **Enter** P05049 **in the first sequence box.**

 P05049 is the accession number of the snake serine protease. If you want to use a sequence that has no accession number, paste it in the window, and make sure that you chose the correct format in Step 7.

9. **Set the second format selector to SWISS-PROT ID or AC.**

10. **Enter** P08246 **in the second sequence box.**

 P08246 is the accession number of the human leucocyte elastase.

11. **Click the Run Lalign button.**

 The computation should be relatively fast (less than one minute). If the program doesn't return any results, check that you have submitted the right parameters.

 Lalign returns the results in the form of an HTML document.

12. **Save your results.**

 Use the File⇨Save As option of your Web browser to keep this result in a file.

Interpreting the Lalign output

Lalign reports the number of local alignments you have specified — as in the preceding Step 3 — sorted according to their score. An interesting property of Lalign is that it only reports *nonoverlapping* alignments.

This means that, in the Lalign output, you can find two amino acids or nucleotides aligned together only once. They can appear in several alignments of course, but they appear in front of each other only once. This prevents Lalign from reporting trivial variations of a very good local alignment.

Figure 8-11 shows you the output of Lalign given the two sequences used in the preceding steps list. Lalign reports the alignments in a BLAST-like format, sorted according to their E-values. On the first line associated with each alignment are the following features:

- **The percent identity:** The proportion of identical residues aligned with one another. For instance, you can see in Figure 8-11 that the best local alignment has a percent identity score of 23.9 percent; the second best 39.1 percent.

- **The local alignment length (Overlap):** This is length of the local alignment.

- **The score:** The score is the sum of cost of the gaps and substitutions, as given by the substitution matrix and the gap penalties. Generally speaking, the higher the score the better the alignment, yet be aware that the absolute value has no clear meaning. It is better to use the E-value as an alignment quality indicator.

- **The E-value:** This value tells you how many times you could have expected to find such a good alignment by chance, given your two sequences. Beware that this E-value is much less meaningful than the one BLAST reports when searching a database. A good E-value must be below 10-4.

TECHNICAL STUFF

Substitution matrices and the cost of mutations

When you make an alignment, you want to find out the rewards and/or costs of aligning a letter with another letter. For instance, if you are aligning DNA, you know that all the mutations are not equal — transvertions that turn pyrimidines into purines are less likely than transition mutations. This means that when you align your DNA sequences, you want to use a substitution matrix that penalizes the alignment of a purine with a pyrimidine more than the alignment of a purine with another — different — purine.

For DNA, there are mathematical formulas that make it possible to precisely establish the cost associated with each possible mutation. In practice these sophisticated models are not so helpful because DNA sequences need to be 70 percent identical to be alignable. In this range of similarity, an identity matrix does just as well as any more sophisticated matrix. You can safely align DNA with a matrix that rewards

identical nucleotides with 1 and different nucleotides (mutations) with -1 or 0.

When it comes to proteins, things are more complicated: there are twenty amino acids and they are more or less different from one another. If you simply look at their chemical formula, you can guess that a mutation turning an isoleucine into a valine is not as bad as a mutation that turns a tryptophan into a cysteine. In order to establish a substitution matrix that would take all these things into account, Margaret Dayhoff created the PAM concept in 1978.

Margaret Dayhoff made alignments of very closely related sequences and analyzed the mutations. For each pair of sequences, she reconstructed the divergence of the sequences and counted every mutation. She also counted the nonmutations.

She then asked a very simple question about the relative mutability of a given amino acid into any

other amino acid. For instance, given the frequency of leucine and valine in my sequences, do I see more mutations of valines into leucines than I would expect by chance? To evaluate this, she simply took the log of the ratio of these two values, where log = measured numbered of mutations/expected number of mutations). This is called a ***log odd***. It has a negative value if a mutation turning a valine into a leucine has been observed less often than we would expect and

therefore is disfavored during evolution. Its value is zero if this mutation has been observed as often as we expected. Finally, it gives a positive value if this type of mutation appears favored during evolution over other mutations.

If you measure this ratio for every possible pair of residues and do a bit of extra mathematical juggling you get a 20*20 matrix like the Pam250 that we wrote below. Note that we only wrote half of it because it is symmetrical.

```
A   2
R  -2   6
N   0   0   2
D   0  -1   2   4
C  -2  -4  -4  -5  12
Q   0   1   1   2  -5   4
E   0  -1   1   3  -5   2   4
G   1  -3   0   1  -3  -1   0   5
H  -1   2   2   1  -3   3   1  -2   6
I  -1  -2  -2  -2  -2  -2  -2  -3  -2   5
L  -2  -3  -3  -4  -6  -2  -3  -4  -2   2   6
K  -1   3   1   0  -5   1   0  -2   0  -2  -3   5
M  -1   0  -2  -3  -5  -1  -2  -3  -2   2   4   0   6
F  -3  -4  -3  -6  -4  -5  -5  -5  -2   1   2  -5   0   9
P   1   0   0  -1  -3   0  -1   0   0  -2  -3  -1  -2  -5   6
S   1   0   1   0   0  -1   0   1  -1  -1  -3   0  -2  -3   1   2
T   1  -1   0   0  -2  -1   0   0  -1   0  -2   0  -1  -3   0   1   3
W  -6   2  -4  -7  -8  -5  -7  -7  -3  -5  -2  -3  -4   0  -6  -2  -5  17
Y  -3  -4  -2  -4   0  -4  -4  -5   0  -1  -1  -4  -2   7  -5  -3  -3   0  10
V   0  -2  -2  -2  -2  -2  -2  -1  -2   4   2  -2   2  -1  -1  -1   0  -6  -2   4
    A   R   N   D   C   Q   E   G   H   I   L   K   M   F   P   S   T   W   Y   V
```

Biologists call this matrix a substitution matrix. As you can see, identities (replacement of an amino acid by itself) have the best scores. For instance, tryptophan is the residue that is the less likely to mutate, hence the very high reward for its conservation (+17). In this matrix, you can also see that the reward/cost for turning a given amino acid into another one depends on their chemical similarity: if you keep the size and the hydropathy the mutation score is often positive or close to 0. This is what happens when you turn a glutamic acid (Q) into an aspartic acid (D). On

the other hand, if the mutation changes the size and the hydropathy, the score is negative. One of the most striking examples is the replacement of the big tryptophan (W) with the small glycine (G): this mutation has a cost of (-7). This is a sure sign that proteins hate it when this happens!

The idea when you make an alignment is to align amino acids or nucleotides so that the sum of their substitution (or conservation) score yields the best possible score for the overall alignment. Local and global alignment programs are designed to do just that.

(continued)

(continued)

The substitution score matrix created by Margaret Dayhoff in 1978 is the **PAM** matrix, where PAM stands for *Point Accepted Mutation*. With a bit of mathematical juggling, it is possible to modify the original matrix and adapt it to distantly related proteins. The PAM index gives you an indication on the evolutionary distance the matrix covers. For instance, PAM 250 is the most commonly used matrix. It is perfectly suited for sequences that are 40-60 percent identical; use PAM with lower indexes for proteins that are more similar.

Later, in 1992 Steve and Georgia Henikoff designed a new set of matrices that they named **BLOSUM** (BLOck Scoring Matrix). The BLOSUM matrices use roughly the same principle as the PAM with the main difference that the BLOSUM matrices are measured on the multiple alignment of conserved regions in distantly related proteins (the BLOCKs). The index that follows a BLOSUM matrix indicates the level of similarity in the sequences on which this matrix has been measured. For instance BLOSUM 62 was measured on pairs of sequences with an average of 62 percent identical amino acids.

It is generally agreed that the BLOSUMs are the best matrices around for making local alignments. They are especially good at identifying the kind of signal that the core of proteins contains.

It is not always clear what to expect when changing matrix, and there are very few examples of dramatic discoveries simply popping out from the use of a new matrix. Ironically, this is quite fortunate because there is an impressive diversity in the matrices to choose from! Few aspects of bioinformatics have attracted as much attention as the design of new matrices, and it seems that over the last 50 years everyone has had her or his idea on a new recipe for assembling the best substitution matrix ever. If you scan the literature, you will find over 50 different types of such matrices. Do not let this impress you: in most cases you can get very sensible results by using a BLOSUM or a PAM.

Be careful if you use a nonstandard matrix: the penalties and the various parameters within the alignment program you use may not be tuned for this new set of substitution scores. For instance BLAST requires that the majority of the scores in the matrix be negative. Thus, using an exotic matrix may generate various artifacts. On the other hand, you can safely assume that any program you use has been tuned to work well with a PAM or a BLOSUM.

The alignment itself contains three types of information:

✔ The residue index on the line above the sequence. The residue corresponding to an index is the one below the last digit of this index.

✔ The alignment, with gaps represented with dashes.

✔ In the line between the two aligned sequences, the (_) symbol means identity, while the (.) symbols mean similarity. Two residues are similar when their substitution score is greater than 0.

```
(A) ./wwwtmp/lalign/.19487.1.seq sp|P05049|SNAK_DROME (SNK)Serine protease snake p - 430 aa
(B) ./wwwtmp/lalign/.19487.2.seq sp|P08246|ELNE_HUMAN (ELA2)Leukocyte elastase pre - 267 aa
 using matrix file: BL50, gap penalties: -14/-4
  23.9% identity in 213 aa overlap; score:  103 E(10,000):  0.028
        220       230       240       250       260       270
sp|P05 CGGALVSELYVLTAAHC-ATSGANHRTWFAWRPQLNETSATQQDIKILIIVLHPKYRSSA
       ::..:..  .::..:::: :. ..          .:.   :.: . . : .. :
sp|P08 CGATLIAPNFVMSAAHCVANVNVRAVRVVLGAHNLSRREPTRQVFAVQRI-FENGYDPVN
          60        70        80        90       100       110
        280       290       300       310       320       330
sp|P05 YYHDIALLKLTRRVKFSEQVRPACLWQCG---APHTTVVAAGWGRTEFLGAKSNALRQVD
       .::..:.. . ..:.: : :    :   . .  .: :::     . ...:....
sp|P08 LLNDIVILQLNGSATINANVQVAQLPAQGRRLGNGVQCLAMGWGLLGRNRGIASVLQELN
         120       130       140       150       160       170

  39.1% identity in 23 aa overlap; score:   47 E(10,000): 2.4e+03

        190       200
sp|P05 IVGGTPTRHGLFPHMAALGWTQG
       :::: .:   .: :..:    :
sp|P08 IVGGRRARPHAWPFMVSLQLRGG
         30        40        50
```

Figure 8-11:
Output of
Lalign.

```
  29.2% identity in 24 aa overlap; score:   42 E(10,000): 5.3e+03

        200       210       220
sp|P05 HMAALGWTQGSGSKDQDIKWGCGG
       .. : :  :.:.   . :: .
sp|P08 QLPAQGRRLGNGVQCLAMGWGLLG
```

The first alignment reported corresponds to the conserved serine protease domain. On its own, this alignment isn't really convincing: It contains less than 24 percent identity over about 200 residues.

To be really convinced, we would need a much higher similarity (at least close to 30 percent) and a better (lower) E-value. The reason we may trust this alignment is because it is consistent with the signal we previously saw on the dot plot (Figure 8-7). The fact that these two different analyses give compatible results (Dotlet and Lalign) makes a good case for the existence of a conserved serine protease domain in our proteins.

If you play with the Lalign parameters, you may find that stronger gap opening or extension penalties may induce Dotlet diagonals to be split by Lalign and turned into two alignments. Lower gapopening or extension penalties, on the other hand, can cause diagonals to fuse. This all depends on the length of the gaps Lalign needs to insert in the local alignments.

When it comes to judging the quality of your alignment, size definitely matters. It isn't uncommon to see short segments with very good scores. This can result from a variety of reasons that include low-complexity segments, small-scale regularity in protein sequences, recurrent patterns of hydrophobic/hydrophilic residues, and so on. For instance, in Figure 8-11, the second and the third local alignments are obviously meaningless because they are too short. However, their E-values are better than the top alignment. This clearly illustrates that a short alignment can more easily get a good score by chance.

Bioinformatics programs aren't perfect: They can make mistakes and deliver wrong results. Unfortunately, right and wrong obey the same rules in bioinformatics as they do in life; they are not absolute concepts.

The top alignment that Lalign reports here illustrates well this limitation. Because of the dot plot information, we trust this alignment to be correct overall. On the other hand, the two sequences are so remotely related and have so little similarity that we must be cautious in our interpretation: Many amino acids could be incorrectly aligned, which means that this alignment is probably right *and* wrong. The only way to really push this analysis further would be to make a multiple sequence alignment or to use structural information. (For more on multiple sequence alignments, see Chapter 9; for more on structures, see Chapter 11.)

Making Global Alignments over the Internet

A global alignment is just what the name implies: an alignment of every amino acid or nucleotide found in your sequences. When you decide to make a global alignment, you implicitly make the hypothesis that your two sequences are related over their entire lengths.

Global alignments are very important for making multiple sequence alignments. However, when done using only two sequences, they offer little interest for biological analysis. Whenever you have two closely related sequences, you can generate a much more informative result by gathering a few more related sequences and making a multiple sequence alignment. Chapter 9 tells you everything you need to know for this purpose.

Global alignments aren't useful at all for discovering similarities between two sequences because the statistical method for evaluating E-values doesn't apply to them. However, beginners often prefer global alignments to their local counterparts because they are simpler to understand.

In a global alignment, no amino acid or nucleotide mysteriously disappears: You find in the output exactly what you put in the box, plus a few extra gaps. There are three main reasons for making a global alignment:

- ✔ Checking minor differences between two sequences. This may happen with data that you've manipulated and possibly altered. The global alignment is the best way to localize potential problems.

- ✔ Analyzing polymorphisms (for example, SNPs) between closely related sequences.

> ✔ Comparing two sequences that partly overlap. In that case you want to make a global pairwise comparison that doesn't penalize misalignments at the extremities of the sequences.

Using Lalign to Make a Global Alignment

In terms of algorithms, there is not much difference between a local and a global alignment. If you were to write the programs for both, you would not find more than one line of difference between these two algorithms!

You can use Lalign to produce global alignments, but there are other choices, some of which we've listed in Table 8-4, at the end of this chapter.

Using Lalign to produce a global alignment is exactly like using Lalign to produce a local alignment. The only thing extra you need to do is to is to check the radio button that says **Global without End-Gap Penalty,** near the top of the Lalign form at ww.ch.embnet.org/software/LALIGN_form.html .

Free Pairwise Sequence Comparisons over the Internet

Many existing servers propose pairwise comparisons. Table 8-4 lists a brief selection of some of the more stable resources.

Table 8-4	Online Pairwise Alignment Programs	
Name	*Address*	*Alignment Type*
lalign	www.ch.embnet.org/software/ LALIGN_form.html	Global/Local
lalign	fasta.bioch.virginia.edu/ fasta_www/plalign.htm	Global/Local
USC	www-hto.usc.edu/software/ seqaln/seqaln-query.html	Global/Local/
alion	fold.stanford.edu/alion/	Global/Local

(continued)

Table 8-4 *(continued)*

Name	Address	Alignment Type
	`genome.cs.mtu.edu/align.html`	Global/Local
align	`www.ebi.ac.uk/emboss/align/`	Global/Local
xenAliTwo	`www.soe.ucsc.edu/~kent/` `xenoAli/xenAliTwo.html`	Local for DNA
Blast2seqs	`www.ncbi.nlm.nih.gov/blast/` `bl2seq/bl2.html`	Local BLAST
Blast2seqs	`web.umassmed.edu/cgi-bin/` `BLAST/blast2seqs`	Local BLAST

Generating an alignment isn't always enough. You may also need to visualize this alignment and to evaluate its statistical significance. Table 8-5 lists a few sites that enable you to do this online.

Table 8-5 Online Pairwise Alignments Analyses

Name	Address	Function
lalnview Visualization	`www.expasy.ch/tools/sim-prot.html`	
prss Evaluation	`www.ch.embnet.org/software/PRSS_form.html`	
prss Evaluation	`fasta.bioch.virginia.edu/fasta/prss.htm`	
graph-align Evaluation	`darwin.nmsu.edu/cgi-bin/graph_align.cgi`	

Chapter 9

Building a Multiple Sequence Alignment

The man with two watches never knows the time, and the man with one watch only thinks he knows.

— A man with multiple watches

*I*n this chapter, we show you how to simultaneously compare many protein or nucleotide sequences. This is something you probably want to do if you have a sequence and you want to elucidate the role of each amino acid or nucleotide that it contains.

To make a multiple sequence comparison, your first job is to select a few sequences and then use them for making a multiple sequence alignment. In this chapter, we show you how to select the most appropriate sequences and how you can generate a multiple sequence alignment. We introduce two different multiple sequence alignment methods that should cover most of your needs: ClustalW and Tcoffee. Because it is very easy to generate bad alignments that look good, we also show you how to evaluate the quality of your alignment.

Sometimes, your multiple sequence alignment isn't good enough to be useful. When this happens, we show you that you have alternatives to sequence alignment and that you can sometimes compare your sequences without aligning them. We introduce two of these methods: the Gibbs sampler for identifying related segments of the same length and Pratt, a motif discovery method for discovering PROSITE patterns.

A multiple alignment is probably the most useful investigative tool you can generate in bioinformatics. In many ways, multiple sequence alignments are to bioinformatics what Swiss knives are to MacGyver; we're hard pressed to think of a situation where such alignments don't help. If you aren't sure whether a multiple alignment can help your work or if you don't see exactly how you can make one of these fit in your experiments, we suggest that you start reading the first section of this chapter, "Finding Out If a Multiple Sequence Alignment Can Help You."

In any case, remember that multiple alignments are useful for predicting protein structures (see Chapter 11), central for predicting the function of proteins, and indispensable for phylogenetic analysis (see Chapter 13). Of course, the better your multiple sequence alignment, the better use you can make of it: better structural models, better functional predictions, and better phylogenetic trees.

Building multiple sequence alignments is far from being an exact science. In fact, it's more art than science, requiring that you use everything you know in bioinformatics and in biology. This chapter is here to give you the secrets of the trade so that you, too, will be able to create the multiple sequence alignment that best suits your needs. We're also here to help you find your own recipes.

If you already have an alignment and you only want to modify it or improve its appearance for publication (highlighting, shading, and so on), go directly to Chapter 10 — that's the chapter entirely dedicated to such fine-tuning techniques.

Finding Out If a Multiple Sequence Alignment Can Help You

Figure 9-1 shows you what a multiple alignment actually looks like: As you can see, it's really only a matter of rewriting your sequences so that similar features end up in the same columns. That's it, plain and simple.

This kind of multiple alignment is ideal if you want to study a sequence family where all sequences share the *same common ancestor*. Don't worry if you have only one member at this stage of the game; we show you later in this chapter how to organize a family gathering!

Figure 9-1:
A multiple
sequence
alignment.

Identifying situations where multiple alignments do not help

Part of knowing when to use multiple sequence alignments involves knowing when *not* to use them. The multiple sequence methods we describe in this chapter don't work well for assembling sequence pieces in a sequencing project. If you have a set of shotgun sequences — short, partly overlapping sequences — or if you want to turn an EST cluster into a gene sequence, multiple sequence alignment methods don't work well, and dumping your sequences into ClustalW or Tcoffee can produce a very disappointing and uninformative result.

If you're facing this particular type of problem, you may want to use specialized sequence assembly tools, such as Phred and Phrap (www.phrap.org). Unfortunately, specialized sequence assembly tools, which we only briefly overview in Chapter 5, are far beyond the scope of this book.

Another reason for not using multiple sequence alignment techniques is if the sequence you're interested in has no homologue in any of the sequence databases. If this happens, you're out of luck; there's no way that you can build a multiple alignment that relates to this sequence. You may try to find a few sequences by using functional criteria and conducting a pattern search with Pratt, but don't count on any miracles.

Helping your research with multiple sequence alignments

With the right sequences in hand, you can build the proper multiple sequence alignment. To be honest, you could build this alignment manually, rather than by using a computer. Manual alignments are like home food: When prepared by a good cook, nothing beats them; for regular folks like us, however, the automatic processing (TV dinner) is generally much better.

The idea behind a multiple alignment like the one in Figure 9-1 is to put amino acids or nucleotides in the same column because they're homologous according to some criterion. You can use four major criteria to build a multiple alignment that all have different properties, as Table 9-1 conveniently shows.

Table 9-1	Main Criteria for Building a Multiple Sequence Alignment
Criterion	*Meaning*
Structural similarity	Amino acids that play the same role in each structure are in the same column. Structure superposition programs are the only ones that use this criterion.
Evolutionary similarity	Amino acids or nucleotides related to the same amino acid (or nucleotide) in the common ancestor of all the sequences are put in the same column. No automatic program explicitly uses this criterion, but they all try to deliver an alignment that respects it.
Functional similarity	Amino acids or nucleotides with the same function are in the same column. No automatic program explicitly uses this criterion, but if the information is available, you can force some programs to respect it or you can edit your alignment manually.
Sequence similarity	Amino acids in the same column are those that yield an alignment with maximum similarity. Most programs use sequence similarity because it is the easiest criterion. When the sequences are closely related, structural, evolutionary, and functional similarities are equivalent to sequence similarity.

The first three criteria have a clear biological meaning — the fourth one doesn't. Yet, when the sequences are similar enough, you can use similarity to produce a multiple alignment that reflects the evolutionary, the structural, and the functional relationships that exist between your sequences. Of course, to do this, you need an alignment that you can trust. Later in this chapter, we show you how to make sure you're putting your money on the right alignment.

Table 9-2 lists the main applications of multiple sequence alignments.

Table 9-2 Main Applications of Multiple Sequence Alignments

Application	Procedure
Extrapolation	A good multiple alignment can help convince you that an uncharacterized sequence is really a member of a protein family.
Phylogenetic analysis	If you carefully choose the sequences to include in your multiple alignment, you can reconstruct the history of these proteins.
Pattern identification	By discovering very conserved positions, you can identify a region that is characteristic of a function (in proteins or in nucleic acid sequences).
Domain identification	It is possible to turn a multiple sequence alignment into a profile that describes a protein family or a protein domain. You can use this profile to scan databases for new members of the family.
DNA regulatory elements	You can turn a DNA multiple alignment of a binding site into a weight matrix and scan other DNA sequences for potential similar binding sites.
Structure prediction	A good multiple alignment can give you an almost perfect prediction of your protein secondary structure for both proteins and RNA. Sometimes it can also help in the building of a 3-D model.
PCR analysis	A good multiple alignment can help you identify the less degenerated portions of a protein family, in order to fish out new members by PCR (polymerase chain reaction). If this is what you want to do, you can use the following site: blocks.fhcrc.org/codehop.html.

Whatever your choice or motivation, always bear in mind that the reason you can use multiple alignments for so many things is that they can give you an instant picture of the forces that shape evolution! When you look at proteins or DNA sequences, the rules are ruthless:

✔ Important amino acids (or nucleotides) are not allowed to mutate.

✔ Less important residues change more easily, sometimes randomly, and sometimes in order to adapt a function.

For you, this has a very simple consequence: When you look at a multiple alignment, you can make the hypothesis that *conserved* positions (columns where all the sequences contain the same amino acid or nucleotide) are more important for the function than *non conserved* positions (columns where the sequences contain different amino acids or nucleotides). Of course, you could tell this from the alignment of just two sequences, but using more sequences makes it easier to discriminate between important and less important positions.

And one more thing: A multiple sequence alignment is a terrific way to present your results. It lets you put lots of information into a single model and makes it incredibly easy to spot inconsistencies or potential problems.

If you want to get an important point across, in a paper or in front of an audience, a good hand-crafted and well-colored multiple alignment says more than three billion nucleotides in bulk.

All this being said, multiple alignments do NOT cure asthma, arthritis, lumbago, or baldness. They won't make you stronger or increase any of your potentials (except maybe your research potential). But at this point, if you need more arguments in order to be convinced of the genius of multiple alignments, chances are that you don't really need to make this kind of alignment.

Choosing the Right Sequences

Anyone who ever worked in a lab knows that molecular biology is very much like cooking: It's all about selecting the right ingredients and putting them into the pot at the right time and in the right order.

Building a multiple alignment obeys the same rule. Before you build your alignment, you must carefully select the sequences you want to align. These sequences are members of the same protein family, and they all share a common ancestor. The family is usually too large to be entirely included in your multiple alignment, and picking the right sequences is an art. If you want to be good at this game, you need to know what you want to show with your alignment and you need to know how the multiple alignment programs work.

Of course, to understand these programs well, you need to see them in action; for that, you need to have a few sequences! This is a typical chicken/egg problem for which no simple solution exists. We suggest that you go through the next section once and come back again after you start playing with the multiple sequence alignments that we describe in the other parts of the chapter. After one or two rounds, things should click into place.

The kinds of sequences you're looking for

Let's assume that you start this procedure with your favorite sequence. You want to thoroughly study this sequence and you know that in order to do this you have to build a multiple alignment. Table 9-3 summarizes most of the things you must take into account when selecting these extra sequences.

Table 9-3	A Few Guidelines for Selecting Sequences
Problem	*Diagnostic*
Proteins or DNA	Use proteins whenever possible. You can turn them back into DNA *after* doing the multiple alignment.
Many sequences	Start with 10-15 sequences and avoid aligning more than 50 sequences.
Very different sequences	Sequences that are less than 30 percent identical with more than half of the other sequences in the set often cause troubles.
Identical sequences	They never help. Unless you have a very good reason to do so, avoid incorporating in your multiple alignment any sequence that's more than 90 percent identical to another sequence in the set.
Partial sequences	Multiple sequence alignment programs prefer sequences that are roughly the same length. Programs often have difficulties comparing a mixture of complete sequences and shorter fragments.
Repeated domains	Sequences with repeated domains cause troubles to most multiple alignment programs, especially if the number of domains is different. When this happens, you may be better off extracting the domains yourself with Dotlet or Lalign (see Chapter 8) and making a multiple alignment of these segments.

Using DNA or protein sequences

If the sequences you're interested in are noncoding sequences, you obviously have no choice — you must use DNA. However, beware that non-coding DNA sequences can be tricky to align. If you cannot generate a proper alignment from sequences that you know are related, you could use a local multiple alignment method, such as the Gibbs sampler, or a pattern extraction motif, such as Pratt. (See "Comparing Sequences That You Can't Align," later in this chapter, for a description of these two methods.)

Multiple sequence alignment methods are at their best when aligning protein sequences. The reason is that protein sequences are three times shorter than the corresponding DNA, and they use a more informative alphabet of 20 amino acids.

If you want to persist in carrying out a phylogenetic analysis on a set of coding DNA sequences, things work better if you do the following:

1. Translate your DNA sequences into proteins.

2. Perform the multiple alignments on the proteins.

3. Thread the DNA back onto the protein multiple sequence alignment framework.

If your proteins are difficult to align because they have few similarities, DNA information does not help. This has to do with the degeneracy of the genetic code and the fact that there are 20 amino acids and only 4 nucleotides. If there is little signal at the protein level, you can be sure that there is NO useful signal at the DNA level.

Choosing the right number of sequences

Of course, there's no absolute answer to this question, such as *42* or *7*. A few years ago, the answer was easy: Get everything you can and go to the lab if there aren't enough sequences in the databases! This is not true anymore. These days, given the sizes of the databases and new complete genome sequences flowing in twice a month, you may easily find hundreds or thousands of sequences that would be suitable for inclusion in your multiple sequence alignment. But that doesn't mean you have to use them all!

In our opinion, you should start with a relatively small number of sequences — between 10 and 15 sequences is suitable for most cases. After you get something interesting happening with this small set, you can always increase its size. In any case, it's hard to see any reason for generating a multiple alignment with more than 50 sequences.

If you start with hundreds of sequences, you immediately hit troubles. There are good reasons for this:

- *Computing* **big alignments is difficult.** Public servers do not have infinite resources. Your job may take a very long time to run (if it runs). For you, this makes it difficult to tune parameters and check alternatives.

- *Building* **big alignments is difficult.** Multiple alignment programs are not very good at handling very large sets of sequences.

✔ *Displaying* **big alignments is difficult.** You can't print them, and they clog your computer when you want to visualize them. If columns are longer than one page, interpretation becomes impossible.

✔ *Using* **big alignments is difficult.** Tree building and structure prediction programs cannot handle them easily.

✔ *Making* **accurate big alignments is difficult.** You want your multiple sequence alignment to be highly reliable so that you can be confident that all the sequences it contains are true members of the family. A major cause for concern is that multiple sequence alignment programs make mistakes. The curse is that these mistakes do not add up, they multiply — making it easy to ruin an entire alignment with a tiny number of bad sequences. Of course, the more sequences you have, the more likely this will happen. The best way to avoid such a disaster is to start small and gradually increase the size of your multiple sequence alignment until it contains all the sequences you're interested in.

Now that you know how many sequences you need, the last question you face is deciding how related these sequences must be. Should you choose sequences that are very similar or very different?

Making the right compromise between similarity and new information

If you think that very similar sequences give very good alignments, you're right! However, a multiple sequence alignment that's correct isn't enough; it must also be useful.

For instance, an alignment that only contains very similar sequences brings little information. You can use it to extrapolate annotations, but you can't do phylogeny, structure prediction, or any of the other useful applications that we list in Table 9-2. These other tasks require being able to observe mutations patterns in every column, which isn't possible if you have only an alignment where most columns are entirely conserved.

Grabbing the most distantly related sequences you can find doesn't work either. Multiple sequence alignment programs cannot use sequences that are too different, even if these sequences are homologous. Two things multiple sequence alignment programs do not like are

✔ Sequences that are very different from every other sequence in the group

✔ Sequences that need long insertions/deletions to be properly aligned

Selecting your sequences for a multiple alignment is a team-building exercise: You want everyone to be a bit special, but you don't want your team members to be so different that they can't communicate. Teams of divas rarely go very far and neither do gangs of sextuplets.

When you select your sequences, the general rule is that you want them to be as distantly related as possible while not requiring too many gaps in order to be properly aligned. These two criteria are mutually exclusive, so finding the right trade-off requires a bit of strategy. The following steps list shows you a general approach that you should have in mind when gathering your sequences:

1. **Select a few sequences.**

 See the section "Gathering your sequences with online BLAST servers," later in this chapter.

2. **Compute a multiple alignment by using one of the servers we introduce in this chapter.**

 See the section "Choosing the Right Multiple Sequence Alignment Method," later in this chapter.

3. **Evaluate visually the quality of your alignment.**

 See the section "Interpreting Your Multiple Sequence Alignment," later in this chapter.

4. **If your alignment looks good, keep the sequences.**

 A good alignment contains nicely conserved blocks separated by regions with insertions and deletions. If you have such an alignment, your sequence set is probably appropriate and you can try to extend it by adding a few new sequences.

5. **If your alignment is difficult to interpret,**

 a. Examine the sequences more closely and try to remove the trouble-makers that are the most distantly related or those that cause long insertions/deletions.

 b. Redo the alignment with the smaller set.

 c. Keep trimming the set until you get a multiple alignment that's easier to interpret.

If you can, make sure that *each sequence is between 30 and 70 percent identical with more than half of the sequences in the set.* This way, you're making a reasonable trade-off between new information and alignment quality.

Before adding a sequence to a multiple alignment, you can try to figure out whether it's a good choice by making pairwise comparisons with some of the tools we describe in Chapter 10. (Pairwise comparisons are methods that let you compare sequences two by two.) However, we don't recommend that you do so exhaustively because it's time consuming.

Naming your sequences the right way

Naming may sound like a trivial issue, but it's not. Multiple sequence alignment programs have no standard way of handling the sequence names. If you stick to the following four rules, you will NEVER get into trouble; if you don't, you're on your own:

- ✔ Never use white spaces in your sequence names.

- ✔ Do not use special symbols. Stick to plain letters, numbers, and the underscore (_) to replace spaces. Avoid ALL other symbols, especially those that are the most tempting for special sequences, such as @, #, _,^ and so on.

- ✔ Never use names longer than 15 characters.

- ✔ Never give the same name to two different sequences in your set. Although some programs accept it, others (such as ClustalW) don't.

If you don't obey these naming rules, some multiple sequence alignment programs may automatically change the name of your sequences, without the courtesy of telling you.

Gathering your sequences with online BLAST servers

There are two types of sequences that you may want to integrate into your multiple alignment:

- ✔ **Characterized sequences:** These are sequences for which you have good annotations and experimental information. You'd definitely want to include these sequences in your alignment because they bring with them biological information and also allow feature propagation.

- ✔ **Uncharacterized sequences:** This category can include your sequence(s) of interest as well as database sequences. Uncharacterized sequences must be members of the family. Your main motivation in including them in your multiple alignment is to distinguish between the conserved positions that cannot mutate and other, less important columns. They help getting some contrast on your sequence of interest.

In this section, we show you how to gather these sequences with the BLAST database search program. If you want to know all the gory details about BLAST, check out Chapter 7. Here, we give you only the bare minimum to get by.

With BLAST, you can search databases for sequences that are homologous (similar) to a query. The original query can be any sequence you are interested in — protein or DNA — and you can use BLAST to search both protein and DNA databases. The main reason for using BLAST is to identify database sequences that are so similar to the query that they probably are homologous. We commonly refer to these sequences as *hits* or *matches*.

There are many BLAST servers around — as becomes abundantly clear in Chapter 7 — and you're free to try any (or all) of them. Nonetheless, they aren't all equally convenient for retrieving the sequences you're interested in. Table 9-4 lists three BLAST servers that are useful for generating a list of sequences in FASTA format (the best format for moving your sequences around, from one program to another) or for sending chosen sequences to a multiple alignment server.

Table 9-4	BLAST Servers Integrating Multiple Alignment Methods
Address	**What You Can Do There**
`www.expasy.ch/cgi-bin/BLASTEMBnet-CH.pl`	Extract entire sequences, Export sequences in FASTA, Submit sequences to ClustalW, Submit sequences to Tcoffee
`npsa-pbil.ibcp.fr/cgi-bin/npsa_automat.pl?page=npsa_blast.html`	Extract entire sequences, Extract sequences fragments, Export sequences in FASTA, Submit sequences to ClustalW
`srs.ebi.ac.uk`	Submit sequences to ClustalW

In the following steps list, we show you how to use the ExPASy server. The SRS server, although sometimes useful, is a bit harder to use than the other two listed in Table 9-4.

Selecting sequences on the ExPASy server

In the following steps list, we select appropriate sequences to make a multiple sequence alignment of calcium-dependent kinase proteins.

You can use this server only to retrieve *protein* sequences. If you're interested in gathering *DNA* sequences, use the European Bioinformatics SRS server (`srs.ebi.ac.uk`).

TIP

The BLAST server of the PBIL site is very similar to ExPASy. If you don't find the database you're interested in on the ExPASy server, try the PBIL.

1. **Point your browser to** www.expasy.ch/cgi-bin/BLASTEMBnet-CH.pl.

 The BLAST page of the ExPASy server appears.

2. **Enter the sequence Accession Number P20472, as shown in Figure 9-2).**

 This is the accession number of the human parvalbumin.

 If you prefer, you can also paste the sequence in raw format. Raw format means using the sequence only, without any header. The program ignores spaces and numbers.

Figure 9-2: Using the BLAST ExPASy server to gather sequences.

> P20472
>
> Output format : HTML
>
> Choose the appropriate BLAST ⊗ program and ⊗ database :
>
> ⊙ **blastp** - query against a protein database SWISS-PROT + TrEMBL + TrEMBL_NEW ▾
> ⊗ Taxonomic groups : [] SWISS-PROT + TrEMBL + TrEMBL_NEW , excluding translated ESTs)
> Enter a species name, a TaxID or the l SWISS-PROT f the OC, OS and OX lines) to
> restrict your search to a particular taxi PDB blons (";"). Example : *Fungi*;
> *Homo sapiens*. Note that this will slow complete microbial proteomes ssing of the results. *To run BLAST*
> *against a complete microbial genome, use the HAMAP pages.* translated EST
>
> ○ **tblastn** - query against the six-frame translation of a nucleotide database All EMBL + GSS (without HTG and ESTs) ▾
> Taxonomic groups : All ▾

3. **Select the BLAST flavor that you're interested in.**

 If you gave a protein sequence in Step 2, select blastp.

 If you gave a coding DNA sequence in Step 2, select tblastn.

4. **Choose SWISS-PROT + TrEMBL + TrEMBL_NEW from the Database pull-down menu.**

 With this choice, you carry out your database search on every known protein sequence. This makes it more likely that you'll find adequate sequences for your multiple alignment.

 If the search reports too many sequences very similar to your sequence of interest, you can decrease the number of identical hits by selecting a smaller database from the Database pull-down menu — SWISS-PROT, for example, or the microbial proteomes.

For a multiple sequence alignment, do not select Translated ESTs from the pull-down menu. These sequences are mostly incomplete protein sequences that may confuse the multiple alignment procedure.

5. **Scroll down to the Options section and set the Number of Best Scoring Sequences to Show option to 1000.**

Doing this makes it more likely that you'll find appropriate sequences in the BLAST result for your multiple alignment.

6. **In the same Options section, set the Number of Best Alignments to Show option to 1000.**

This choice makes it possible to judge the quality of the alignment before selecting a sequence.

7. **Click the Run BLAST button.**

After a brief pause, a Results page appears.

8. **Scroll down the page to select the sequences you want.**

You select a sequence by selecting the box to its right.

This is the most delicate part of the process. There is no absolute rule to selecting your sequences, but you can use the following guidelines:

- **Select the top sequence.** This top sequence is usually your sequence of interest. If your sequence of interest is not on the top, you may have to add it later to the list.

- **For a first analysis, you want to select ten sequences or less.** Ideally, the ten sequences to select should be evenly spaced between the very good E-values (10^{-40}) and less good E-values (10^{-5}). Figure 9-3 gives you an idea of what such a selection looks like.

 For the purpose of this steps list, choose (from top to bottom): P20472, P80079, P02626, P02619, P43305, P32930, Q91482, P02620, P02622.

- **Before selecting a sequence, check that it is similar to the query along its entire length.**

The alignment section is at the bottom of the BLAST output. You must be especially careful with hits that have E-values higher than 10^{-10}. They are equally likely to correspond to a good partial match, a global overall match, or a match between a protein fragment and your sequence. Inspecting the alignment is the only way to distinguish between these situations.

9. **Choose the method you want to use to export your sequences from the Send Selected Sequences To pull-down menu, as shown in Figure 9-4.**

Figure 9-3:
Selecting
sequences
on the
BLAST
output.

```
□ tn   BAB23980   Pva protein [Pva] [Mus musculus]                            163 2e-40
□ tn   BAB22033   Pva protein [Pva] [Mus musculus]                            163 2e-40
□ sp   P32848     PRVA_MOUSE Parvalbumin alpha [PVALB] [Mus musculus]         163 2e-40
☑ sp   P80079     PRVA_FELCA Parvalbumin alpha [PVALB] [Felis silvestris...   162 7e-40
□ sp   P80026     PRVM_CHICK Parvalbumin, muscle [Gallus gallus]              151 1e-36
□ sp   P51434     PRVA_CAVPO Parvalbumin alpha (Fragment) [PVALB] [Cavia...   130 2e-30
□ tn   CAC83047   Parvalbumin alpha [Rana sp. CH-2001]                        124 1e-28
□ tn   P18087     PRVA_RANCA Parvalbumin alpha (PA 4.97) [Rana catesbeiana]   121 1e-27
□ tn   CAC83046   Parvalbumin alpha [Rana esculenta]                          120 3e-27
□ sp   P02627     PRVA_RANES Parvalbumin alpha [Rana esculenta]               120 3e-27
□ sp   P02628     PRVA_ESOLU Parvalbumin alpha [Esox lucius]                  110 3e-24
☑ sp   P02626     PRVA_AMPME Parvalbumin alpha [Amphiuma means]               109 6e-24
□ sp   P30563     PRVA_TRISE Parvalbumin alpha [Triakis semifasciata]         105 6e-23
□ sp   P02614     PRVB_GRAGE Parvalbumin beta [Graptemys geographica]         105 8e-23
□ sp   P02629     PRVA_LATCH Parvalbumin alpha [Latimeria chalumnae]          105 8e-23
□ sp   P19753     PRVT_CHICK Parvalbumin, thymic (Avian thymic hormone) ...   100 3e-21
□ tr   Q90WX7     Parvalbumin 3 [Rana catesbeiana]                             97 2e-20
□ sp   P02618     PRVB_CYPCA Parvalbumin beta [Cyprinus carpio]                96 8e-20
☑ sp   P02619     PRVB_ESOLU Parvalbumin beta [Esox lucius]                    95 1e-19
□ sp   Q91483     PRV2_SALSA Parvalbumin beta 2 (Major allergen Sal s 1)...    94 2e-19
□ tr   Q8UUS2     Parvalbumin [CYP C 1.02] [Cyprinus carpio]                   93 5e-19
□ sp   P02616     PRVB_AMPME Parvalbumin beta [Amphiuma means]                 92 7e-19
□ tn   CAC95152   Parvalbumin beta protein [PRVB] [Rana esculenta]             92 9e-19
□ sp   P02617     PRVB_RANES Parvalbumin beta [Rana esculenta]                 92 9e-19
☑ sp   P43305     PRVU_CHICK Parvalbumin, thymic CPV3 (Parvalbumin 3) [G...    92 1e-18
□ tr   Q8UUS3     Parvalbumin [CYP C 1.01] [Cyprinus carpio]                   90 4e-18
□ sp   P05941     PRVB_OPSTA Parvalbumin beta (Parvalbumin IIIF) [Opsanu...    90 5e-18
```

Figure 9-4:
Output
selector on
the ExPASy
BLAST
page.

```
Retrieve sequences (FASTA format)   ▼
Clustal W (multiple alignment)
T-COFFEE (multiple alignment)
Retrieve entries (SWISS-PROT format)
Retrieve sequences (FASTA format)
```

There are three ways to export your sequences:

- **FASTA:** Generates a file that contains your sequences in FASTA format. You can save this file with the File⇨Save As option of your browser. When you need to, you can reopen this file with your browser, in order to cut and paste its content into another server form.

- **ClustalW:** Sends your sequences to the EMBnet ClustalW server. This causes a prefilled ClustalW form to appear in your browser. Click the Run ClustalW button to obtain a default alignment, or see the ClustalW steps list later in this chapter to know how to proceed.

- **Tcoffee:** Another multiple sequence alignment program. It's more accurate than ClustalW, but it can't align as many sequences. When the form appears, you can click the Run Tcoffee button to obtain a default alignment, or see the Tcoffee steps list in the chapter.

Gathering a known collection of sequences from SWISS-PROT

If you already know the name or accession number of every sequence you want to include in your multiple alignment and if these sequences are in SWISS-PROT or in TrEMBL, you can directly access them by using a special online ExPASy facility.

This can be handy if you're using a BLAST server that does not have sequence extraction facilities. You can simply copy the accession number of the sequences you are interested in and extract them here.

This procedure works only if the sequences that you're interested in belong to the SWISS-PROT or TrEMBL databases. The following steps list shows you how to use this server:

1. **Point your browser to** `www.expasy.ch/sprot/sprot-retrieve-list.html`.

 The SWISS-PROT/TrEMBL Retrieve a List of Entries page appears.

2. **Leave the File Name field blank so that the browser returns the sequences directly.**

3. **On the Format line, select the FASTA check box.**

4. **Enter the accession numbers of your sequences in the Sequence window, as shown in Figure 9-5.**

 Enter one accession number (or sequence name) per line. For our example, we entered P20472, P80079, P02626, P02619, P43305, P32930, Q91482, P02620, and P02622.

 The Upload a File Field is convenient if you have all your accession numbers in a file. You can generate this file with Microsoft Word, but it must be in text form. You can enter the filename in the box or use the Browse button and choose your file for upload.

5. **Underneath the field where you entered your accession numbers, click the Create FTP file button.**

 This submits your sequence request to the ExPASy server.

6. **Save the results into a text file with the File⇨Save As option of your browser.**

 This server gives the sequences names that are longer than 15 characters. These can give you trouble on some servers.

Format: ○ SWISS-PROT or ◉ FASTA

Note: If you do not enter a filename, the sequences will be sent directly to your browser.

- **Upload a file from your computer which contains a list of SWISS-PROT/TrEMBL/TrEMBL_new accession numbers (AC) or entry names (ID) or protein_ids, one per line:**

 [] Browse...

 [Create FTP file] or [Reset]

Figure 9-5:
Gathering
several
sequences
from the
ExPASy
server.

- **Or enter a list of SWISS-PROT/TrEMBL/TrEMBL_new accession numbers (AC), entry names (ID) or protein_ids, one per line:**

```
P20472
P80079
P02626
P02619
P43305
P32930
Q91482
P02620
P02622
```

Choosing the Right Multiple Sequence Alignment Method

Before you start making multiple sequence alignments, you must know that none of the methods available today are perfect. They all use approximations.

Building a multiple alignment that lets you make a real discovery requires some practice. The usual strategy requires comparing several alternative results and looking for robustness and stability.

In this section, we show you how to use ClustalW, the most commonly used multiple sequence alignment package. It is a simple, no-fuss, no-questions-asked kind of tool. We also show you how to use Tcoffee, one of the latest multiple sequence alignment packages that you can use. With Tcoffee, you can combine sequences and structures, evaluate an alignment, or merge several alternative multiple alignments into a single unified result.

Using ClustalW

ClustalW is by far the most commonly used program for making multiple sequence alignments. If you see a multiple alignment in a scientific publication, you can safely bet that the authors used ClustalW to generate it.

ClustalW is the latest and the most famous of the long series of Clustal programs. The conception of this software has been directed by Des Higgins who, according to legend, wrote the first Clustal on the back of an envelope while sipping pints of Guinness in a Dublin pub.

ClustalW uses a progressive method to build its alignments. Instead of aligning all the sequences at the same time, it adds them one by one. If you want to get the best out of ClustalW, understanding its underlying principle helps a lot.

Many ClustalW servers are around. They usually run the same version of this program, but their interfaces give you access to different options. At the end of this chapter, we give you a list of servers that run ClustalW. Shopping around to find a ClustalW server that's both fast *and* reliable is worth your time.

Running the EBI ClustalW server

ClustalW is a typical example of a program that can produce a reasonable output with its default settings. You should probably worry about the settings mostly only when you want to change the output format of the program.

Before you head off to a ClustalW server, you must do a little spade work ahead of time. Specifically, you need to gather together all the sequences you want to work with. If you're not quite sure how to do this, check out the section "Gathering a known collection of sequences from SWISS-PROT," earlier in this chapter.

The most convenient way to use ClustalW is to feed it sequences in FASTA format. However, you can also give to ClustalW a variety of formats, including SWISS-PROT and PIR (the Protein Information Resource format), as well as sequences in the most common multiple alignment formats.

If you give ClustalW a set of sequences already aligned, it does not remove the existing gaps. This means that the alignment you input influences the alignment ClustalW will produce.

The order in which you give sequences to ClustalW sometimes influences the alignment. If you change this order, even with the same sequences, the alignment may change.

With your sequences in hand, it's now time to fire up the ClustalW server:

1. **Point your browser to the EBI ClustalW server page at**
 `www.ebi.ac.uk/clustalw/index.html`.

 The ClustalW page dutifully appears.

2. **Paste the sequences you collected in the Sequence window.**

3. **Choose Fast from the Alignment pull-down menu (Figure 9-6).**

ClustalW behind the scenes

ClustalW is the latest of the Clustal software series. Clustal was the first multiple sequence alignment program that could run on almost any platform. Legend has it that Des Higgins made the original design for Clustal on the back of an envelop, in a smoky Dublin pub, in the early nineties. A few years later, when Des moved to the European Molecular Biology Laboratory in Heidelberg, he took Clustal along and kept it alive, turning it into one of the most successful series of bioinformatics programs: ClustalV, ClustalW and ClustalX, an X-window implementation.

Des Higgins did not truly invent the Clustal algorithm, Paula Hogeweg had already described it in the early eighties. What Des did was to put together all the right ingredients so that his program could be used by anybody on almost any computer. He also packaged and arranged everything so that it would be very easy to use.

These are some of the reasons why Clustal became an instant hit.

ClustalW uses a progressive algorithm. This means that it builds the alignment progressively, adding the sequences little by little until the complete multiple sequence alignment is finished. The reason for doing this is that state-of-the art sequence alignment programs find it hard to align more than two sequences at a time. Since these pairwise alignment programs are the only tools we've got in on hand to produce a multiple alignment, the only solution is to cheat a little. This is what the progressive algorithm does.

The trick is simple. You start comparing all your sequences two by two, so that you can cluster them by similarity (Clustal does this for you). The clustering looks like a phylogenetic tree. (It is the file with a .dnd extension that ClustalW outputs.) This clustering is named a dendrogram. If we were to align four sequences A, B, C and D, the dendrogram may look like this:

The topology of this dendrogram tells us a simple story: it says that A and B are more similar to each other than C and D. This means that if we align A with B, we are less likely to make a mistake than if we align A with C or D.

To make the progressive alignment, ClustalW follows the dendrogram topology: it starts aligning A with B. After this it aligns C with D. When

this is done, Clustal has two small multiple alignments (AB and CD). This is where Clustal pulls out its main trick: it aligns the two alignments as if each of them was a single sequence! It is not as complicated as it seems and there are many ways to do this. For instance, you could replace each alignment with a single consensus sequence. Clustal uses a slightly more sophisticated method but the idea is essentially

(continued)

(continued)

the same: it treats multiple alignments like single sequences and aligns them two by two.

Now you may ask, where do we cheat then? The answer is simple: we make a multiple alignment with ALL the sequences in our set, but we do not use ALL the information they contain. For instance, when Clustal aligns A and B, it does not use C and D. This could be a problem. if A and B are very different, we will produce a wrong pairwise alignment. This is a real waste if the two other sequences contain some useful information we did not use. Imagine, for instance, that A, B, C and D all contain a very short but important motif. This motif does not look so important when you only compare A and B, it only shows up when you look at all the sequences simultaneously. However, with a progressive strategy, C and D come too late to rescue the incorrect alignment of A and B.

Unfortunately, using all the information is much too expensive in terms of computation. This is why we cheat a little and use the progressive alignment. This should not worry you too much though. Even if it is greedy and approximate, Clustal often delivers pretty good alignments.

When you ask yourself which type of sequences are best suited for ClustalW, imagine that your sequences are like big stones spread across a shallow river. Making a multiple alignment is like crossing this river by jumping from stone to stone: it does not matter how large the river is, as long as you always find a stone to jump to. Similarly, it does not matter how many sequences you have and how far apart they are from each others, as long as a chain of correct alignments exists that can take you across the entire set. Of course, if a sequence or a small group of sequences is very different from the rest, you fall in the rapids!

It may sometimes be the case that your sequence set contains many identical or similar sequences. This is usually a problem because it makes it harder for sequences that belong to a minority subgroup to get properly aligned. If you can, you want to avoid this situation by removing the sequences yourself, but if you have no choice, it can be reassuring to know that Clustal W is equipped to deal with redundancy.

In case you were wondering, the W in ClustalW, does stand for Dubya, but here it only stands for *W*eights, not the U.S. President. ClustalW uses a sophisticated scheme so that very similar sequences do not end up dominating the multiple sequence alignment. In fact every sequence is supposed to receive a weight proportional to the amount of new information it contributes.

4. **Use the Output Format pull-down menu to set the selection of your choice.**

 Output formats have various pros and cons. See Chapter 10 for a discussion on this. It is safe to use Aln Without Numbers, the default ClustalW format.

 It is never too late to change a format. If you didn't generate your multiple alignment in the format that suits you best, DON'T recompute it! You can easily reformat alignments by using an online reformat utility such as ftmseq at www.bimcore.emory.edu/Pise/. (For more on reformatting, see Chapter 10.)

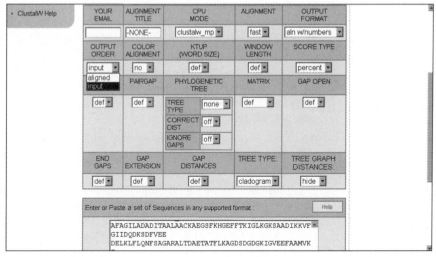

Figure 9-6:
The EBI
ClustalW
server.

5. Choose Input from the Output Order pull-down menu (Figure 9-6).

When you set the output order to Input, ClustalW outputs your sequences in their original order.

If you set the output order to Aligned, the sequences appear in the order they were aligned. This order depends on the guide tree, and it puts closely related sequences next to each other.

Although the Aligned output is more informative than Input, Aligned output makes it difficult to compare alignments generated with different methods or different parameters. Usually, it's better to prearrange your sequences in the most informative way and select the Input option.

6. Click the Run button at the bottom of the page.

An intermediate page appears. Wait until your browser displays the Results page.

7. Save your results.

Your results come in three sections:

- **Pairwise scores:** This is the default output of ClustalW. You can safely ignore the content of this section. It has mostly to do with the pairwise comparisons that ClustalW makes to figure out the guide tree.

- **The Multiple Alignment:** This section is in the middle of the output. It contains your alignment. You can display this alignment and save it as a text file if you click the hyperlink that comes just before the alignment.

- **The guide tree:** This section contains the tree that ClustalW used to guide its progressive alignment strategy. This tree is displayed in the phylip format (text and parenthesis) and with a graphic representation.

 You can save the Phylip tree format by clicking the hyperlink above it and then using the File⇨Save As feature of your browser. If you want to keep a trace of the tree graph, you need to print it or make a screen capture.

 When you mouse over the tree graph and click the left button, several options appear that enable you to control the appearance of your graph.

 The guide tree is NOT a phylogenetic tree. Confusing the ClustalW guide tree with a true phylogenetic tree is a major source of error. Producing a phylogenetic tree with ClustalW is possible but, if you want to do so, you must use ClustalW in a slightly different manner, which we explain in Chapter 13.

Changing ClustalW parameters

There are three parameters in ClustalW that can change your alignment: substitution matrices, gap-opening penalties, and gap-extension penalties. (Refer to Figure 9-6 to see the pull-down menus associated with these parameters.) Table 9-5 lists these three types of parameters along with potential consequences when you change them.

Table 9-5 Controlling the Effect of Parameter Tuning in ClustalW

Parameter	Effect
Substitution matrix	Substitution matrices control the cost of mutations in sequence alignments. (For more on cost, see Chapter 8.) If you select a category of matrices like PAM or BLOSUM, ClustalW automatically chooses the most adapted index. Predicting the effect of changing matrices is difficult, and there is no such thing as an ideal matrix. If your sequences are closely related, such a change has no effect. If your alignment is difficult to interpret, it may be worth changing from BLOSUM to PAM.
Gap-opening penalty	Gap-opening penalties (GOP) control the cost of opening gaps in your alignment. The higher the value, the more difficult it is to insert a gap in your alignment. The gap-opening penalty is applied once for the opening of each gap. Tuning has little effect because ClustalW readjusts these values automatically.
Gap-extension penalty	Gap-extension penalties control the size of the gaps. It is impossible to predict the optimal couple GOP/GEP, but it is clear that an optimal value exists for almost every protein family and that the only way to find this value is empirically.

Never change parameters in order to force ClustalW to produce an alignment that you know is right. If you know the correct alignment, edit the ClustalW alignment with one of the editors we show you in chapter 10. The only good justification for changing the parameters is if you want to see whether slight changes can improve the overall appearance of your alignment in terms of blocks and weakly conserved positions.

Making and evaluating alignments with Tcoffee

Tcoffee is one of the most recently developed methods for conducting multiple sequence alignments. It uses a principle that's a bit similar to ClustalW but yields more accurate alignments at the cost of a slightly longer running time. Tcoffee builds a progressive alignment like ClustalW, but it compares segments across the entire sequence set. Table 9-6 lists the main applications of Tcoffee.

Table 9-6	Possible Usages of the IGS Tcoffee Server
Usage	*Description*
Multiple alignment	Given a set of sequences, Tcoffee can produce a multiple sequence alignment.
Evaluation	Evaluate the reliability of an existing multiple alignment.
Using structures	If some of your sequences have a known structure, Tcoffee can use them to help the alignment. This can be useful if you're trying to align distantly related sequences.
Combining alignments	If you have several alignments of the same sequences produced with different methods (ClustalW and Tcoffee), you can use Tcoffee to combine these alignments into a single one. Tcoffee also shows you the regions where your alignments agree most.

Making a multiple alignment with Tcoffee

The following steps list shows you how you can make a multiple alignment with the IGS Tcoffee server. As was the case with ClustalW, you're going to have to gather together all the sequences you want to work with. Again, if you're not quite sure how to do this, check out the section "Gathering a known collection of sequences from SWISS-PROT," earlier in this chapter.

With your sequences in hand, it's Coffee Time!

1. **Point your browser to the Tcoffee server home page at** igs-server. cnrs-mrs.fr/Tcoffee.

2. Hover the mouse over Make a Multiple Alignment in the table and click Regular.

The Build a Multiple Alignment page appears (Figure 9-7).

3. Enter your e-mail address.

If your job times out, the program can return the results to you by e-mail.

4. Paste your sequences in the large window.

You can use most formats. If your sequences are in a text file, you can upload this file by using the Browse button.

5. Click the Run Tcoffee button at the top or the bottom of the page.

Tcoffee can be slow. Look at the real world through the window till your results return. . . .

6. Examine your results.

Tcoffee returns a table that contains hyperlinks to your results, as shown in Figure 9-8.

Results of T-Coffee

Your data will remain available on this server over the next 3 days. It will then be deleted.
Do not forget to bookmark this URL (http://193.50.234.246/~paralle/Tcoffee/tmp/232094154_result.html) or save it for further reference.

Multiple Alignment	aln	html	pdf
Phylogenetic tree	dnd	ph	png

log of t_coffee
Command line

3-D coffee?

If some of your sequences have a known 3-D structure, you can force Tcoffee to use these structures. To do so, modify the steps list in the section "Making a multiple alignment with Tcoffee" thusly:

1. **In your list of sequences, replace the name of the sequences whose structure you want to use with their PDB identifiers.**

 A *PDB identifier* is the name of the structure of your sequence in the PDB (a database of protein structures). If you don't know whether your sequence corresponds to a structure, the easiest thing to do is to make a BLAST between this sequence and PDB (see Chapter 11).

 If the sequence corresponds to only one chain in the PDB file, append the chain index to the PDB identifier. For instance, use 1gofB for chain B of the PDB structure 1gof.

In order to get the best from your structures, you must use at least two structures. A single structure has a limited effect.

2. **Point your browser to** `igs-server.cnrs-mrs.fr/Tcoffee`, **hover your mouse over Make a Multiple Alignment, and then click Advanced.**

3. **Enter your e-mail address and paste your sequences in the appropriate text fields.**

4. **Scroll down the page to the Parameters section and deselect the Fast(pair) option.**

5. **Select the Mixed(pair) option.**

 Mixed(pair) aligns each pair of sequences with an appropriate method. It does structure-structure alignment if both sequences have a known structure; otherwise, it does a standard sequence-sequence alignment.

6. **Run Tcoffee as you normally would and enjoy the results!**

The first row of the table is dedicated to multiple sequence alignments and includes

- **aln:** A text file in the same format as ClustalW alignments. Keep this file if you want to use your alignment as input for another program.

- **html:** A colorized alignment where every residue appears on a background that indicates the quality of this alignment. Red indicates high-quality segments while blue indicates regions of your alignment that you have no reason to trust.

- **pdf:** Same as the html but in PDF. It can be easier to display and print, depending on your browser's settings.

 These two files are only meant for display; you can't use them as an input for other sequence analysis programs.

The second row is dedicated to phylogenetic trees:

- **dnd:** the guide tree or dendrogram generated by Tcoffee in Newick format (see Chapter 13). You should not use it in place of the true phylogenetic tree.

- **ph:** The true phylogenetic tree in Newick format, generated from the Tcoffee multiple alignment by using the Neighbor Joining method (see Chapter 13). This is not a guide tree but a real phylogenetic tree.

- **png:** The gif picture of the phylogenetic tree that corresponds to the ph file.

Evaluating the quality of an alignment with Tcoffee

If you want, you can give Tcoffee a multiple alignment that you generated with your favorite method (or by hand if you are a specialist), and you can ask Tcoffee to evaluate the quality of this multiple sequence alignment for you. This can give you an idea on the portions of your alignment that you can trust and those that it is safer to ignore.

You can easily do such an evaluation on the Tcoffee server (`igs-server.cnrs-mrs.fr/Tcoffee`). Simply hover the mouse over the Evaluate a Multiple Alignment option, click Regular, and then follow the instructions. You can use any of the most common formats (MSF, ALN, FASTA, and PIR).

These evaluations are only empirical, and they do not replace E-values. Nonetheless, it is useful to know that residues with a yellow/orange/red background have an index above 5 (out of 10) and have more than an 80 percent chance of being correctly aligned.

The nice thing about the libraries is that they can contain whatever you want: pairwise alignments, multiple alignments, global or local, structure-based sequence alignments, or even alignments you made yourself by using experimental information. Tcoffee also makes it possible to measure the local agreement between your multiple sequence alignment and any library.

Tcoffee and sequence democracy

Tcoffee is a progressive alignment method and, in many ways, is a proud relative of ClustalW. The main difference between Tcoffee and ClustalW is that Tcoffee doesn't directly use substitution matrices to align sequences; it is much more lazy and simply relies on other methods to work for it.

When you give sequences to Tcoffee, it starts making pairwise comparisons. To do so, it takes every possible pair of sequences and makes a global alignment with ClustalW. It also makes a Lalign comparison for each of these pairs. (Lalign is a local alignment method developed by Huang and Miller to do pairwise local alignments.) Given two sequences, Lalign produces the ten best local alignments. The complete collection of local and global alignments is a library.

After the library is finished, Tcoffee builds a multiple alignment that has the highest possible agreement with all the pairwise alignments in the library. It does so by using a progressive algorithm like ClustalW. This process looks very much like a general election, where all the bits of information contained in the library are competing to find their way into the final alignment.

Interpreting Your Multiple Sequence Alignment

Interpreting an alignment is a bit of an art. E-values — the scores that tell you how reliable your database search is — do not (yet) exist for multiple sequence alignments, which means that deciding whether your alignment is correct or not still involves some educated guesswork.

DNA alignments are by far the most difficult to interpret. If you're analyzing this type of sequence, you want a very high level of conservation, knowing that single conserved columns are likely to be meaningless. A DNA block is only informative when it contains several identical columns in a cluster. Even with the DNA of closely related sequences, obtaining such an alignment is still difficult. This is why most biologists (when they have a choice) much prefer protein alignments.

Recognizing the good parts in a protein alignment

The most convincing evaluative grid we have for a protein multiple alignment stems from our knowledge of protein structures (see Chapter 11). We know that structures contain surface loops that evolve rapidly. Loops are softer portions of the protein that connect its more rigid portions. Protein structures also contain core regions inside the protein that act as support walls for the protein. These support walls evolve less rapidly than the loops on the surface (see Chapter 11).

In your multiple alignment, you can expect to find nice, gap-free blocks that correspond to the core regions and gap-rich regions that correspond to the loops. The alignment shown in Figure 9-9 (which appears in the following section, "Taking your multiple alignment further") is a good illustration of this principle.

Now, how can you tell whether a block is a good block or not? When you look at a ClustalW or a Tcoffee alignment, you can see that the last line contains seemingly cabalistic signs such as (*), (:), or (.). These three symbols have very precise meanings:

- (*) A star indicates an entirely conserved column.
- (:) A colon indicates columns where all the residues have roughly the same size and the same hydropathy.
- (.) A period indicates columns where the size *or* the hydropathy has been preserved in the course of evolution.

After you get used to them, these indications are really quite priceless. For instance, your average *good block* is a unit at least 10-30 amino acids long exhibiting at least one to three stars (*), five to seven colons (:), and a few periods (.) sprinkled here and there. This is exactly what you see in the alignment shown in Figure 9-9.

The magic thing about multiple sequence alignments is that 4 or 5 conserved positions over 50 amino acids can be enough to convince us that we're looking at a genuine signal. This is less than 10 percent identity! If you remember that we require at least 25 percent identity to consider a pairwise alignment interesting, you can easily understand why multiple alignments are so powerful.

Another criterion when looking at a multiple alignment is knowing the type of amino acids you can expect to see conserved. Amino acids aren't equal and they all have very characteristic patterns of mutation/conservation in a multiple sequence alignment. Table 9-7 lists the most common features associated with some conserved columns you may come across.

Conserved columns in a multiple sequence alignment are meaningful only when the surrounding columns are not conserved.

Table 9-7	Patterns of Conservation in Multiple Sequence Alignments
Amino Acid	*Characteristic*
W, Y, F	It is common to find conserved tryptophans. Tryptophan is a large hydrophobic residue that sits deep in the core of proteins. It plays an important role in their stability and is therefore difficult to mutate. When tryptophan mutates, it is usually replaced by another aromatic amino acid, such as phenylalanine or tyrosine. Patterns of conserved aromatic amino acids constitute the most common signatures for recognizing protein domains.
G, P	It is common to find conserved columns with a glycine or a proline in a multiple alignment. These two amino acids often coincide with the extremities of well-structured beta strands or alpha helices. (For more on these structures, see Chapter 11.)
C	Cysteines are famous for making C-C (disulphide) bridges. Conserved columns of cysteines are rather common and usually indicate such bridges. Columns of conserved cysteines with a specific distance provide a useful signature for recognizing protein domains and folds.
H, S	Histidine and serine are often involved in catalytic sites, especially those of proteases. Conserved histidine or a conserved serine are good candidates for being part of an active site.

Amino Acid	Characteristic
K, R, D, E	These charged amino acids are often involved in ligand binding. Highly conserved columns can also indicate a salt bridge inside the core of the protein.
L	Leucines are rarely very conserved unless they're involved in protein-protein interactions such as a leucine zipper.

Taking your multiple alignment further

One sentence summarizes what you really want from your multiple alignment: You want to identify important positions! You want to find the amino acids that are not allowed to mutate, those that you find conserved even when aligning distantly related proteins.

Consider, for instance the alignment in Figure 9-9. You can use any ClustalW or Tcoffee server to generate this multiple sequence alignment using any server you fancy. It is a good alignment: It contains distantly related proteins, and it is beginning to tell us a nice story about the various components of our protein family. For instance, we can clearly see that the N-terminus region seems to be more conserved than the C-terminus region. In the N-terminus, we can see a short stretch of highly conserved amino acids that make this region a good candidate for being a binding or an active site.

```
! Sequence: BLASTP:temp_job1_swissprot_34_PRV1_SALSA
! Sequence: BLASTP:temp_job1_swissprot_36_PRVB_MERME
! Sequence: BLASTP:temp_job1_swissprot_38_PRVB_GADCA
CLUSTAL W (1.82) multiple sequence alignment

temp_job1_swissprot_1_PRVA_HUM     SMTDLLNAEDIKKAVGAFSATDSFDHKKFFQMVGLKKKSADDVKKVFHML
temp_job1_swissprot_7_PRVA_FEL     SMTDLLGAEDIKKAVEAFTAVDSFDYKKFFQMVGLKKKSPDDIKKVFHIL
temp_job1_swissprot_13_PRVA_AM     SMTDVIPEADINKAIHAFKAGEAFDFKKFVHLLGLNKRSPADVTKAFHIL
temp_job1_swissprot_19_PRVB_ES     -----LKDADVAAALAACSAADSFKHKEFFAKVGLASKSLDDVKKAFYVI
temp_job1_swissprot_23_PRVU_CH     SLTDILSPSDIAAALRDCQAPDSFSPKKFFQISGMSKKSSSQLKEIFRIL
temp_job1_swissprot_26_ONCO_HU     SITDVLSADDIAAALQECQDPDTFEPQKFFQTSGLSKMSANQVKDVFRFI
temp_job1_swissprot_34_PRV1_SA     ----LCKEADIKTALEACKAADTFSFKTFFHTIGFASKSADDVKKAFKVI
temp_job1_swissprot_36_PRVB_ME     ----ILADADITAALAACKAEGSFKHGEFFTKIGLKGKSAADIKKVFGII
temp_job1_swissprot_38_PRVB_GA     ----ILSNADIKAAEAACFKEGSFDEDGFYAKVGLDAFSADELKKLFKIA
                                   : :.    *:  *    :*.   *   *:  *  ::.. *  .

temp_job1_swissprot_1_PRVA_HUM     DKDKSGFIEEDELGFILKGFSPDARDLSAKETKMLMAAGDKDGDGKIGVD
temp_job1_swissprot_7_PRVA_FEL     DKDKSGFIEEDELGFILKGFYPDARDLSVKETKMLMAAGDKDGDGKIDVD
temp_job1_swissprot_13_PRVA_AM     DKDRSGYIEEEELQLILKGFSKEGRELTDKETKDLLIKGDKDGDGKIGVD
temp_job1_swissprot_19_PRVB_ES     DQDKSGFIEEDELKLFLQNFSPSARALTDAETKAFLADGDKDGDGMIGVD
temp_job1_swissprot_23_PRVU_CH     DNDQSGFIEEDELKYFLQRFECGARVLTASETKTFLAAADHDGDGKIGAE
temp_job1_swissprot_26_ONCO_HU     DNDQSGYLDEEELKFFLQKFESGARELTESETKSLMAAADNDGDGKIGAE
temp_job1_swissprot_34_PRV1_SA     DQDASGFIEVEELKLFLQNFCPKARELTDAETKAFLKAGDADGDGMIGID
temp_job1_swissprot_36_PRVB_ME     DQDKSDFVEEDELKLFLQNFSAGARALTDAETATFLKAGDSDGDGKIGVE
temp_job1_swissprot_38_PRVB_GA     DEDKEGFIEEDELKLFLIAFAADLRALTDAETKAFLKAGDSDGDGKIGVD
                                   *:* ..::: :** :*  *     *:  ** ::  .* **** *. :
```

Figure 9-9: A good multiple sequence alignment.

This is interesting, but it isn't enough to make a big story. This alignment still contains too many conserved positions for a detailed analysis. At this point, what we could do is add a few distantly related sequences, one by one, and carefully check the effect of these sequences on the overall alignment quality. More specifically, we want to make sure that these distantly related sequences actually enhance existing patterns rather than completely destroy blocks.

Here is a possible strategy to further reveal the evolutionary constraints within our protein. This strategy relies on the integration within the multiple alignment of precisely those sequences that BLAST reported as marginal hits when we first scanned SWISS-PROT for homologues of the human parvalbumin. (If this all sounds a bit unfamiliar, take a look at the "Selecting sequences on the ExPASy server" section, earlier in this chapter.) You don't actually need to rerun BLAST to use this example. We give you the info you need to know. First and foremost, go ahead and gather your sequences as follows:

1. **Point your browser to** www.expasy.ch/sprot/sprot-retrieve-list.html.

 The SWISS-PROT/TrEMBL: Retrieve a List of Entries page appears.

2. **In the Format line, select the FASTA radio button.**

3. **Enter the accession number of your sequences in the Sequence window.**

 Enter one accession number per line. For our example, we entered P20472, P80079, P02626, P02619, P43305, P32930, Q91482, P02620, P02622, **P02586**.

 P02586 is the TPCS_RABIT, the troponin C of rabbit. In the BLAST of the human parvalbumin against SWISS-PROT that we used to select the other sequences, BLAST reported this hit with an E-value of 5!

 On its own, this result is not interesting, but now that we have a multiple sequence alignment, we can see if this rabbit can tell us something about our human protein.

4. **Click the Create FTP File button.**

5. **Copy the sequences onto the Clipboard.**

Now that you've got all sequences in your basket (well, on the Clipboard, at least), get ready to do some alignment work:

1. **Point your browser to the EBI ClustalW server home page at** www.ebi.ac.uk/clustalw/index.html.

2. **Paste the sequences you gathered in the preceding steps list into the Sequence window.**

3. **Choose Fast from the Alignment pull-down menu.**

4. **Use the Output Format pull-down menu to choose the output format you want.**

5. **Choose Input from the Output Order pull-down menu.**

6. **Click the Run button at the bottom of the page.**

7. **Save your results.**

You can see in Figure 9-10 that the new sequence respects the blocks that already existed (Figure 9-9), while shunting some conserved positions. It also reveals regions where insertion and deletions are likely to occur. These regions are good candidates for being loops. (Remember that loops are portions of the protein structure that are a bit more flexible than the rest of the structure and act as connectors; see Chapter 11 for more details.)

Figure 9-10:
Making a multiple alignment of distantly related proteins.

Figure 9-10 also reveals the existence of two conserved regions in the N-terminus of the human parvalbumin. This is an interesting result that narrows down the range of possibilities if we were looking for the amino acids responsible for the function of our protein.

At this point, going a bit further and adding another distantly related protein is a good idea. Our aim is to check that these few highly conserved regions are indeed conserved across the whole protein family, even when we compare distantly related cousins.

To do this, you can go once again through the two previous steps lists by using an extra sequence:

P20472, P80079, P02626, P02619, P43305, P32930, Q91482, P02620, P02622, P02586, **P19123**

P19123 is TPCC_MOUSE, the mouse troponin C. It is a very remote homologue of the human parvalbumin. Figure 9-11 shows the result of the inclusion of this new protein in our multiple sequence alignment. It clearly shows that most conserved columns remain.

With such a result, we can safely bet that these two conserved regions are surely involved in some biological function. We know that most of these proteins bind calcium, so we can safely bet that the calcium-binding site involves some of these conserved positions. This is indeed the case, and if you look at the SWISS-PROT annotation, you can see that these regions correspond to the portion of the human parvalbumin involved in the binding of calcium.

Figure 9-11:
Including more distant relatives in the multiple alignment.

```
sp|P20472|PRVA_HUMAN    -----------------------------------------------SMTDLLNAEDIK
sp|P80079|PRVA_FELCA    -----------------------------------------------SMTDLLGAEDIK
sp|P02626|PRVA_AMPME    -----------------------------------------------SMTDVIPEADIN
sp|P02619|PRVB_ESOLU    -----------------------------------------------SFAGL-KDADVA
sp|P43305|PRVU_CHICK    -----------------------------------------------MSLTDILSPSDIA
sp|P32930|ONCO_HUMAN    -----------------------------------------------SITDVLSADDIA
sp|Q91482|PRV1_SALSA    -----------------------------------------------MACAHLCKEADIK
sp|P02620|PRVB_MERME    -----------------------------------------------AFAGILADADIT
sp|P02622|PRVB_GADCA    -----------------------------------------------AFKGILSNADIK
sp|P02586|TPCS_RABIT    -TDQQAEARSYLSEEMIAEFKAAFDMFDADGG-GDISVKELGTVMRMLGQTPTKEELDAI
sp|P19123|TPCC_MOUSE    MDDIYKAAVEQLTEEQKHEFKAAFDIFVLGAEDGCISTKELGKVMRMLGQNPTPEELQEM
                                                                    :

sp|P20472|PRVA_HUMAN    KAVGAFSATDSFDHKKFFQMVG------LKKKSADDVKKVFHMLDKDKSGFIEEDELGFI
sp|P80079|PRVA_FELCA    KAVEAFTAVDSFDYKKFFQMVG------LKKKSPDDIKKVFHILDKDKSGFIEEDELGFI
sp|P02626|PRVA_AMPME    KAIHAFKAGEAFDFKKFVHLLG------LNKRSPADVTKAFHILDKDRSGYIEEEELQLI
sp|P02619|PRVB_ESOLU    AALAACSAADSFKHKEFFAKVG------LASKSLDDVKKAFYVIDQDKSGFIEEDELKLF
sp|P43305|PRVU_CHICK    AALRDCQAPDSFSPKKFFQISG------MSKKSSSQLKEIFRILDNDQSGFIEEDELKYF
sp|P32930|ONCO_HUMAN    AALQECQDPDTFEPQKFFQTSG------LSKMSANQVKDVFRFIDNDQSGYLDEEELKFF
sp|Q91482|PRV1_SALSA    TALEACKAADTFSFKTFFHTIG------FASKSADDVKKAFKVIDQDASGFIEVEELKLF
sp|P02620|PRVB_MERME    AALAACKAEGSFKHGEFFTKIG------LKGKSAADIKKVFGIIDQDKSDFVEEDELKLF
sp|P02622|PRVB_GADCA    AAEAACFKEGSFDEDGFYAKVG------LDAFSADELKKLFKIADEDKEGFIEEDELKLF
sp|P02586|TPCS_RABIT    IEEVDEDGSGTIDFEEFLVMMVRQMKEDAKGKSEEELAECFRIFDRNADGYIDAEELAEI
sp|P19123|TPCC_MOUSE    IDEVDEDGSGTVDFDEFLVMMVRCMKDDSKGKSEEELSDLFRMFDKNADGYIDLDELKMM
                        ::.  *          *   *..  .::; :**   :

sp|P20472|PRVA_HUMAN    LKGFSPDARDLSAKETKMLMAAGDKDGDGKIGVDEFSTLVAES----
sp|P80079|PRVA_FELCA    LKGFYPDARDLSVKETKMLMAAGDKDGDGKIDVDEFFSLVAKS----
sp|P02626|PRVA_AMPME    LKGFSKEGRELTDKETKDLLIKGDKDGDGKIGVDEFTSLVAES----
sp|P02619|PRVB_ESOLU    LQNFSPSARALTDAETKAFLADGDKDGDGMIGVDEFAAMI-----KA
sp|P43305|PRVU_CHICK    LQRFECGARVLTASETKTFLAAADHDGDGKIGAEEFQEMVQS-----
sp|P32930|ONCO_HUMAN    LQKFESGAREITESETKSLMAAADNDGDGKIGAEEFQEMVHS-----
sp|Q91482|PRV1_SALSA    LQNFCPKARELTDAETKAFLKAGDADGDGMIGIDEFAVLV-----KQ
sp|P02620|PRVB_MERME    LQNFSAGARALTDAETATFLKAGDSDGDGKIGVEEFAAMV-----KG
sp|P02622|PRVB_GADCA    LIAFAADLRALTDAETKAFLKAGDSDGDGKIGVDEFGALVDKVGAKG
sp|P02586|TPCS_RABIT    FRASG---EHVTDEEIESLMKDGDKNNDGRIDFDEFLKMMEGVQ---
sp|P19123|TPCC_MOUSE    LQATG---ETITEDDIEELMKDGDKNNDGRIDYDEFLEFMKGVE---
                        :     .:: :   :: .* :.** *. :** ::
```

Comparing Sequences That You Can't Align

Sometimes you need to compare sequences that don't necessarily have a common ancestor or are so distantly related that considering them as homologous is difficult.

Multiple sequence alignment programs are usually hopeless in these situations. They'll include proteins that contain homologous domains but that are otherwise unrelated or segments of DNA that seem to contain similar regulating features although it is impossible to see anything common between them in a multiple alignment.

When a situation like this arises, do not expect miracles from bioinformatics. In most cases, your first intuition that there is nothing to get out of these sequences proves to be the right one. Nonetheless, before giving up completely, it is worth trying the two techniques we only briefly introduce in this section. The first one, the Gibbs sampler, simultaneously looks into all your sequences for short, partially conserved gap-free segments. Pratt, the second method, looks for flexible patterns, a special category of segments that can contain gaps and only needs to be conserved at certain positions.

Making multiple local alignments with the Gibbs sampler

The Gibbs sampler is a stochastic method: It first scrambles your sequences, aligns them randomly until a good solution appears, and then keeps scrambling the sequences to improve this good solution. The interesting thing about using chance is that you can solve complicated problems without having to explore all the possible solutions.

Gurus name this type of method *stochastic* because it contains an element of chance. Stochastic methods are by far the most powerful in bioinformatics. Unfortunately, they do not have the popularity they deserve because they make it difficult to reproduce the same results twice. For instance, if you run the Gibbs sampler twice on the same set of sequences, you may not get exactly the same solution. For most biologists, reproducibility is paramount, and — let's face it — we all hate it when computers change their minds from one minute to the next.

Nonetheless, if you're ready for a bit of fuzzy logic, you may find that the Gibbs sampler can offer very sensible solutions to extremely complicated problems. For instance, the Gibbs sampler is very good at identifying HTH (Helix Turn Helix) domains across a protein family. It is also a nice way to search for regulatory elements shared by otherwise unrelated DNA sequences. The Pasteur institute maintains a Gibbs sampler server that you can access at `bioweb.pasteur.fr/seqanal/interfaces/gibbs-simple.html`.

When using it, bear in mind that, to be accurate, the Gibbs sampler needs as many sequences as possible. You should not use it with less than 20 sequences.

Searching conserved patterns

The Gibbs sampler is useful only when the segments you're looking for have exactly the same length, like Helix-Turn-Helix domains. However, the vast majority of the interesting motifs you may come across when doing bioinformatics have different lengths and the most standard situation is to find motifs that are poorly conserved. When this happens, your only hope is to find a few highly conserved amino acids or nucleotides that anchor the motif.

If your sequences are poorly related but contain such motifs, the only way to analyze them is to use a pattern-finding motif. If you want to know more about patterns (biologists often call them PROSITE patterns), you can go to Chapter 7. Several tools exist for identifying these conserved patterns in a set of unaligned sequences. Showing you how to use these tools and to interpret their result is beyond the scope of this book. However, if you want to experiment with them, you should know that Pratt is one of the most powerful because it allows some flexible spacing between the conserved positions. You can also use TEIRESIAS, MEME, or SMILE. Table 9-10 lists some of these resources.

Internet Resources for Doing Multiple Sequence Comparisons

The amount of resources for making multiple alignments online is almost overwhelming. The usual words of caution apply to the resources that we list in the following tables: Use stable resources and always make a few simple tests to make sure that the service you're using does what it's supposed to do. Never depend blindly on a resource you don't control. As they say, "Good servers go bad and bad servers go down."

If you want to do many multiple sequence alignments — or if your company does not authorize you to send your sequences over the Internet — you may have to install these programs on your own machine. We give you addresses where you can download the source code or the executable files of some of these programs.

Making multiple alignments with ClustalW around the clock

ClustalW is the most common online program. Table 9-8 gives you a list of available servers, starting with the site where you can download ClustalW to install it on your own machine.

Note: The last link in Table 9-8 is the location where you can download the executable to install it on your own machine. It is *not* a Web server.

Table 9-8	A List of ClustalW Servers	
Name	*Location*	*Address*
EBI	Europe	www.ebi.ac.uk/clustalw/index.html
EMBnet	Europe	www.ch.embnet.org/software/ClustalW.html
PIR	USA	pir.georgetown.edu/pirwww/search/multaln.html
BCM	USA	searchlauncher.bcm.tmc.edu/multi-align/multi-align.html
GenomeNet	Japan	clustalw.genome.ad.jp/
DDBJ	Japan	www.ddbj.nig.ac.jp/E-mail/clustalw-e.html
Strasbourg	Europe	ftp://ftp-igbmc.u-strasbg.fr/pub/ClustalW/

Finding your favorite alignment method

If ClustalW isn't suited for the sequences you want to align, Table 9-9 saves the day by giving you the addresses of a few other servers.

Table 9-9	Multiple Sequence Alignment Resources over the Internet
Method	*Address*
Tcoffee	igs-server.cnrs-mrs.fr/Tcoffee www.ch.embnet.org/software/TCoffee.html www.cmbi.kun.nl/bioinf/tools/T_COFFEE/
Dialign	bibiserv.techfak.uni-bielefeld.de/dialing/bioweb.pasteur.fr/seqanal/interfaces/dialign2-simple.html
DCA	bibiserv.techfak.uni-bielefeld.de/cgi-bin/dca_submit

Searching for motifs or patterns

If your sequences are too distantly related for making a multiple sequence alignment, your last chance is trying to find a conserved pattern across them. Table 9-10 lists some of the online methods you can use for this purpose.

Table 9-10	Motif-finding Methods Available Online
Method	**Address**
Gibbs Sampler	bioweb.pasteur.fr/seqanal/interfaces/ gibbs-simple.html, bayesweb.wadsworth.org/ gibbs/gibbs.html
Pratt	www.ebi.ac.uk/pratt/index.html
eMotif	dna.stanford.edu/emotif/
MEME	meme.sdsc.edu/meme/website/
TEIRESIAS	cbcsrv.watson.ibm.com/Tspd.html
Bioprospector	http://bioprospector.stanford.edu/
Improbizer	www.soe.ucsc.edu/~kent/improbizer/ improbizer.html
BLOCK-Maker	blocks.fhcrc.org/blocks/blockmkr/ make_blocks.html

Chapter 10

Editing and Publishing Alignments

- -

In This Chapter

▶ Reformatting your multiple sequence alignment

▶ Manually editing a multiple sequence alignment with Jalview

▶ Shading a multiple sequence alignment with Boxshade

- -

It looked so good that I thought it just had to be genuine.

— Everyone's favorite secret thought

*T*his chapter is all about using and modifying an existing multiple sequence alignment. If you don't have one ready yet, don't worry; we start by showing you how to generate a dummy multiple sequence alignment that you can use throughout the entire chapter.

The first real section in this chapter — as opposed to the kind-of-real section about making a dummy multiple sequence alignment — tells you what gurus assume we all know about alignment formats. We also show you how to decide which format you want to use (or avoid!) and, generally speaking, how to find your way across the format jungle.

If you have generated your multiple sequence alignment with any kind of automatic method, you probably need to arrange (edit) it manually before you can really use it. When you stop to think about it, editing a multiple sequence alignment isn't trivial: It requires inserting gaps in an entire sub-group of sequences or shifting several sequences all at once, and so on. When you do this with a standard word processor, pretty soon you're going to feel as if you're playing with a Rubik's Cube. No matter how slowly and carefully you start, after three minutes of moving things about, everything is out of control! In order to avoid this kind of confusion, we show you some powerful editing tools in this chapter that you can use over the Internet.

After you have the alignment of your dreams, there's a good chance that you may want to prepare it for publication. Nothing looks nicer than a beautiful multiple alignment with a well-chosen color scheme and a few pieces of annotation where it matters. In fact, good-looking alignments are so convincing that you should convince yourself of the quality of your alignment *before* beautifying it! Several online tools are available to do this, and we introduce them in this chapter.

Finally, we show you a couple of tools that can help you extract information from your multiple sequence alignment. These include methods such as the protein Logos, for identifying the most conserved positions.

All in all, this is a small chapter, but if you decide to go through it, you'll have no more excuse for attaching simple raw text alignments to your reports, publications, or lab books!

If you do not have an alignment ready, you can download one for your crash tests from igs-server.cnrs-mrs.fr/Tcoffee/dummy_aln.html. It is an alignment in ALN format (see Figure 10-1).

Figure 10-1:
A multiple
sequence
alignment in
ALN or
ClustalW
format.

```
CLUSTAL W (1.74) multiple sequence alignment

CYC1_YEAST       -------------------TEFKAGSAKKGATLFKTRCLQCHTVEKGGP-------HKVG
CYC_ORYSA        ----------------ASFSEAPPGNPKAGEKIFKTKCAQCHTVDKGAG-------HKQG
CY2_RHOVI        ----MRKLVFGLFVLAASVAPAAAQDAASGEQVFK-QCLVCHSIGPGAK-------NKVG
CY2_RHOGL        -------------------GSAPPGDPVEGKHLFHTICILCHTDIKGR--------NKVG
CY2_RHORU        MKKGFLAAGVFAAVAFASGAALAEGDAAAGEKVSK-KCLACHTFDQGGA-------NKVG
C550_PARDE       -MKISIY-ATLAAITLALPAAAQDGDAAKGEKEFN-KCKACHMIQAPDGTDII-KGGKTG
CY2_RHOCA        -MKISLTAATVAALVLAAPAFA--GDAAKGEKEFN-KCKTCHSIIAPDGTEIV-KGAKTG
CY2_RHOSH        -MKFQVKALAAIAAFAALPALAQEGDPEAGAKAFN-QCQTCHVIVDESGTTIAGRNAKTG
CYC4_PSEST       ---MNKVLVSLLLTLGITGMAHAAGDAEAGQGKVA-VCGACHGVDGNSPAPNFPKLAGQG
                                          .. *      *  **               *

CYC1_YEAST       PNLHG------IFGRHSGQAEGYS-YTDANIKKN-----VLWDENNMSEYLT--------
CYC_ORYSA        PNLNG------LFGRQSGTTPGYS-YSTANKNMA-----VIWEENTLYDYLL--------
CY2_RHOVI        PVLNG------FLGRHSGTIEGFA-YSDANKNSG-----ITWTEEVFREYIR--------
CY2_RHOGL        PSLYG------VVGRHSGIEPGYN-YSEANIKSG-----IVWTPDVLFKYIE--------
CY2_RHORU        PNLFG------VFENTAAHKDDYA-YSESYTEMKAKG--LTWTEANLAAYVK--------
C550_PARDE       PNLYG------VVGRKIASEEGFK-YGEGILEVAEKNPDLTWTEADLIEYVT--------
CY2_RHOCA        PNLYG------VVGRTAGTYPEFK-YKDSIVALGASG--FAWTEEDIATYVK--------
CY2_RHOSH        PNLYG------VVGRTAGTQADFKGYGEGMKEAGAKG--LAWDEEHFVQYVQ--------
CYC4_PSEST       ERYLLKQLQDIKAGSTPGAPEGVG--RKVLEMTGMLDPLSDQDLEDIAAYFSSQKGSVGY
                                          .             : *.

CYC1_YEAST       -NP-------------------------KKYIPGTKMAFGGL-------------KK
CYC_ORYSA        -NP-------------------------KKYIPGTKMVFPGL-------------KK
CY2_RHOVI        -DP-------------------------KAKIPGTKMIFAGV-------------KD
```

Getting Your Multiple Alignment in the Right Format

When using online servers — or even local programs — you do not always have full control of what goes in and what comes out of the program you use. For instance, many multiple sequence alignment programs commonly output only one format: MSF (Multiple Sequence Format). What can you do if you want to analyze this multiple sequence alignment with a program that only reads FASTA formatted alignments? The answer is simple: Use a reformatting program just like the ones we introduce in this chapter.

When it comes to multiple sequence alignments, there is no such thing as a perfect format. Anyone used to dealing with the many services available online knows that each of them tends to have its favorite format. (See the "A format for everyone: Democracy or anarchy?" sidebar if you want proof of this fact.) Understanding everything there is to know about formats is an ability you only acquire with some practice, yet in this chapter we give you some ideas, and we tell you how to be on the lookout for potential problems.

Unfortunately, nothing is standardized yet, and the number of programs able to use XML is still limited. To be honest, you can expect the ancient formats we describe in this section to still be around when your grandchildren retire in their early nineties.

When you choose to store your data in a specific format, you must ask yourself four questions:

- ✔ Do most programs support this format?
- ✔ Will my collaborators be able to use it?
- ✔ Can I store all the information I need with this format?
- ✔ Is it easy to manipulate?

Today no format is so perfect that you would answer yes to each of these questions; any choice relies on the right trade-off between what you really need and what's available. If you can, avoid using a mixture of formats when running a project.

The taxonomy of multiple sequence alignment formats is rather complex. Table 10-1 lists the most common formats you may come across when manipulating multiple sequence alignments.

Table 10-1	A Classification of Multiple Sequence Alignment Formats	
Name	*Type*	*Usage*
post-script, pdf, html	Graphic	Terminal formats suitable for printing only
FASTA (Figure 10-2)	Text	Easy to manipulate Noninterleaved (not readable by humans) Easy for renaming sequences Supported by most programs Cannot incorporate extra annotation
PIR	Text	Similar to FASTA but with an extra line Can incorporate limited extra annotation
MSF (Figure 10-3)	Text	Most standard multiple alignment format Difficult to modify manually Interleaved (easy to read for humans) Can include extra annotation, such as weights Supported by most programs, but some programs only support it partially!
Selex	Text	Extended version of MSF Can include extra annotation Supported by few programs
ALN (Figure 10-1)	Text	Simplified version of MSF Default output of ClustalW Interleaved (easy to read for humans) Supported by many programs
Phylip	Text	Variant of ALN Useful for doing phylogenetic analysis Supported by most phylogenetic packages

Recognizing the main formats

The alignments you see in publications are usually in a graphic format that isn't very useful for doing analyses. The last section of this chapter shows you resources that you can use for generating these alignments. You've already seen the ALN format in this chapter (refer to Figure 10-1) and FASTA is another format (see Figure 10-2) that has cropped up a number of times in previous chapters. In the lab, MSF, as shown in Figure 10-3, is probably the most common format. Note that MSF and ALN are *interleaved* (meaning easier for humans to read but not for machines) whereas FASTA is not.

A format for everyone: Democracy or anarchy?

The variety of formats is a major curse in bioinformatics. During a long period of time, the tradition was that anyone developing a new program would design his or her own house format. Each new format was slightly different from the others and especially appealing to a particular category of biologists. For instance, people doing phylogeny became very attached to the Phylip format, whereas the users of GCG, a popular bioinformatics package, preferred MSF, and so on. Whether we like it or not, the weight of history has made each of these formats totally acceptable, and today they all live in perfect harmony, side by side on our computer keyboards — or do they?

Another source of confusion in the field was the communication wars between biologists and computer scientists. For many years, biologists have complained about the computer scientist

formats, basically saying, "We cannot make sense of this gibberish." In retaliation (and because they needed to), biologists also created formats that they could use, to the contempt of computer scientists who dubbed these formats "amateurish and ambiguous." Both the biologists and the computer scientists were probably right in their respective evaluations.

Things may be about to change with the XML language coming to the fore. XML (eXtensible Markup Language) is an extended version of HTML — the language of the Web. XML makes it possible to simply describe your data with keywords everyone can agree on. Today, biologists and computer scientists consent — albeit weakly — that XML could be the solution and the main bioinformatics programs are now able to produce output in XML.

Figure 10-2:
Multiple
sequence
alignment in
FASTA
format.

```
>CYC1_YEAST, 216 bases, B0EA875C checksum.
...................TEFKAGSAKKGATLFKTRCLQCHTVEKGGP.
......HKVGPNLHG......IFGRHSGQAEGYS.YTDANIKKN.....V
LWDENNMSEYLT.........NP...........................
KKYIPGTKMAFGGL..............KKEKDRN...DLITYLKKACE.
.................
>CYC_ORYSA, 216 bases, 9C7091DB checksum.
................ASFSEAPPGNPKAGEKIFKTKCAQCHTVDKGAG.
......HKQGPNLNG......LFGRQSGTTPGYS.YSTANKNMA.....V
IWEENTLYDYLL.........NP...........................
KKYIPGTKMVFPGL..............KKPQERA...DLISYLKEATS.
.................
>CY2_RHOVI, 216 bases, 3BC15531 checksum.
....MRKLVFGLFVLAASVAPAAAQDAASGEQVFK.QCLVCHSIGPGAK.
......NKVGPVLNG......LFGRHSGTIEGFA.YSDANKNSG.....I
TWTEEVFREYIR.........DP...........................
KAKIPGTKMIFAGV..............KDEQKVS...DLIAYIKQFNAD
GSKK............
>CY2_RHOGL, 216 bases, 8127F66D checksum.
...................GSAPPGDPVEGKHLFHTICILCHTDIKGR..
......NKVGPSLYG......VVGRHSGIEPGYN.YSEANIKSG.....I
VWTPDVLFKYIE.........HP...........................
QKIVPGTKMGYPGQ..............PDPQKRA...DIIAYLETLK..
.................
>CY2_RHORU, 216 bases, 9393D345 checksum.
MKKGFLAAGVFAAVAFASGAALAEGDAAAGEKVSK.KCLACHTFDQGGA.
......NKVGPNLFG......VFENTAAHKDDYA.YSESYTEMKAKG..L
TWTEANLAAYVK.........DPKAF.......................V
LEKSGDPKAKSKMT..............FKLTKDDEIENVIAYLKTLK..
.................
>C550_PARDE, 216 bases, 9EBA85A1 checksum.
```

```
PileUp

  MSF:  216  Type: P    Check:  1457   ..

 Name: CYC1_YEAST oo  Len:  216  Check:  3274  Weight:  0.100
 Name: CYC_ORYSA oo  Len:  216  Check:  6625  Weight:  0.100
 Name: CY2_RHOVI oo  Len:  216  Check:   273  Weight:  0.100
 Name: CY2_RHOGL oo  Len:  216  Check:  3372  Weight:  0.100
 Name: CY2_RHORU oo  Len:  216  Check:  9999  Weight:  0.100
 Name: C550_PARDE oo  Len:  216  Check:  3882  Weight:  0.100
 Name: CY2_RHOCA oo  Len:  216  Check:  6091  Weight:  0.100
 Name: CY2_RHOSH oo  Len:  216  Check:  4400  Weight:  0.100
 Name: CYC4_PSEST oo  Len:  216  Check:  3541  Weight:  0.100

//

CYC1_YEAST       .......... ........T EFKAGSAKKG ATLFKTRCLQ CHTVEKGGP.
CYC_ORYSA        .......... ......ASFS EAPPGNPKAG EKIFKTKCAQ CHTVDKGAG.
CY2_RHOVI        ....MRKLVF GLFVLAASVA PAAAQDAASG EQVFK.QCLV CHSIGPGAK.
CY2_RHOGL        .......... ........G SAPPGDPVEG KHLFHTICIL CHTDIKGR..
CY2_RHORU        MKKGFLAAGV FAAVAFASGA ALAEGDAAAG EKVSK.KCLA CHTFDQGGA.
C550_PARDE       .MKISIY.AT LAAITLALPA AAQDGDAAKG EKEFN.KCKA CHMIQAPDGT
CY2_RHOCA        .MKISLTAAT VAALVLAAPA FA..GDAAKG EKEFN.KCKT CHSIIAPDGT
CY2_RHOSH        .MKFQVKALA AIAAFAALPA LAQEGDPEAG AKAFN.QCQT CHVIVDESGT
CYC4_PSEST       ...MNKVLVS LLLTLGITGM AHAAGDAEAG QGKVA.VCGA CHGVDGNSPA
```

Figure 10-3:
A multiple sequence alignment in MSF format.

To make an interleaved format, the idea is to take the complete multiple sequence alignment and chop it into blocks of 60 residues. The file displays the blocks one after the other.

If you want to further investigate the intricacies of the world of formatting, here are two extensive resources for keeping your formats straight: www.hgmp.mrc.ac.uk/Software/EMBOSS/Themes/SequenceFormats.html and www.molbiol.ox.ac.uk/help/sequenceformats.htm.

Working with the right format

Sometimes when a program offers several outputs, keeping only the graphic output is tempting because it seems — with its flashy colored annotation — to be the most complete. This graphic output usually comes in post-script, .pdf, .html, .gif, or .jpeg format. If you keep only this graphic output, though, you can be badly stuck.

Graphic outputs are useful for displaying your information. Using graphic output with any sequence analysis program, however, is virtually impossible. You can request a graphic output if you want the pretty picture, but you must also keep the text version. Remember that turning a text into a graphic is always easier than the other way round!

Text representations are much more civilized; they are the default output of most multiple sequence alignment programs and servers. Their main

purpose is to be easy to display, easy to manipulate, and easy to transfer from one program to the next. Of course, they're much less flashy than their graphic counterparts!

MSF is the most standard format for multiple sequence alignments. This is the most common format if you download an alignment from the Web, for example. FASTA or PIR are a bit less common. If you're using phylogenetic programs, Phylip is fairly common. ALN is also fairly common because it is the default output of ClustalW, the most commonly used multiple sequence alignment program. ALN offers a good trade-off between most formats.

Converting formats

If the program you're using doesn't produce alignments in the format you need, it is possible to use a third-party conversion tool to get to the format that you want. In the following steps list, we show you how to convert an alignment in ALN format (refer to Figure 10-1) into a FASTA format multiple alignment.

1. **Point your browser to** `bioweb.pasteur.fr/seqanal/interfaces/` `fmtseq.html`.

 The Institut Pasteur's FMTSEQ Sequence Conversion page appears, as shown in Figure 10-4.

2. **Enter your e-mail address in the field provided.**

fmtseq : sequence conversion (J. R. Knight)

| Reset | Run fmtseq | ● | bioinformatics4dummies@wiley.com | your e-mail |

(● = required, ◐ = conditionally required)

● Sequence File : please enter either :

1. the name of a file: [] [Browse...]

```
CLUSTAL W (1.74) multiple sequence alignment

CYC1_YEAST         -------------------
```
2. *or* the **actual data** here: `TEFKAGSAKKGATLFKTRCLQCHTVEKGGP-------HKVG`

[9. FASTA ▼] Output Sequence Format

Input parameters

Figure 10-4:
The
FMTSEQ
server.

3. **Paste your alignment in the Sequence window.**

FMTSEQ automatically recognizes most formats. If you don't have any spare sequences lying around, use the dummy multiple sequence alignment we created at the beginning of the chapter.

If your sequence names are longer than 15 characters, the rest of the name gets incorporated IN the sequence. This is a major source of confusion.

4. **Select the format you want to output from the Output Sequence Format drop-down menu.**

For our example, we chose FASTA.

5. **Under the Input Parameters heading, select the format you want to input from the Input Sequence Format drop-down menu.**

If you leave this selector blank, as shown in Figure 10-5, FMTSEQ tries to guess your format automatically. If it fails to do so, you can select your input format here (assuming you know it!).

6. **In the Output Parameters section, make sure that the Convert to Lowercase and the Convert to Uppercase options are deselected.**

When you deselect these two options, FMTSEQ retains the original case of each symbol in your alignment. Some programs, such as Dialign, use case to indicate the local reliability of the alignment.

The pretty print option gives you access to a menu that makes it possible to control every detail in your output format, such as the number of characters per line, the indentation and other minute details.

Input parameters

Set the gap symbol for the input (-gapin) ? ○ [default] ○ '.' ○ '-'

[▮▮▮▮▮▮▮▮▮▮▮ ▾] Input sequence format (-inf[ormat])

[Return to the main part with your favorite browser's Back function]

Output parameters

☐ convert to lowercase (-c)
☐ convert to upper case (-C)
☐ Remove gaps from sequences (-degap)
Set the gap symbol for the output (-gapout) ? ○ [default] ○ '.' ○ '-'
☐ Only list sequence information (-li)
☐ Long form conversion (input header included as comment) (-long)
Gaps in sequences (-raw) ? ○ [default] ○ leave gaps ○ do not leave gaps
☐ Reverse-complement each sequence (-re)

Pretty-print options

Figure 10-5:
Output
parameters
of FMTSEQ.

7. **Click the Run Ftmseq button at the top or the bottom of the page.**

 Wait until your browser displays the result page. In most cases, the results come back interactively in a few seconds, but if the server is very busy it may send the output using the e-mail you provided in Step 2.

8. **Click the fmtseq.out link to see your alignment reformatted.**

9. **Save the reformatted file with the File⇨Save As option of your browser.**

 You can save this file as a .txt or .html file.

Sequence reformatting is one of the most common facilities on the Web. Table 10-2 lists some of these sequence reformatting servers and what you can do using them.

Table 10-2	Sequence Text Conversion on the Web	
Name	*Address*	*Description*
FMTSEQ	tofu.tamu.edu/Pise/5.a/ fmtseq.html,bioweb.pasteur. fr/seqanal/interfaces/ fmtseq-simple.html,www. bimcore.emory.edu/Pise/	Converts most formats
READSEQ	bimas.dcrt.nih.gov/molbio/ readseq/,analysis.molbiol. ox.ac.uk/pise_html/ readseq.html,dot.imgen. bcm.tmc.edu/seq-util/ readseq.html,iubio.bio. indiana.edu/cgi-bin/ readseq.cgi,bimas.dcrt. nih.gov/molbio/readseq/, www.ebi.ac.uk/readseq/ index.html	A very popular tool for reformatting; does not recognize ALN
SEQCHECK	www.swbic.org/products/ bioinfo/seqcheck/ seqcheck.php	Cleans your FASTA sequences

Watching out for lost data

No matter what they tell you, there's always the danger that, when changing the format of your sequences, you can lose information. Each type of format supports different types of information. The problem is that, in most cases, you only realize this when it's too late. Table 10-3 lists the kinds of features formatting can destroy.

Along the same lines, do not take for granted that similar online servers do the same thing (even if they have the same name and the same interface). Two servers running *readseq* may run different versions of this program, or the same version with different default parameters. As a consequence, a problem that doesn't occur with one server may occur with the next server. It really pays to keep your eyes peeled and to keep backup copies of your original files.

Table 10-3	Information You Can Lose When Reformatting
Information Type	*Nature of the Loss*
Sequence name	Long names can be truncated when switching formats. Special characters may also be modified. This happens when converting from FASTA to ALN. The effect of the truncation is unpredictable. Sometimes a portion of the name is added to the sequence!
Upper/lowercase	Case sometimes contains useful information. Strictly speaking, FASTA only supports uppercase. Some programs hate to receive an input with a mixture of cases.
Gap type	Some formats, such as MSF, use different symbols for different types of gaps (., -, ~). These are often turned into (-) symbols after reformatting.
Annotation	MSF can support weight values for the sequences. This information is lost in most conversions. The extra line of annotation in PIR often disappears when changing the format of these alignments. Any annotation that comes after the sequence name in FASTA is bound to disappear after a conversion.
Special amino acid or nucleotides	Some reformatting programs support the code for ambiguities such as X for undetermined amino acids or N for nucleotides. If your sequences travel across many programs, these symbols may disappear. This can be a problem if you rely on the offset of some residues within the sequences. If you can, stick to the standard 4 nucleotides alphabet and the 20 amino acid alphabet.

Using Jalview to Edit Your Multiple Alignment Online

Multiple sequence alignment methods aren't perfect. The people who make them know this, and the people who use them must also consider this fact.

To produce an alignment that you know is right, you can torture a program (and yourself) until it spits out exactly what you want. This may take time and may never lead to the desired result. You can also bite the bullet and edit this alignment yourself. The best recipe for instant insanity is trying to modify your multiple sequence alignment with a standard word processor, such as Microsoft Word or any similar product. It is an experience that has shattered some of the smartest people. The chances of never recovering are high, so don't do it!

Because editing a multiple sequence alignment is so complicated, biologists have developed text editors that are specific for multiple sequence alignment. They make it easy for you to see exactly what's going on. They also make it possible to group sequences so that you can realign two subgroups without having to modify the entire alignment. Most of these editors require that you install something on your computer. This should not deter you because some of these programs — Seaview, for instance — are really easy to install on a PC.

However, if you want to stick to your browser, you can use Jalview, a Java applet that you simply need to load into your Web browser for instant action.

Jalview is a Java applet and uses the same language as Dotlet (see Chapter 8) and has the same general properties:

- ✔ It runs on your own computer.
- ✔ When you load a sequence in Jalview, your sequence does not travel over the Internet; it stays in your computer.

If you want to ensure that none of your data travels across the Internet, choose the File➪Work Offline option on your browser as soon as Jalview is loaded. Do not load confidential sequences in Jalview BEFORE doing this. The Web interface is NOT secure.

Starting Jalview

A convenient way to use Jalview online is to use it in conjunction with the EBI ClustalW server. You can also use the interface provided by the Spanish EMBnet node. (See Table 10-5 for a listing of such interfaces.)

This does NOT mean that you can edit only the multiple alignment generated with ClustalW. After you've loaded Jalview into your computer, you can load it with any alignment you fancy.

1. **Point your browser to** `www.ebi.ac.uk/clustalw/index.html`.

 Steps 1-3 show you how to reach Jalview on the EBI server; as an alternative, you can access Jalview directly from the Spanish EMBnet server at `www.es.embnet.org/Services/MolBio/jalview/index.html` and proceed to Step 4.

2. **Paste your multiple sequence alignment into the Sequence window.**

 You can paste either an alignment or unaligned sequences. If your alignment is rather large and you do not need ClustalW to align it, simply paste a few dummy sequences so that the server can run.

 For the purpose of this steps list, you can use your own alignment or the dummy alignment we propose in the first section of this chapter.

3. **Click the Run button.**

 An initial Results page appears.

4. **Wait until your browser returns a page similar to Figure 10-6.**

 This page loads the Java applet into the memory of your computer. You must wait until the word *Jalview* appears inside the gray button, as shown in Figure 10-6. If you are on a slow connection, this can take a few minutes.

Figure 10-6:
The Jalview applet ready for launching.

5. **Click the Jalview button.**

 The Jalview applet window appears, as shown in Figure 10-7. It contains the alignment that you have just computed with ClustalW, but it is possible to replace this alignment with any other alignment you're interested in.

Figure 10-7:
The Jalview
Applet.

Working offline with Jalview

If your sequences are confidential, you may want to use Jalview offline. You can use the following procedure to make sure that your data does not travel unprotected over the Internet:

1. **Load Jalview with unimportant sequences.**

 Follow the steps list in the preceding section until you obtain a Jalview Applet. (Refer to Figure 10-7.)

2. **Select the File⇨Work Offline option of your Internet browser.**

3. **In the Jalview window, select File⇨Input Alignment Via Text Box.**

 The Input Alignment via Text Box dialog box duly appears.

4. **Paste your multiple sequence alignment (or your sequences) in the text box, as shown in Figure 10-8.**

5. **Select the format that corresponds to your multiple alignment from the Alignment Format drop-down menu.**

6. **Click the Apply button.**

 A new window should appear, containing your sequences ready to be safely edited.

Jalview does not conveniently let you know — by aural or visual clues — if you give it the wrong format; it just sits there and does nothing after you click the Apply button. If you notice a certain lack of activity after clicking the Apply button, you can safely assume that you have a format problem.

```
Input alignment via text box                                          [X]

CLUSTAL W (1.74) multiple sequence alignment

CYC1_YEAST      --------------------TEFKAGSAKKGATLFKTRCLQCHTVEKGGP-------HKVG
CYC_ORYSA       ----------------ASFSEAPPGNPKAGEKIFKTKCAQCHTVDKGAG-------HKQG
CY2_RHOVI       ----MRKLVFGLFVLAASVAPAAAQDAASGEQVFK-QCLVCHSIGPGAK-------NKVG
CY2_RHOGL       --------------------GSAPPGDPVEGKHLFHTICILCHTDIKGR-------NKVG
CY2_RHORU       MGKGFLAAGVFAAVAFASGAALAEGDAAAGEKVSK-KCLACHTFDQGGA-------NKVG
C550_PARDE      -MKISIY-ATLAAITLALPAAAQDGDAAKGEKEFN-KCKACHMIQAPDGTDII-KGGKTG
CY2_RHOCA       -MKISLTAATVAALVLAAPAFA--GDAAKGEKEFN-KCKTCHSIIAPDGTEIV-KGAKTG
CY2_RHOSH       -MKFQVKALAAIAAFAALPALAQEGDPEAGAKAFN-QCQTCHVIVDESGTTIAGRNAKTG
CYC4_PSEST      ---MNKVLVSLLLTLGITGMAHAAGDAEAGQGKVA-VCGACHGVDGNSPAPNFPKLAGQG
                                      .  *     *  **        *

CYC1_YEAST      PNLHG------IFGRHSGQAEGYS-YTDANIKKN-----VLWDENNMSEYLT--------
CYC_ORYSA       PNLNG------LFGRQSGTTPGYS-YSTANKNMA-----VIWEENTLYDYLL--------
CY2_RHOVI       PVLNG------LFGRHSGTIEGFA-YSDANKNSG-----ITWTEEVFREYIR--------
CY2_RHOGL       PSLYG------VVGRHSGIEPGYN-YSEANIKSG-----IVWTPDVLFKYIE--------
CY2_RHORU       PNLFG------VFENTAAHKDDYA-YSESYTEMKAKG--LTWTEANLAAYVK--------
C550_PARDE      PNLYG------VVGRKIASEEGFK-YGEGILEVAEKNPDLTWTEADLIEYVT--------
CY2_RHOCA       PNLYG------VVGRTAGTYPEFK-YKDSIVALGASG--FAWTEEDIATYVK--------
CY2_RHOSH       PNLYG------VVGRTAGTQADFKGYGEGMKEAGAKG--LAWDEEHFVQYVQ--------
CYC4_PSEST      ERYLLKQLQDIKAGSTPGAPEGVG--RKVLEMTGMLDPLSDQDLEDIAAYFSSQKGSVGY
                                       .              : *.

CYC1_YEAST      -NP---------------------KKYIPGTKMAFGGL-------------KK
```

Alignment format CLUSTAL ▼

Status:

Apply Close

Attention : fenêtre d'applet

Figure 10-8:
Alignment
Input box.

Changing the color scheme in Jalview

Choosing a color scheme is really a matter of taste, and most multiple sequence alignment specialists can spend hours explaining to you why they prefer a special color scheme to every other alternative in the galaxy. Whatever your choice, anytime you show a color alignment, you must be ready to answer the most common question in the world of biology:

"What is the color scheme in this alignment?"

To know more about the color schemes available in Jalview, you can read the excellent online documentation of this program at www.es.embnet.org/Doc/jalview/contents.html#colour. In Jalview, you'll find all the available color schemes under the Colour menu. Clustalx Colors is one of the most readable options.

Editing a group of sequences

When you edit an alignment, you usually want to keep some of your sequences aligned the way they are relative to one another. What you want to do is *collectively* modify their alignment. To do this, you need to define them as a group, as follows:

1. **Using the Jalview output we got for our dummy multiple sequence alignment, click to select the sequences CYC1_YEAST, CYC_ORYSA, and CY2_RHOVI, as shown in Figure 10-9.**

 Note that depending on the version of ClustalW you use to produce this alignment, the order of the sequences may slightly vary from what you see in Figure 10-9.

 Selecting the sequences on a phylogenetic tree is possible:

 a. Select Calculate⇨Neighbor Joining Tree Using PID.

 b. On the tree that pops up, you can select sequences individually, or you can draw a line that defines the groups as subtrees.

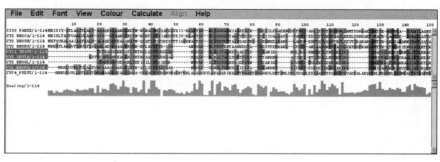

Figure 10-9: Selecting sequences in Jalview.

2. **Choose Edit⇨Groups from the main menu.**

3. **In the pop-up window, first click the Add New Group button and then the Add Selected IDs button. Click the Apply and Close buttons to finish (see Figure 10-10).**

 This creates a new group that contains the sequences you have selected. You can create as many groups as you want. You can also use this menu to edit an existing group (Add and Delete Buttons).

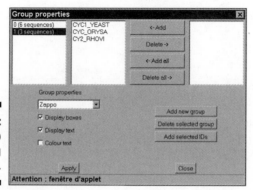

Figure 10-10: The group editing window.

4. **Choose Edit⇨Group Editing Mode from the main menu.**

 This command toggles group editing on.

5. **Click anywhere on a sequence and drag to the left or to the right to insert or remove gaps.**

 All sequences in the same group as the sequence you've chosen move in step with your sequence, as shown in Figure 10-11.

 When you select a group, you're actually defining two groups: the sequences within the group and the sequences outside the group. Sequences outside the group behave as a second uniform group.

Figure 10-11: Effect of dragging a group to the right.

Useful features of Jalview

Although explaining all the possibilities of Jalview is clearly beyond the scope of this chapter, we encourage you to explore them. Jalview is like all the programs we introduce here: You can't break anything, so experiment away!

This being said, remember that it is easy to ruin a good alignment. Save intermediate results. There is no Undo button in Jalview, and even if you *can't* break anything, you *can* lose quite a bit of time.

Table 10-4 lists some of the features we find especially useful.

Table 10-4	Some Useful Features of Jalview
Command	*Description*
Calculate⇨Autocalculate Consensus	Automatic update of the graph below the alignment. This graph — looking like a city skyline — indicates the level of conservation within the alignment. If you set this option, the graph is updated automatically while you're editing.
Calculate⇨Remove Redundancy	Makes sure that no pair of sequences is more than P percent similar.
Calculate⇨Neighbor Joining Tree Using PID	Computes and displays a phylogenetic tree in graphic format on which you can select sequences for group editing.

Saving your alignment in Jalview

If you work on Windows, it is not very convenient to save a colored version of your alignment. The best thing to do is to save your edited alignment into a file and to use some of the tools we introduce in the next section of this chapter to color it and generate a graphic display.

To save the multiple alignment you have edited in Jalview:

1. **Choose File⇨Output Alignment Via Text Box from the Jalview main menu.**

 This makes the Text Box appear.

 Do not use the File⇨Save Alignment to Local File option of Jalview.

2. **Select the alignment format. (We recommend MSF or ALN.)**

3. **Click the Apply button.**

 A properly formatted alignment appears in the window.

4. **Open a Microsoft Word document.**

5. **Select, copy (Ctrl+C), and paste (Ctrl+V) your alignment from the Jalview Text Box to the Word Document.**

6. **Save the Word document as a** .txt **or a** .doc **document.**

7. **Click the Close button on the Jalview Text Box.**

Preparing Your Multiple Alignment for Publication

Showtime has finally come: You have the multiple alignment you want, and you're determined to show the world! You want to show this alignment to your colleagues, you want to include it in publications, on posters, in mails, and maybe on T-shirts, mugs, and other merchandise — you're even mulling over the idea of a tattoo. In short, you need a high impact picture to convince people that your research is going gangbusters. In this section, we list a few online utilities that you can use to beautify your multiple sequence alignments and make them look sharp as a tack.

Using Boxshade

Boxshade is a utility that allows you to put some life into your alignment. It shades columns according to their level of conservation and produces files that you can easily manipulate for inclusion in reports or articles. We give you other powerful tools of the same kind in Table 10-7.

To get the ball rolling with Boxshade, do the following:

1. **Point your browser to** www.ch.embnet.org/software/BOX_form.html.

 The Boxshade page of ch.EMBnet.org appears, part of which you can see in Figure 10-12.

Figure 10-12: Part of the Boxshade home page.

2. **Choose RTF_new from the Output Format drop-down menu.**

 Most of these formats are difficult to use on a PC. The most convenient is RTF New (Rich Text Format). It generates a file that most word processors (such as Microsoft Word) can read.

3. **Select the font size you want.**

 If your alignment is long, select a small font size. If you have selected RTF_new as output, you can change this later.

4. **Choose Add a Consensus Line with Letters from the Consensus Line drop-down menu.**

 This adds a consensus sequence that contains the most common amino acid for each column.

5. **Select the fraction of sequences.**

 Boxshade only shades columns that have a certain level of conservation. This insures that conserved columns show up in your alignment. If you request 0.5, it means you want half of the amino acids (or nucleotides) to be conserved for some shading to occur.

 Conservation doesn't necessarily mean identity in Boxshade. Similar residues, such as isoleucine and valine, also account for conservation. Two types of shading exist:

 - *Black*: Identical amino acids or nucleotides

 - *Grey*: Similar amino acids

6. **Select the format of the multiple sequence alignment you want to use.**

 If you're planning to use the dummy multiple alignment from igs-server.cnrs-mrs.fr/Tcoffee/dummy_aln.html, select ALN.

 Other possible formats include anything that the readseq program can read, such as FASTA or PIR alignments.

7. **Paste your multiple alignment in the Sequence window.**

8. **Click the Run Boxshade button.**

 An intermediate page appears, as shown in Figure 10-13.

 If this page does not contain an active hyperlink, like the one shown in Figure 10-13, it means that Boxshade had troubles reading your alignment. When this happens, double-check the alignment format. If this doesn't help, use a different format.

9. **Click the <u>here is your output</u> link and save to a local file; open the local file with Word.**

 Figure 10-14 shows you what your multiple sequence alignment looks like after proper shading.

BOXSHADE result for

BOXSHADE has now created the ouput file that can be downloaded.
The command line used was:

ulimit -t 30; box -def -numdef -out=wwwtmp/1.BOX.24948.7736.ps -in=wwwtmp/1.BOX.24948.7736.ali -par=pt10.par -type=2 -dev=1 -thr=0.5 >/dev/null

here is your output number 1

Important note: If you get an error message when clicking on the link above Please read this page

Figure 10-13:
The Boxshade intermediate page.

Figure 10-14:
Boxshade output.

Logos

Logos are a terrific way of generating high impact pictures from multiple sequence alignments. They have been around in sequence analysis for quite some time. Figure 10-15 shows a logo representation of the alignment in Figure 10-1. Notice how the conserved amino acids (like cysteines) stick out to indicate regions of potential biological importance.

When looking at a sequence logo, you can consider the following elements:

✔ Each position corresponds to a column in the multiple alignment.

✔ The total height of a logo position depends on the conservation in the corresponding multiple alignment column. Very conserved alignment columns give high logo positions. Positions that contain a very hetero-geneous mixture of symbols yield low logo positions.

✔ The size of each letter in a logo position depends on how frequent this letter is in the column.

✔ The top letter is always the most frequent in the column.

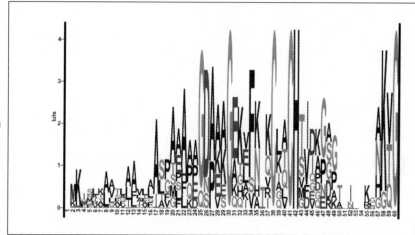

Figure 10-15:
Logo representation of a multiple sequence alignment.

It is possible to build logos with protein or nucleotide multiple sequence alignments. Of course, the better the alignment, the more informative the logo. The most complete logo server around is the one from Tom Schneider at `www.bio.cam.ac.uk/cgi-bin/seqlogo/logo.cgi`, but an alternative that is much simpler to use can be found at `www.cbs.dtu.dk/~gorodkin/appl/plogo.html`. To run this server, use the following steps list.

1. **Reformat your alignment in FASTA.**

2. **Copy and paste your FASTA alignment into a word processing program.**

3. **For each sequence, replace the name with a space.**

 In other words

 >*name*
 XXXXXXX—XXXXX

 must become

 > XXXXXX—XXXXX

 Make sure that your sequences do not contain any line returns.

4. **Point your browser to** `www.cbs.dtu.dk/~gorodkin/appl/plogo.html`.

 The Protein Sequence Logos page of the Center for Biological Sequence Analysis (Technical University of Denmark) duly appears.

5. **Paste your reformatted alignment in the Sequence window.**

6. **Click the Make Logo button.**

 Hopefully, you get something neat looking (refer to Figure 10-15). If nothing happens, make sure that your reformatting is right, check that you have removed the sequence names, and make sure that a single space appears between the (>) sign and the first sequence symbol.

Logos make sense only if you have a nice block with a few highly conserved positions surrounded by highly degenerated positions. There is a utility on the Web that identifies blocks within your multiple alignments and turns each of them into a logo. You can find this utility on the Block server at `blocks.fhcrc.org./blocks/process_blocks.html`.

Editing and Analyzing Multiple Sequence Alignments for Free over the Internet

This last section briefly introduces or recapitulates lists of resources available on the Internet for editing, analyzing, or visualizing your multiple sequence alignments. If you need to reformat your multiple sequence alignment, you can use the resources indicated in Table 10-2.

Finding multiple sequence alignment editors

The only tool available for editing multiple sequence alignments online is Jalview. If you're ready to install some program locally on your machine, a vast selection of high quality packages exists. Table 10-5 lists some of these; most come with extensive documentation for installation and usage.

Table 10-5	Multiple Sequence Editing Packages	
Name	*Address*	*Description*
Jalview	`www.ebi.ac.uk/clustalw/index.html,www.es.embnet.org/Doc/jalview/help.html,www.ebi.ac.uk/~michele/jalview/download.html`	Java package, available online
CINEMA	`www.bioinf.man.ac.uk/dbbrowser/CINEMA2.1/`	Java Package
Seaview	`pbil.univ-lyon1.fr/software/seaview.html`	X-Window
Belvu	`www.cgr.ki.se/cgr/groups/sonnhammer/Belvu.html`	Useful to remove redundancy

Name	Address	Description
Bioedit	`www.mbio.ncsu.edu/BioEdit/` `bioedit.html`	Adapted for RNA
DCSE	`rrna.uia.ac.be/dcse/`	Adapted for RNA
Review	`bioweb.pasteur.fr/cgi-bin/` `seqanal/review-edital.pl`	A complete list

Finding tools to interpret your multiple sequence alignment

The interpretation of a multiple alignment depends very much on its appearance. Some tools on the Net can help you make sense of your multiple alignments by extracting blocks or singling out special positions. Table 10-6 lists some of these resources.

Table 10-6	Extracting Information from a Multiple Sequence Alignment	
Name	**Address**	**Description**
Logo	`www-lmmb.ncifcrf.gov/~toms/` `sequencelogo.html, www.cbs.` `dtu.dk/~gorodkin/appl/plogo.html`	Logos
Blocks	`blocks.fhcrc.org/blocks/` `process_blocks.html`	Identifies blocks
Blockgap	`www.bork.embl-heidelberg.de/` `Alignment/blockgap.html`	Measures blocks
Lama	`blocks.fhcrc.org/blocks-bin/` `LAMA_search.sh`	Compares your multiple alignment with the BLOCKs database
Amas	`www.compbio.dundee.ac.uk/` `servers/amas_server.html`	Identifies important features in the multiple alignment

Finding tools for publishing multiple alignments

Boxshade — which we describe earlier in this chapter — is one of the simplest tools for improving the aspect of a multiple sequence alignment. It isn't the only server that you can use for this purpose, however. Table 10-7 lists some alternative resources that are available online.

Table 10-7	Multiple Alignment Beautifying Tools	
Name	*Address*	*Description*
ESPript	prodes.toulouse.inra.fr/ ESPript/cgi-bin/ nph-ESPript_exe.cgi	Shading and coloring
Boxshade	www.ch.embnet.org/software/ BOX_form.html	Shading in black and white
Mview	bioweb.pasteur.fr/seqanal/ interfaces/ mview_blast-simple.html	Can process BLAST alignments

Part IV

Becoming a Specialist: Advanced Bioinformatics Techniques

The 5th Wave By Rich Tennant

Hmm – Nice phylogenetic tree structure.

In this part . . .

This part is dedicated to a major slice of bioinformatics: prediction. Accurate predictions are the ultimate dream of biologists. It is a very practical dream because computer predictions are cheap whereas experiments are expensive!

In this part, we introduce one of biology's Holy Grails: the prediction of protein structure. We also present you with its more successful nucleic acid counterpart: the prediction of RNA structure. We call this sort of stuff advanced bioinformatics because the results of these methods only make sense if you know how to interpret them and if you have a good understanding of their limitations. These aren't methods that you can trust blindly and, in this part, we tell you when to keep your eyes peeled!

Chapter 11

Working with Protein 3-D Structures

. .

. .

There is no foreseeable limit to the number of proteins whose structure is worth analyzing, since each will have its own unique function which demands explanation in structural terms.

— John C. Kendrew, Nobel Lecture 1962

*W*hen you study a protein, you are usually interested in its function. One thing all biologists know these days is that there is a tight relationship between the structure and the function of a protein. As soon as you find an interesting feature at the level of the protein sequence — perhaps a motif or an evolutionarily conserved segment — the right thing to do is to ask what this feature looks like within the 3-D structure.

Typical questions you want to ask are

✔ Do the amino acids at given sequence positions contribute to the stability of the protein structure?

✔ Why is this sequence segment conserved (or variable)?

✔ Are these amino acids located at the surface of my protein molecules?

✔ Are these amino acids directly involved in the function of the protein?

✔ Are these amino acids involved in the binding of another molecule?

In this chapter, we show you how to look at your sequence from a 3-D perspective.

From Primary to Secondary Structures

Biologists refer to the sequence of a protein as its *primary structure* as opposed to the 3-D or *tertiary* structure that is the final shape of the protein. No prize for guessing that an intermediary level of structure exists that is known as the *secondary* structure!

When the first crystallographers started looking at protein structures, they discovered (and predicted) that there was a hierarchy in the way that amino acid sequences fold onto themselves to become a biologically active molecule. Amino acids first look to their immediate neighbors in the sequence to form regions of regular, periodical conformations (backbone shapes). After this is done, the chain collapses further by folding those preshaped regions onto each other (or onto unstructured regions), leading to the final 3-D structure in which residues that are far apart in the sequence come in direct contact with each other.

There are three types of local segments:

- **Helices:** Where residues seem to be following the shape of a spring. The most common are the so-called alpha helices
- **Extended or Beta-strands:** Where residues are in line and successive residues turn their back to each other
- **Random coils:** When the amino-acid chain is neither helical nor extended

Within the latter category, biologists like to distinguish those cases where the chain makes a sharp turn (90° or more) by labeling them *loops*.

Predicting the secondary structure of a protein sequence

Predicting the secondary structure of proteins was one of the hottest goals of the 1990s. It is fair to say that this is one of the great successes of that decade. Nowadays, fairly good servers are available that use Hidden Markov Models and neural networks to accurately predict the secondary structure of any protein that may interest you. In this section, we show you how to use PSIPRED, one of the most accurate of these servers.

If your protein has enough homologues in the current databases, you can expect its secondary structure prediction to be close to 80 percent accurate. However, bear in mind that this is *only* a prediction; as with all predictions, they can be more or less inaccurate.

To get started using PSIPRED, do the following:

1. **Prepare your favorite protein in a FASTA-formatted .txt file.**

 We used GenPep NP_360043, Rickettsia conorii sequence TolB for this example. You can fetch it from the NCBI Web server (www.ncbi.nlm.nih.gov); see Step 1 in the later section "Looking at sequence features in 3-D."

2. **Point your browser to** bioinf.cs.ucl.ac.uk/psipred/.

 This displays the home page of the Protein Structure Prediction server, which is maintained by the Bioinformatics Unit of University College in London (UK).

 This server offers access to several structural prediction programs in addition to PSIPRED.

3. **Scroll down the page and click the <u>Access the Server Here</u> link.**

 A one-page input form appears in a new page.

4. **Enter your e-mail address in the top field.**

 This server returns its results by e-mail, usually in less than 30 minutes (sometimes more, depending on its load).

5. **Copy your sequence and paste it into the Paste PROTEIN Sequence Here window.**

 When you paste the sequence, DO NOT include the FASTA header (first line with the name).

6. **If you like, enter a short name for your sequence in the field beneath the sequence window.**

7. **Click the Predict button.**

 By default, the server runs PSIPRED, the secondary structure prediction method.

 Alternatively, you can select three other prediction methods: one for transmembrane segments and two for fold recognition. Feel free to explore these if you have time.

If everything is fine, a new page pops up and informs you that your prediction job has been submitted and that the result will be sent to the e-mail address you provided in Step 4. (Check to make sure that this is correct!) It also provides an estimated time for completion (usually largely overestimated). At this point, there's little that you can do but wait for your e-mail to show up. Go and have a coffee!

When you come back, look for an e-mail from `psipred¢.ucl.ac.uk`, similar to what you see in Figure 11-1. The output is a simple text file in which each line of the sequence (AA) is now printed side by side with two other lines:

- The *prediction* line (Pred), consisting of H, E, and C characters, denoting the predicted conformation for each residue (H = *Helical*, E = *Extended*, and C = *Random Coil*).
- The *confidence* (Conf) line, consisting of digits from 9 to 0, indicating the reliability of the prediction for each position (9 = high, 0 = poor).

PSIPRED is very popular because of its accuracy (over 80 percent correctly predicted positions, on average) and also because it produces a nice graphic output that comes in handy for publication. To retrieve the graphic files, look at the very end of your PSIPRED e-mail. It reads like this:

Calculate PostScript, PDF and JPEG graphical output for this result using `http://bioinf.cs.ucl.ac.uk/cgi-bin/psipred/graphics/nphview.cgi?id=xxxxxxx`.

Clicking this URL gets you to a PSIPRED output page that offers you the opportunity to download the graphic files in three different formats:

- **PostScript File:** `http://bioinf.cs.ucl.ac.uk/psiout/xxxxxx.ps`
- **PDF File:** `http://bioinf.cs.ucl.ac.uk/psiout/xxxxxx.pdf`
- **JPEG Page 1:** `http://bioinf.cs.ucl.ac.uk/psiout/xxxxx_1.jpg`
- **JPEG Page 2:** `http://bioinf.cs.ucl.ac.uk/psiout/xxxxx_2.jpg`

Figure 11-1: Typical PSIPRED e-mail: Partial output for R. conorii TolB (NP_360043).

```
                     Random Coil                              Helical
                          |                                      |
Conf: 9137789999873764553278774015653465046516888751111 57789999998
Pred: CCHHHHHHHHHHCCCCCCCEEEEEEECCCCCCCCCEEEECCCCCCCCCCHHHHHHHHHHHH
  AA: MRNIIYFILSLLFSVTSYALETINIEHGRADPTPIAVNKFDADNSAADVLGHDMVKVISN
           10        20        30        40        50        60

                                                 Extended
                                                   |
Conf: 643025612266422432455676566620022157149999999978998607899997
Pred: HHHHCCCCCCCCCCCCCCCCCCCCCCCCCCCCCHHHHCCEEEEEEEEEECCCCCEEEEEEEE
  AA: DLKLSGLFRPISAASFIEEKTGIEYKPLFAAWRQINASLLVNGEVKKLESGKFKVSFILW
           70        80        90       100       110       120

Conf: 145510000013430511202233200012210234455566883899982589888636
Pred: CCCCCCCEEEEEEECCHHHHHHHHHHHHHHHHHHCCCCCCCCCCCCEEEEEECCCCCCCCCE
  AA: DTLLEKQLAGEMLEVPKNLWRRAAHKIADKIYEKITGDAGYFDTKIVYVSESSSLPKIKR
          130       140       150       160       170       180
```

If your computer is already set up with Adobe Acrobat, select the graphic file in PDF format by simply clicking the appropriate URL. That should open up a display similar to what's shown in Figure 11-2. You may also download the JPEG version if you want to incorporate it into a PowerPoint presentation.

The graphic output is organized along the same lines as the text output, with the confidence line replaced by a confidence *profile*, and the Prediction line completed by explicit symbols (Helical: *cylinder*, Extended: *arrow*, Coil: *nothing*).

Predicting additional structural features

PSIPRED is ideal if you need a quick answer to a basic question, such as *what is the secondary structure of my protein?* If you want to know more, you must turn to other sites capable of predicting a much broader variety of features in your protein.

Figure 11-2:
Typical PSIPRED graphics: Partial output for R. conorii TolB (NP_360043).

The PredictProtein server is probably the most comprehensive site for protein structure analysis. Unfortunately, the original site (based at Columbia University) is also very busy and response times can be more than a day or two. Luckily, the PredictProtein server has numerous mirror sites throughout the world, including Europe, the United States, Asia, and Australia (beware, some of the URLs listed on the server's home page are obsolete). Here is a list of up-to-date URLs:

Europe:	www.embl-heidelberg.de/predictprotein/
	www.cmbi.kun.nl/bioinf/predictprotein/
USA:	cubic.bioc.columbia.edu/predictprotein/
	www.sdsc.edu/predictprotein/

Your basic PredictProtein server

To check out a PredictProtein server, have your protein sequence handy in another browser window or in a local file, and follow these steps:

1. **Point your browser to** www.sdsc.edu/predictprotein/.

 The PredictProtein Server page dutifully appears.

 We found (for now!) that the San Diego site tends to be the more responsive.

2. **Click the <u>Submit</u> a Protein Sequence for Prediction link (the first bullet).**

 A default submission form appears.

3. **Enter your e-mail address in the top text field.**

4. **Select the output format by checking a box.**

 In the default mode (Results on Our Web Site), the server sends you an e-mail indicating the URL of your result file on the PredictProtein site. This file remains available for *one day*. It can be in HTML or Text format. In this mode, a graphic processing of the results is also offered.

 If you deselect the default mode, the results are sent to you as a regular, rather long e-mail. To keep the e-mail size reasonable, we advise you *not* to use the HTML format.

 So, in all cases, the server sends you an e-mail; with an address where to find your results or with the results themselves.

5. **Copy your sequence from another browser window or your local file and paste it into the Paste, or Type Your Sequence box.**

 The server only accepts a single protein sequence, in one-letter code, in the raw (plain) format. No FASTA header.

6. **If you like, you can enter a short name for your sequence in the One-Line Name of Protein text field.**

7. **Click the Submit/Run Prediction button.**

Now all you have to do is check your e-mail from time to time. To estimate how long it will take, click the yellow Wait button at the bottom-left of the form to display the status of the PredictProtein queue.

A default analysis returns the following:

✔ A secondary structure prediction on the three conformational states (H = *Helical*, E = *Extended*, and C = *Random Coil*)

✔ A prediction of the solvent accessibility of the various residues

✔ A prediction of transmembrane helices and their topology

✔ A prediction of globular regions in your protein

✔ A prediction of the coiled/coil regions of your protein

✔ A description of the PROSITE motifs matching your sequence (for more on PROSITE motifs, see Chapters 4 and 6)

✔ A description of the putative domain structure for your protein (Prodom domains)

✔ A prediction of bound cysteines (disulphide bonds) in your sequence

✔ A description of the composition-biased regions in your sequence

These various features are reported only if they occur in your query sequence. If you need the details on any of these predictions, you can run additional programs by using the Advanced or Expert input form.

The PredictProtein Meta server

The PredictProtein site also runs a META server, an interface that lets you submit your sequence to many servers at once and then automatically retrieves the results for you. Here's how it's done:

1. **Point your browser to** www.sdsc.edu/predictprotein/.

 You're back at the UC-San Diego PredictProtein Server page.

2. **Click the <u>Submit</u> Requests to META-PP link (the second bullet).**

 This displays a one-page input form.

3. **Enter your e-mail address in the top text field.**

 This server returns its results by e-mail.

4. **Copy your sequence from another browser window or from a local file and paste it into the Paste, or Type Your Sequence box.**

 A one-letter code sequence given in the raw (plain) format is fine.

5. **Scroll down to find the table of options available and choose the methods that you're interested in.**

 Most of these services are alternative methods for doing secondary structure prediction or threading. The threading methods try to fit your sequence into a known 3-D structure. We describe threading methods later in this chapter.

6. **Click the Submit/Run Prediction button.**

Although it may seem convenient to submit all these analyses at once, our experience is that it is quicker and less confusing to go to the relevant sites and use them without the META intermediary.

Finding the best server around

There are many servers around the world and, as is to be expected, many use different kind of algorithms — which usually means that some perform faster than others. Of course, their authors keep updating them. If you want to know which one is currently the best server, you can check out EVA, the secondary structure server monitoring system, at `cubic.bioc.columbia.edu/eva/`. (EVA is short for *EV*aluation of *A*utomatic protein structure prediction.)

From the Primary Structure to the 3-D Structure

In the previous section, we show you how to get some preliminary information about what your protein sequence may look like when folded into a biologically active molecule. Such information would involve helical and extended regions, loops, residues exposed at the surface, and so on.

If you have nothing else at hand, this may be useful. For instance, such information may help you predict the potential effect of a mutation or choose the right protein fragment to make antibodies. Unfortunately, you'll soon find out that secondary structure predictions are far less useful than we would like them to be. They fall short of being the real thing: the detailed spatial representation of your molecule. In this section, we show you how to gather and use 3-D information to better understand what goes on in your protein sequence.

The good news for biologists is that plenty of experimental 3-D structure information is available on the Internet. As was the case when the molecular biologists of the world agreed to centralize their sequence data into the GenBank/EMBL/DDBJ databank repository, all structural biologists have agreed to deposit their 3-D structure coordinates into a single database: the Protein Data Bank. Everybody refers to this database by its acronym: PDB.

Retrieving and displaying a 3-D structure from a PDB site

Like other data repositories, the Protein Data Bank (PDB) offers a rather daunting interface that wasn't particularly designed with the nonspecialist in mind. Yet, in those rare cases where you know precisely what you're looking for — and even know what you're doing! — you may want to directly retrieve a protein 3-D structure data set from one of the PDB sites. Before you query the PDB, be sure to collect some precise information about the structure you're looking for, such as the exact protein name or, even better, its PDB identifier. You can usually obtain this identifier from such user-friendly sources as the ExPASy/SWISS-PROT server or by using the various NCBI query tools. (See Chapter 4 for more on how to use these tools.)

Here's how to obtain and display a PDB structure. For now, let's assume that we are looking for the structure of the E. coli protein named TolB, with PDB ID code 1CRZ.

1. **Point your browser to** www.rcsb.org/pdb/.

 This takes you to the PDB home page.

2. **Enter the PDB ID code** 1CRZ **in the text box to the left of the Find a Structure button.**

3. **Click the Find a Structure button.**

 Figure 11-3 shows the resulting output form. The top of the form gives some bibliographical information and describes the content of the entry. More technical data follows, describing obscure crystallographic parameters.

 For now, we want to see only what the protein molecule looks like in 3-D.

4. **Click the small molecular icon at the top right of the form.**

 A View Structure page appears, offering to display your molecule in a variety of Interactive 3-D display formats (VRML, RasMol, and so on).

 Obtaining simple still images is also possible. Because the interactive displays require installing specialized software, let's go for simple things first: the static view (already quite spectacular).

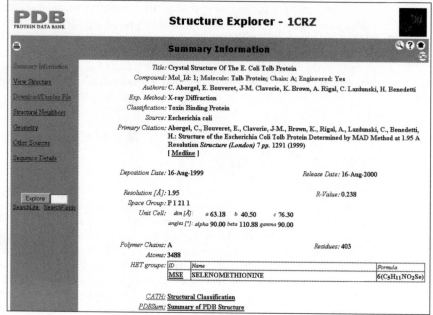

Figure 11-3:
Top of the
PDB output
form for the
query 1CRZ.

5. **Under the Still Images heading, click the <u>Ribbons (500x500)</u> link.**

 Your browser displays an image like the one shown in Figure 11-4. You can also use the Cylinders option that gives a different representation of the helical regions of your protein.

6. **Right-click the picture.**

 What you do next depends on whether you're using Netscape or Internet Explorer.

 If you're using Netscape:

 a. Choose Save Image (1CRZx500.jpg) from the context menu.

 b. In the Save File dialog box, select the filename (1CRZx500.jpg) and the proper location on your disk.

 If you're using Internet Explorer:

 a. Choose Save Picture as from the context menu.

 b. In the Save Picture dialog box, select the file format (JPEG or Bitmap), the filename (1CRZx500), and navigate to the proper location on your disk.

7. **Click Record.**

This figure is now suitable for inclusion in reports and presentations. This is a great way to impress your boss or your professor!

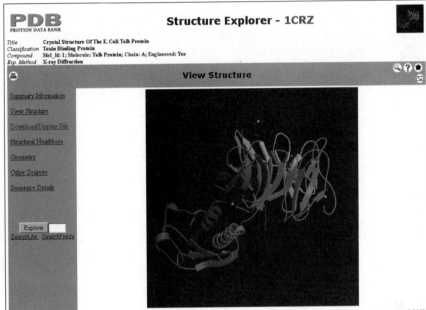

Figure 11-4:
Still view of
the 1CRZ
PDB entry.

A unique worldwide protein structure repository

Since the early '60s, protein structures have accumulated at an increasing speed. Approximately 20,000 of them are publicly available from the Protein Data Bank (PDB), the worldwide repository for the processing and distribution of 3-D biological macromolecular structure data. The PDB is maintained by a consortium of laboratories at the following sites:

✔ www.rcsb.org/pdb/ (San Diego, USA)

✔ rutgers.rcsb.org/pdb/ (Rutgers University, USA)

✔ nist.rcsb.org/pdb/ (NIST, USA)

And perfectly mirrored throughout the world at the following:

✔ pdb.bic.nus.edu.sg/pdb/ (Singapore)

✔ pdb.protein.osaka-u.ac.jp/pdb/ (Osaka, Japan)

✔ pdb.ccdc.cam.ac.uk/pdb/ (Cambridge, UK)

✔ www.pdb.ufmg.br/pdb/ (Minas Gerais, Brazil)

The PDB consortium ensures a minimal quality control of the deposited sets of coordinates, guarantees their public availability (requested for publication by most scientific journals), and promotes common distribution formats.

This figure is a still figure. The angle was chosen for you, and there is nothing you can do to change it! If you need a different angle, you must use a different display tool. See "Displaying a PDB structure the interactive way," later in this chapter.

The preceding steps list shows you how to jump directly from the PDB entry code to a visual representation of the molecule. This may not be enough and you may also want to keep the PDB entry for further study on your own computer. Read on to find out how.

PDB files aren't meant to be read by nonspecialists. They are rather long because they contain a vast amount of data. For instance, the 1CRZ contains the detailed x, y, z coordinates of 3,488 atoms as well as additional information about their connectivity!

If you're sure that you need to download a PDB file, you can simply stop after Step 3 in the previous steps list and look for the various links provided on the left column of the screen. (Refer to Figure 11-3.) The possibilities are

✔ View Structure

✔ Download/Display File

✔ Structural Neighbors

✔ Geometry

✔ Other Sources

✔ Sequence Details

Follow the second link (Download/Display File) to get your own copy of the coordinates. Feel free to explore the other links. For instance, the Sequence Details link gives you the actual secondary structure along with the sequence. You can compare it to the prediction that we made earlier in this chapter.

Guessing the 3-D structure of your protein

The previous section shows how to retrieve and display a PDB structure in a static way. This is definitely useful but falls short of addressing the questions we listed at the beginning of this chapter. Basically, we still do not know what the interesting bits of our sequence look like in their 3-D context.

In the remaining sections of this chapter, we follow a realistic scenario to show you how you can sometimes get a fairly good idea of the 3-D structure of your protein by doing a handful of simple operations.

We assume that you have determined the sequence and the secondary structure of your protein and that you are now asking this basic question: *How does it fold?* A simple way to answer this question is to look for a homologous protein with a known 3-D structure in the PDB. The next steps list shows you how.

Imagine, if you will, that we've just determined the sequence of the TolB gene of the bacteria *Rickettsia conorii* and we are curious about its structure. To satisfy our curiosity, we do the following:

1. **Fetch the protein sequence from NCBI:**

 a. Open a window in your browser and go to www.ncbi.nlm.nih.gov.

 b. Choose Protein from the Search drop-down menu.

 c. Type the identifier **NP_360043** in the query window and click Go.

 d. When the answer is back, change the Display format to FASTA, and then click Text.

 The *Rickettsia conorii* TolB protein sequence is now ready for use.

2. **Open a new browser window and point it to the NCBI BLAST server at** www.ncbi.nlm.nih.gov/BLAST/.

3. **Click the <u>Standard Protein-Protein BLAST [blastp]</u> link.**

 The blastp input page appears.

4. **Select pdb from the Choose Database drop-down menu.**

5. **Copy the TolB sequence from the other browser window and paste it in the blastp query window.**

6. **Deselect the Do CD-search box.**

7. **Click the BLAST! button.**

 An intermediate page appears, telling you that your request has been successfully submitted.

8. **Click the Format! button on the intermediate page and wait for the Blastp search to finish.**

Figure 11-5 shows an example of the kinds of results you can expect. There are two strong matches with E-values much lower than 1. They are identified as PDB entries 1C5K and 1CRZ. This is good news. If you click these matches, their identifiers reveal that they both correspond to the TolB protein from E. coli, the structure of which has been determined by two different laboratories.

Two relevant structures

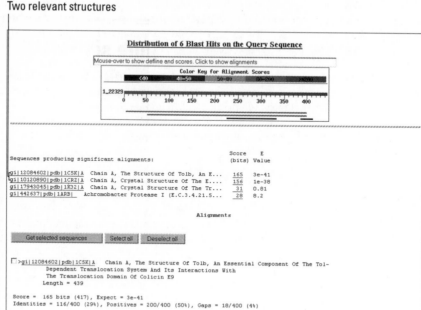

Figure 11-5:
Blastp
output of the
Rickettsia
conorii TolB
sequence
against
PDB.

This result is very interesting for two important reasons:

1. The homologous PDB sequences have 29 percent identical residues with our query.

2. The homology region covers the entire sequences.

By experience, we know that, given such a percentage of identical residues, the 3-D structure of our query protein (the Rickettsia TolB) is probably quasi-identical to the 3-D structure of its E. coli homologue. We thus have a nearly perfect 3-D model to work with. The next section shows you how to use this model to further interpret the TolB protein sequence.

Looking at sequence features in 3-D

On its own, a protein sequence is an important piece of information, but it becomes more revealing and valuable when you compare it with others from a variety of different species. Chapter 9 insists on the role of multiple alignments to identify the most significant regions of a sequence: Conserved residues (or alternatively highly variable ones) are often key to predicting or understanding a protein function. Going further, by precisely locating these conserved residues in space, provides additional clues as to their biological roles: Such residues can delineate a cavity at the active site of an enzyme; they can be found at the surface and interact with other molecules, and so on.

TIP

Sequence versus structure similarity

It is a fact of life that the 3-D structures of homologous proteins are much more invariant (conserved) than the corresponding amino-acid sequences. Proteins exhibiting down to 50 percent identical residues (once optimally aligned over most of their lengths) usually exhibit virtually identical 3-D structures. When you have between 25 and 50 percent identical residues over an entire domain, the 3-D structures are still superimposable, with most of the significant variations located in loops at the surface of the protein. Thus, down to approximately 20 percent residue identity, determining the 3-D structure of one member of a protein family also provides a very good model for all the others. Below the 20 percent threshold, however, pairwise sequence alignments become unreliable, where

the 3-D-structures may or may not remain highly similar: This is the so-called *twilight zone*. The stability of protein folding with respect to sequence variation is a well-accepted (and used) notion, even though it is not truly understood. One explanation frequently cited is that evolution would have selected natural sequences, so their mutations are the least likely to result in overall structural changes. Another one is that the intrinsic properties of the peptidic chain might strongly limit the number of distinct folding patterns, compared to the immense combinatorial diversity of sequences of a given length (20^{100} for a 100-residue protein). Similar sequences thus had no choice but to fold in similar structures.

We can use the example of the TolB protein family (of relatively unknown function) to illustrate the interplay between multiple alignments and structural analysis. First, fetch several TolB homologue sequences from various bacterial species:

1. **Fetch the protein sequences from NCBI:**

 a. Open a window in your browser and go to www.ncbi.nlm.nih.gov.

 b. Choose Protein from the drop-down Search menu.

 c. In the query window, type in the following identifiers: NP_360043 (R. conorii), P19935 (E. coli), NP_404737 (Y. pestis), NP_249663 (P. aeruginosa), and NP_438543 (H. influenzae); click Go.

 d. When the answer is returned, change the Display format to FASTA and click Text.

 Five TolB protein sequences are now ready for use. Now we can build a multiple alignment of these sequences.

2. **Open a new browser window and go to** igs-server.cnrs-mrs.fr/ Tcoffee.

3. **Hover the mouse over Make a multiple alignment and click Regular.**

4. **Copy the five TolB sequences from the other browser window and paste them into the Tcoffee input window.**

 Be sure to include their FASTA headers.

5. **Click Run Tcoffee (at the bottom of the form).**

 The Results of Tcoffee form appears.

6. **Choose HTML from the Multiple alignment output choices.**

 This displays the multiple alignment in color with the most reliable regions in red. This part of the alignment is shown in Figure 11-6. It clearly reveals a segment of strictly conserved (invariant) residues shared by all these TolB proteins. Our challenge in the remaining pages of this chapter is to figure out WHERE in the TolB structure these residues are located.

Figure 11-6: An intriguing segment of strictly conserved residues in the TolB proteins.

Displaying a PDB structure the interactive way

To interpret the significance of an invariant segment in our TolB proteins, we must be able to display its 3-D structure and *turn this structure around interactively* so that we can look at it *from any angle*. We must also be able to *highlight* the residues we find interesting.

In short, we need a protein model that we can rotate around and analyze in parallel with its sequence. Not long ago, this kind of business was still the privilege of protein crystallographers working on expensive computers. Not anymore! If you can bear with us until the end of this chapter, you'll be doing it on your own PC in just a few minutes. Just follow our instructions:

1. **Point your browser to** www.ncbi.nlm.nih.gov/Structure/.

 The structure server of the NCBI appears.

2. **In the Search bar at the top of the page, leave Structure in the first drop-down menu and enter the PDB code of your relevant model in the accompanying text field.**

 For our example, we'll use 1CRZ.

 If you have the choice of several structures, a simple rule is to choose the one with the best resolution; for our example, the best resolution is 1CRZ (1.95 Angstrom).

3. **Click the Go button.**

 Wait for a new page to appear.

4. **Click the 1CRZ link.**

 You may have to change the *display format* to Structure and click the Display button again.

 At this point, you should be looking at the MMDB structure card, as shown in Figure 11-7.

Click here to view. Click here to install.

MMDB
Structure Summary

NCBI

| PubMed | BLAST | Structure | Taxonomy | OMIM | Help? | Cn3d |

Description: Crystal Structure Of The E. Coli Tolb Protein.
Deposition: C.Abergel, E.Bouveret, J-M.Claverie, K.Brown, A.Rigal, C.Lazdunski & H.Benedetti, 16-Aug-99
Taxonomy: Escherichia coli
Reference: PubMed **MMDB:** 14149 **PDB:** 1CRZ

View 3D Structure of Best Model ▼ with Cn3D ▼ Display ▼ NEW *Get Cn3D 4.0!*

Figure 11-7:
MMDB
structure
card for
PDB entry
1CRZ.

This MMDB form exhibits a prominent View 3-D Structure button. What will happen next depends on the software already installed on your computer.

5. **Click the View 3-D Structure button.**

 Doing so downloads a file with the atomic coordinates of the 1CRZ protein structure. This file is formatted for the 3-D structure viewer Cn3D program. If nobody ever used your computer to display a protein structure in 3-D, a dialog box will pop up asking you what to do with this file. This is telling you that your system doesn't contain the Cn3D program required to display the structure. In order to continue, you have to install it. Here's how

a. **Click the <u>Get Cn3D!</u> link.**

This takes you to the Cn3D home page, as shown in Figure 11-8. From here you can access a complete tutorial and/or continue the installation process.

b. **Click <u>Download</u> Cn3D for PC, Mac, or Unix link.**

c. **In the window that appears (see Figure 11-9), select the version suitable for your computer: PC, Mac, or Unix.**

d. **Whatever platform you select, a new page appears with an FTP address, similar to the Windows page shown in Figure 11-10.**

e. **Click the FTP address link.**

For our Windows system, the link is

`ftp://ftp.ncbi.nih.gov/cn3d/Cn3D-4.0.0.msi`

This downloads the CnD3 installer, which takes just a few seconds.

The system may also require you to install a newer version of the Windows Installer. Just follow the instructions. This also only takes a few extra seconds.

Download Cn3D.

Learn how to use it.

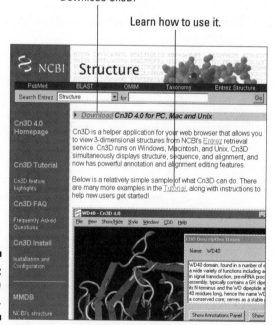

Figure 11-8:
The Cn3D
home page.

Select your system.

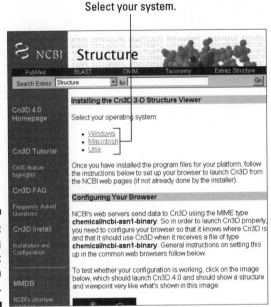

Figure 11-9:
Installing
Cn3D:
System
selection.

Download the Cn3D installer.

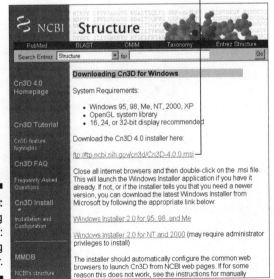

Figure 11-10:
Installing
Cn3D:
Installing
the installer.

f. **After everything is done, locate the Cn3D-4.0.0.msi icon on your desktop and click it twice.**

Cn3D is being installed on your computer, and then the Cn3D installation wizard welcomes you.

g. **Follow its instructions and fill out the forms; click the Next button when you're done.**

You have to provide a name, an organization, and select a destination folder (default is fine). After the last step, the wizard confirms the successful installation.

Now that Cn3D is in your system, go back to NCBI and redo the sequence of steps that we describe in the previous section.

6. **Click the View 3D Structure button again.**

Now that Cn3d is installed, the miracle occurs and your screen should look like Figure 11-11. At the top of the MMDB page, the 3-D protein structure that corresponds to the 1CRZ PDB entry is displayed in a live window.

FSPDG

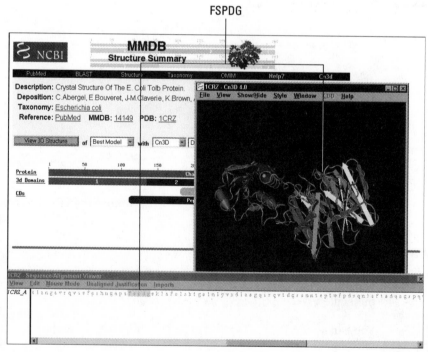

Figure 11-11:
Interactive
3-D display
of the E. coli
TolB protein
structure
(1CRZ).

7. **Spin your protein around.**

 You can turn your protein around by moving your mouse and spin it around like you would do with your partner when dancing salsa! This is impressive, fun, and tremendously useful!

8. **Prepare your protein for taking a picture.**

 Zoom in and out with the drop-down menus on the top bar of the window.

 Change the appearance of the model (wire, tube, ball and stick, or even solid space filling).

9. **Save a picture by choosing Export PNG from the File drop-down menu in the Cn3D window.**

 Saving your protein as a still picture in a file enables you to include it in your presentation or report.

This spectacular show runs locally on your computer, independently of your Internet connection. Terminate your browser, and the live molecule stays on your screen! Neat! It also means that your data doesn't travel over the Internet. This is important if you work for a paranoid boss.

Relating sequence and structure interactively

Last but not least, you probably noticed another window displayed below the live molecule. This is the sequence viewer. By default, this window displays the protein sequence associated with the PDB entry. Here we have the sequence of the E. coli TolB molecule.

The time has come to find out where in the structure that curious region of invariant residues that read FSPDG is located. With Cn3D, it could not be easier.

1. **In the sequence viewer top bar, click the View button.**

2. **From the pop-up menu that appears, choose the Find Pattern option.**

3. **In the pattern input box that pops up, type** FSPDG.

The corresponding segment is immediately highlighted in bright yellow in the sequence AND in the structure. To locate it in the sequence, scroll along the sequence by using the bottom cursor. To locate it in the structure, turn the molecule around.

You can select any residue or group of residues in the sequence viewer by clicking it.

What can we tell about our invariant segment from all this? It is quite an interesting story actually. Figure 11-11 shows that our segment occurs in a loop, well exposed at the surface of the molecule. Surface loops are notorious for being the most variable parts of proteins. (They are more free to mutate.) The strict conservation of these residues suggests that they may have a specific role in the protein function, such as participating in an interaction with another protein or a small molecule.

Beyond This Chapter

In this chapter, we've provided you with a very general idea of what biologists do when they analyze the structure of their proteins with their computers. The applications shown in this chapter probably cover 90 percent of what nonspecialists do when they study the relationships between their sequences and its potential structure.

However, we haven't even mentioned the many other tools and sophisticated approaches that exist for studying protein structures. So here we give you a quick list of what we have *not* talked about, with some related URLs if you're curious about delving deeper into 3-D structures.

Finding other PDB viewers

If all you want to do is spin a few molecules around, you should be aware that, apart from Cn3D, two other very powerful PDB viewers are available over the Internet:

- **RasMol:** www.rasmol.org
- **SwissPDBviewer:** www.expasy.ch/swissmod/SWISS-MODEL.html

Classifying your PDB structure

The classification of protein structures in categories of prototype folds and domain topologies is a hot topic. This is related to the problem of identifying and quantifying structural homologies between very distantly related proteins, as well as the threading problem (see the upcoming section "Threading sequences onto PDB structures"). The three main structure classification sites are

- **The CATH classification:** www.biochem.ucl.ac.uk/bsm/cath_new/
- **The Dali 3D neighbor finding server:** www2.ebi.ac.uk/dali/
- **The SCOP classification:** scop.mrc-lmb.cam.ac.uk/scop/

Doing homology modeling

The detailed modeling of the 3-D structure of a protein with significant sequence similarity to a homologous protein of known structure can go much beyond the simple example we have presented here. Specialists use energy minimization techniques to replace the residues in the original model and refine their positions. A couple of good entry points to find out more about this homology modeling approach are

> ✔ **Modeller:** guitar.rockefeller.edu/modeller/modeller.html
> ✔ **SWISS-MODEL:** www.expasy.ch/swissmod/SWISS-MODEL.html

Folding proteins in a computer

Some groups are making constant progress in their attempts to generate a structural model for a protein solely from its sequence by simulating its folding process. This is referred to as the *ab initio* (or *in silico*) folding approach. The successes are still limited to a certain type of small protein sequences. Find out more about it at the Folding@home project Web site (folding. stanford.edu/).

Threading sequences onto PDB structures

Other groups have taken up the inverse problem of finding out which of the known structures might be best suited as a folding mold to a given protein sequence. This research field is known as *threading*. In the context of a growing database of structures, threading methods might perform increasingly well and expand the range of *homology modeling* to the case of sequences without any significant similarities in the PDB. To find out more about threading, check out these sites:

> ✔ **NCBI Structure threading:** www.ncbi.nlm.nih.gov/Structure/ RESEARCH/threading.html
> ✔ **The PROSPECT server:** compbio.ornl.gov/structure/prospect/

Looking at structures in movement

Some specialists are developing techniques to simulate the movements known to occur in protein structures, when stably folded. Understanding molecular dynamics is an important part of understanding the details of protein-protein interaction, or enzymatic reactions. Find out more about molecular dynamics and protein movements at

- ✔ **The Brooks group:** www.scripps.edu/brooks/
- ✔ **The database of Macromolecular Movements:** molmovdb.mbb.yale.edu/MolMovDB/

Predicting interactions

Finally, researchers are tackling the difficult problem of predicting if, and how, two different protein molecules (or a protein and a small molecule) might interact with each other. This is technically referred to as *molecular docking*. The top performance in this field is still very inadequate. In most cases, the 3-D structures of the complex of two interacting proteins cannot yet be computationally derived from their individual structures. Visit the following Web sites to find out more about docking:

- ✔ **Macromolecular Interactions:** www.bio.vu.nl/nvtb/Docking.html
- ✔ **Hex's home page:** www.biochem.abdn.ac.uk/hex/

TIP

Looking for paradoxical sequence features

The rationale we followed for the analysis of the TolB sequences is of general use. Functional clues mostly arise from the observation of paradoxical features in protein sequences and structures. Evolutionarily conserved residues are uninformative when they are located deep inside a protein structure or in segments of well-defined secondary structures (alpha helices, beta sheets). In those cases, their invariance is most likely due to structural constraints. To identify residues involved in the protein function, look for surface amino acids or loop residues that should have complete freedom to vary but do not. Also, isolated residues, far apart in the sequence but close in the structure, may be involved in enzymatic catalysis or ligand binding if they are strictly conserved. Other paradoxical situations may arise such as hydrophobic residues exposed at the protein surface, or highly variable regions in conserved secondary structure elements.

Chapter 12

Working with RNA

. .

In This Chapter

▶ Searching databases for RNA molecules

▶ Predicting the secondary structure of an RNA molecule

▶ Publishing an RNA secondary structure

▶ Finding non-coding RNAs over the Internet

. .

*I*n this chapter, we show you that it is possible to use a computer to ana-
lyze the most fascinating molecule of them all: the RNA molecule. RNA is
incredible because it does almost anything you can think of, including trans-
porting and transmitting genetic information — just like DNA does — and
performing catalytic functions, just like proteins do. (See the sidebar "RNA
secondary structures.") In fact, the more that biologists study RNA, the more
they wonder why nature bothered inventing DNA and proteins!

There are sound arguments that support the belief that RNA is the true
ancestral support of life, but we don't plan to go so far back in time in this
chapter! We want to concentrate on another really exciting fact about RNA:
It is one of the most clear-cut successes in bioinformatics. If you use them
well, RNA computer-based methods can tell you as much as any million-
dollar experiment, simply because RNA is much more predictable than DNA
or proteins and thus lends itself well to computer analysis.

This chapter is not here to give you an accurate view of the RNA world. This
is simply too complex a subject. (You'll have to go back to your textbooks if
you want the full story.) What we show here are the few things that you can
easily do on a computer if you know what you're looking for and if you want
to find out whether detailed RNA analysis could play a role in your project.

In the first section of this chapter, we show you how you can use mfold, a
famous online server to predict and display the secondary structure of your
RNA sequence. We also show you how to make sure that you have the correct
structure and how to check for this.

The second section focuses on database search methods that are RNA specific. If you know the secondary structure of an RNA molecule, you can use this information to scan databases for sequences that have a similar fold, even if their sequence similarity is very low. This is a very powerful technique that you can use to discover new members of an established gene class or new members of an RNA family that you've discovered yourself.

Sadly, the big players, such as NCBI or the EBI, haven't put as much effort on the RNA side of things as they have in other fields. The good news is that the RNA community is very active and open. On the Internet, you can find a wealth of high quality resources available to anyone who needs them, including specialized databases and powerful software. We introduce some of them in the last section of this chapter.

Predicting, Modeling and Drawing RNA Secondary Structures

A major biological advance of the seventies was the discovery that RNA can have complex 2-D and 3-D structures, as shown in Figure 12-1. The very exciting thing about these structures is that they obey laws that seem much simpler and much more predictable than the laws that reign in the world of proteins. Many of the RNA short- and long-range secondary structures rely on standard Watson and Crick base pairing, just like DNA does. (See the sidebar "RNA secondary structures.")

Figure 12-1:
Typical secondary structures in RNA.

RNA secondary structures

A single-stranded RNA molecule is exactly like sticky tape. It will not be stable unless some of its exposed bases are protected from water. A good way to protect an RNA base from the solvent is to pair it with another RNA base. Of course, all the base pairs are not as good as one another. Pairing a guanine with a cytosine is more stabilizing than pairing an adenine with a uracile.

The pairing of these bases forms the *RNA secondary structure*. When a molecule contains two long stretches that are complementary, this gives a nice stable stem. (Refer to Figure 12-1.) The unpaired bases between the stem strands make up a loop. Stems do not have to be perfect; they can also contain unpaired residues that RNA gurus name *bulges.*

We assume that the natural tendency of the RNA molecule is to reach its most stable conformation by assembling a nice collection of pairwise interactions that gives the molecule the highest stability it can have. This is what we call the *lowest energy model*. Such a stable RNA structure always has a negative energetic value, such as -70 Kcal/mol. So if you want to unfold it, you need to provide some energy (heat).

Just like for proteins, we don't know exactly how the RNA molecule finds this lowest energy form, but we know that this happens very rapidly in the cell. The stability of an RNA fold doesn't depend only on the number of GC pairs it contains. Many parameters, such as the stacking of base pairs — making some stems much more stable than others — and the loop sizes, influence it. Sophisticated algorithms (like the one of mfold) can take these subtle effects into account when computing an optimal fold.

Tertiary interactions also play a role in the overall stability. These include pseudo-knots — again, refer to Figure 12-1 — that are usually long-range interactions between a loop and another portion of the RNA molecule. The interaction between the RNA molecule and other chemical elements, such as ions, proteins, or other RNAs, also plays an important role in its stabilization.

Unfortunately, we have difficulties predicting tertiary interactions or the effect of the proteins on the folding of an RNA molecule. So when you predict the secondary structure of your RNA, this prediction usually depends on the assumption that this RNA folds on its own in the cell. Because this is almost never true, there is always a chance for your prediction to be partly or totally incorrect. The general rule is that the most energetically stable features tend to be reasonably close to the truth.

Using Mfold

Mfold is an implementation of the Zuker algorithm that makes it possible to predict the energetically optimal secondary structure of an RNA molecule. It uses a sophisticated model that can take into account many realistic physical parameters that affect the RNA folding, such as pH, temperature, and local composition bias of your RNA.

However, the energetic model that mfold uses ignores 3-D interactions or protein-RNA interactions that could stabilize a suboptimal fold. There is always a chance that a prediction from mfold does not correspond to the true biological fold of your RNA.

To help you make a decision, mfold returns suboptimal models so that you can judge the stability of the optimal fold. It is also possible to force mfold to respect an interaction that you know is correct.

You cannot force pseudo-knots in mfold.

To try mfold, you'll need to do a couple of things. First and foremost, you'll need to fetch an RNA sequence. "Where?" you might ask. When it comes to non-coding RNA, standard DNA databases are poorly equipped. A few specialized databases exist, but most of them are dedicated to Ribosomal RNA. The last section of this chapter gives you the Internet address of a few of these resources (Table 12-2).

What's the best length for your RNA sequence? Good question. The time mfold needs to analyze sequences increases very rapidly with their length. For instance, if it takes 1 second to fold a molecule that's 100 bases long, it takes about 10 seconds to fold a sequence that's 200 bases long.

The current limit of the server we're going to use for our example is 3,000 bases. This is convenient for most non-coding RNAs except the largest Ribosomal ones.

With the preliminaries out of the way, it's time to do some mfolding. We're going to use a sample RNA sequence, but you're free to use one of your own if you want:

1. **Point your browser to** rna.wustl.edu/GtRDB/Hin/Hin-seqs.html.

 This returns a list of tRNA sequences predicted by Sean Eddy's tRNA scan server of the Haemophilus influenza genome. The sequences are ordered according to the reliability of their predictions. Sequences at the top of the list are almost 100 percent sure.

2. **Select the first tRNA sequence, as shown in Figure 12-2, and copy it to the Clipboard.**

3. **Point your browser to** www.bioinfo.rpi.edu/applications/mfold/.

 The Applications page of the Center for Bioinformatics (Rensselaer Polytechnic Institute) duly appears.

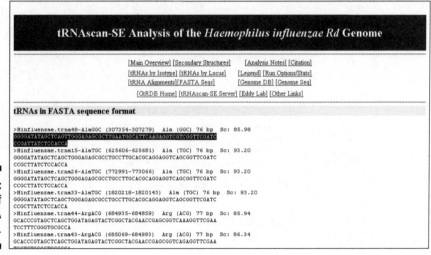

tRNAscan-SE Analysis of the *Haemophilus influenzae Rd* Genome

[Main Overview] [Secondary Structures] [Analysis Notes] [Citation]
[tRNAs by Isotype] [tRNAs by Locus] [Legend] [Run Options/Stats]
[tRNA Alignments][FASTA Seqs] [Genome DB] [Genome Seq]
[GtRDB Home] [tRNAscan-SE Server] [Eddy Lab] [Other Links]

tRNAs in FASTA sequence format

```
>Hinfluenzae.trna48-AlaGGC (307354-307279)  Ala (GGC) 76 bp  Sc: 85.98
GGGGATATAGCTCAGTTGGGAGAGCGCTTGAATGGCATTCAAGAGGTCGTCGGTTCGATC
CCGATTATCTCCACCA
>Hinfluenzae.trna15-AlaTGC (625606-625681)  Ala (TGC) 76 bp  Sc: 93.20
GGGGATATAGCTCAGCTGGGAGAGCGCCTGCCTTGCACGCAGGAGGTCAGCGGTTCGATC
CCGCTTATCTCCACCA
>Hinfluenzae.trna26-AlaTGC (772991-773066)  Ala (TGC) 76 bp  Sc: 93.20
GGGGATATAGCTCAGCTGGGAGAGCGCCTGCCTTGCACGCAGGAGGTCAGCGGTTCGATC
CCGCTTATCTCCACCA
>Hinfluenzae.trna33-AlaTGC (1820218-1820143)  Ala (TGC) 76 bp  Sc: 93.20
GGGGATATAGCTCAGCTGGGAGAGCGCCTGCCTTGCACGCAGGAGGTCAGCGGTTCGATC
CCGCTTATCTCCACCA
>Hinfluenzae.trna44-ArgACG (684935-684859)  Arg (ACG) 77 bp  Sc: 85.94
GCACCCGTAGCTCAGCTGGATAGAGTACTCGGCTACGAACCGAGCGGTCAAAGGTTCGAA
TCCTTTCGGGTGCGCCA
>Hinfluenzae.trna43-ArgACG (685069-684993)  Arg (ACG) 77 bp  Sc: 86.34
GCACCCGTAGCTCAGCTGGATAGAGTACTCGGCTACGAACCGAGCGGTCAGAGGTTCGAA
TCCTCTCGGGTGCGCCA
```

Figure 12-2:
Selection of
a tRNA
sequence.

At the time of this writing, a new mfold server is being announced. Most of what we are saying here will hold for the new server. However, if you want to follow this steps list, you can use as an alternative one of the two European mirrors of mfold:

`bibiserv.techfak.uni-bielefeld.de/cgi-bin/mfold_submit`

`bioweb.pasteur.fr/seqanal/interfaces/mfold-simple.html`

or its Australian counterpart:

`mfold.burnet.edu.au`

4. Scroll down the page and click the RNA Folding link.

The mfold server page appears, as (partially) shown in Figure 12-3.

5. Enter a title for your sequence in the Name field.

For our example, we named our sequence tRNA.

6. Paste your tRNA sequence in the Sequence window.

You can safely keep all the other default parameters.

See the next steps list in the "Forcing interaction in mfold" section if you want to provide the program with some predefined constraints.

Figure 12-3:
mfold server
home page.

7. **Scroll down the page and click the Fold RNA button near the bottom of the form.**

 The Output page (eventually) appears, as shown in Figure 12-4. The output itself looks rather complicated because it contains links to your results in almost any format you can think of. It also contains energetic information concerning your model.

Figure 12-4:
Output of
the mfold
server.

8. Display the jpg picture of the best-scoring model.

 a. Scroll down the Output page until you find the section entitled View Individual Structures. (Refer to Figure 12-4).

 The structures are ranked according to their stability. For instance, Structure1 is the most stable model, Structure2 is the second best, and so on.

 b. Look for the listing of different file formats and then click the <u>jpg that corresponds to Structure1</u> link.

9. Save your secondary structure prediction by right-clicking it and then choosing the Save As option.

Interpreting mfold results

Given your sequence, mfold looks for the arrangement that yields the secondary structures with the lowest possible energy. Of course, the correctness of this secondary structure depends on the correctness of the energy model. If the model is wrong, the structure is wrong. (See the "RNA secondary structures" sidebar.)

The general rule is that the more stable the conformation mfold predicts, the more likely this interaction is to be correct. To help you get an idea of the nature of this stability, mfold returns several results related to the stability of your RNA. Among these, you may give further attention to the following:

✔ **The Energy Dot Plot:** The energy dot plot shows the stems that are part of the optimal fold of your sequence. A stem shows up as a black diagonal perpendicular to the main diagonal, as shown in Figure 12-5.

 Distal interaction (that is, pairs that involve nucleotides very far from each other) appears as diagonals close to the top-right (or bottom-left) corner.

 Colored diagonals indicate suboptimal secondary structures. The color ranges from red (the best) to yellow (the worst).

✔ **The View Individual Structures section:** This section contains data relative to the various structures mfold has computed. The optimal structure is the top one.

 This section also indicates the free energy of your structure. This energy must be as low as possible. For a molecule that's around 200 bases long, you should expect a free energy lower than -50 Kcal to consider the reported fold as plausible.

✔ **The Dot Plot Folding Comparisons section:** This very interesting section makes it possible to compare the fold of several suboptimal solutions. In order to do so, follow these steps:

1. **Choose .jpg from the Image Format drop-down menu.**

2. **Select all the alternative structures you want to compare from the Compare Selected Foldings line.**

3. **Click the Do the Comparison button, as shown in Figure 12-6.**

 A graph similar to the one displayed in Figure 12-5 appears. It indicates the elements of predicted secondary structures that are common to the selected suboptimal solutions.

 Colored elements correspond to stems that occur only in suboptimal secondary structure predictions.

Figure 12-5:
Energy dot plot of an RNA secondary structure prediction.

Figure 12-6:
Comparing alternative secondary structures with mfold.

Finding a high consistency between the optimal solution and the suboptimal ones makes it more likely that the optimal solution is in fact biologically correct.

The predictions you make on a single sequence can yield interesting results. RNA is at its best, however, when you make the prediction on a multiple alignment rather than on a single sequence, thanks to what is known as the *covariance phenomenon*. (See the "RNA bases evolve in pairs" sidebar for more on this phenomenon.)

Unfortunately, multiple sequence-based predictions are not available for free over the Internet, but if you are ready to install some software — and you have a powerful machine — you can use a handy bit of software called Cove, available for download at

```
www.genetics.wustl.edu/eddy/software/#cove
```

or some of the methods available through registration in Robin Gutell's laboratory at

```
www.rna.icmb.utexas.edu
```

RNA bases evolve in pairs

Covariance analysis is the identification of the positions that evolve together in an RNA molecule. This type of analysis relies on multiple sequence alignments. You need to consider the columns two at a time and check whether the nucleotides they contain always mutate simultaneously. Simultaneous mutations in two columns constitute what we name co-evolution. Usually this co-evolution is meant to keep a Watson and Crick base pair, but it can also be a more complex interaction. If you find that your two columns co-evolved, you can safely bet that the nucleotides they contain interact. It is *almost* as simple as it sounds!

This technique is so powerful that it made it possible to identify many of the interactions that occur in the huge ribosomal RNA. Covariance analysis also made it possible to discover new types of non-canonical base pair interactions.

The recent publication of the complete Ribosome structure has confirmed that most of the prediction made with a cheap PC and a lot of patience by Robin Gutell and others were indeed mostly correct — and had been obtained for a fraction of the price of an x-ray structure!

There is no example of such a dramatic success in the protein world! This makes RNA structure prediction one of the great victories of bioinformatics.

Forcing interaction in mfold

If you have some experimental data and you know that specific base pairs occur in your RNA sequence, you can force mfold to use this information. For example, you can tell mfold that nucleotides 10, 11, and 12 interact respectively with nucleotides 22, 21, and 20. To do this you need to write a line of instruction in the interaction form of mfold that says:

F, 10, 22, 3

This means that you (F)orce an interaction to start with nucleotide 10 — paired with nucleotide 22 — and that you want this interaction to be the beginning of a stem that is 3 nucleotide pairs long (nucleotide 10 with 22, 11 with 21, and 12 with 20).

Figure 12-7 shows you what the interaction form of mfold looks like when it is filled out correctly.

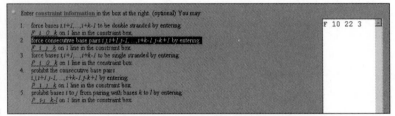

Figure 12-7:
Forcing
secondary
interaction
in mfold.

Searching Databases and Genomes for RNA Sequences

In a cell, all genes are transcribed into RNA molecules so that the cell can use them. One thing that we all forget sometimes is that there are two kinds of RNAs: those that the cell turns into a protein using the ribosome machinery and those that the cell uses directly. The latter are known as *non-coding RNAs.*

Non-coding RNAs constitute a rich family with new members discovered every year. The more we understand the basic mechanisms that govern the life of a cell, the more we realize that non-coding RNAs are everywhere. They help with transcribing, splicing, replicating, and probably many other things we aren't yet aware of. It seems that every basic mechanism of life heavily relies on these little guys, and arguing that a big chunk of the biology future belongs to them is probably the safest bet we'll make in this book.

Unfortunately, with the exception of the ribosomal RNA, these small non-coding RNAs tend to be rather short and poorly conserved. Their most common feature is often their secondary structure. This makes it difficult to find them using any of the homology-based methods we routinely use for finding proteins. (For more on such methods, see Chapter 7.)

The methods we show you in this section initiate database searches using not only the sequence of RNAs but also information related to the known secondary structures. If you have just used mfold to predict the fold of the RNA family you're interested in, you can now use the fold you've discovered to look for new members of the family. In this section, we first show you how to use tRNAscan to look only for tRNAs, and we also introduce PatScan, a method that allows you to look for RNAs with a secondary structure that you can specify yourself.

Finding tRNAs in a genome

tRNAs are small non-coding RNAs that the cell uses to assemble proteins. They constitute one of the best examples of important non-coding genes that are difficult to localize in a genome. All the bacterial and eukaryotic genomes contain tRNAs. Unfortunately, these ubiquitous genes are just as difficult to identify as any small non-coding RNA. They require database search techniques much more sophisticated than BLAST.

The good news here is that there are some very efficient programs out there for predicting tRNA genes in a eukaryotic or a prokaryotic genome. The state of the art method is RNAscan-SE that you can access from the Washington University server at www.genetics.wustl.edu/eddy/tRNAscan-SE.

Using PatScan to look for RNA patterns

tRNAscan is perfect if you're interested only in tRNAs. However, if your interest lies in a specific RNA family, you want to use a tool that's more general and enables you to use your own knowledge of the secondary structure of your RNA family. PatScan is ideal for this purpose.

PatScan lets you search databases with patterns that can accurately describe RNA secondary structures. From a computational point of view, this type of search is quite expensive, which is why PatScan requires your e-mail address and returns a result after a few hours. (In our experience, however, PatScan has never taken more than half an hour to return the result.)

If PatScan is not adapted to your needs because you don't want your sequences traveling across the Internet, you can use RNAmot, an alternative program that's similar to PatScan. To use the program, however, you'll need to install the software on your own computer. If you want to do this, beware that this is not really a tool for novices. RNAmot is available for download at www.esil.univ-mrs.fr/~dgaut/download/.

Designing a PatScan pattern is the trickiest step when using PatScan. It is something that may require a bit of thought. If you want to know the exact rules that govern the PatScan patterns, you can read the excellent online documentation of this program, available at www-unix.mcs.anl.gov/compbio/PatScan/HTML/matching_nucleotides.html.

To help design your pattern, start by identifying the elements you're interested in. Use the following checklist as a guide:

- ✔ On the primary structure, report the secondary structure elements.
- ✔ Identify the corresponding stems and name them. The name of the left side of the first stem is "p1" and the name of its right side is "~p1".
- ✔ Count the length of each of these elements.
- ✔ In each stem element, identify non-canonical Watson and Crick base pairs.
- ✔ Write this structure with your own vocabulary.

 For instance:

 [stem p1 of length 8 to 9] [Loop of length 3 to 8] [stem ~p1 of length 8 to 9, complementary to the left stem p1]

- ✔ Turn your secondary structure into a PatScan Pattern.
- ✔ p1=8...9 defines the stem p1 that contains 8 to 9 nucleotides. [Note the 3 periods (...) between the 8 and 9.]
- ✔ 3...8 defines a pattern of 3 to 8 nucleotides.
- ✔ ~p1 defines a stem reverse complementary to p1.

 You can then write your pattern: p1=8...9 3...8 ~p1.

 If you want to define several stems, give them different names, such as p1, p2, p3, and so on.

 If you need to, you can have nested stems, like p1=8...9 3...8 p2=5...7 4...5 ~p2 3...8 ~p1.

- ✔ You can accommodate in PatScan almost anything you need to describe complex RNA structures. The ability to describe pseudo-knots — long-range interactions between a loop and another portion of the RNA molecule, as shown in Figure 12-1 — makes PatScan a very powerful program. Table 12-1 lists some of its possibilities.

Table 12-1	RNA-Specific Pattern Rules in PatScan	
Rule	**Example**	**Description**
Pairing rule	r1={au,gc} p1=10...12 3...8 r1~p1	Indicates that the p1 pairing only involves au and gc pairs
Mismatches	p1=10...12 3...8 r1~p1[1,4,2]	The pairing between the two p1 stems can involve at most: 1 mismatch in p1 Vs ~p1 4 unpaired symbols in p1 2 unpaired symbols in ~p1
Specify patterns	p1=10...12 AUGC~p1	Specifies that the loop MUST contain only AUGC. Note that this pattern must be in upper case, as opposed to the pairing rules.

With these guidelines, take PatScan on a trial run:

1. **Prepare your PatScan pattern.**

2. **Point your browser to** `www-unix.mcs.anl.gov/compbio/PatScan/HTML/patscan.html`.

 The PatScan server page appears, as shown in Figure 12-8.

Figure 12-8: Home page of the PatScan server.

3. **Choose the database you want to search from the Database drop-down menu.**

In our example, we chose DNA: Human.

The default database on this server is SWISS-PROT. In most cases, this database is *not* appropriate for searching RNA secondary structure patterns.

You must change this database to a DNA database.

4. **Paste or enter your pattern into the Pattern window.**

We entered p1=8...9 3...8 ~p1.

5. **Enter your e-mail address into the appropriate field.**

6. **Click the Submit button.**

Depending on the size of the database you have selected and on the length and complexity of your pattern, your search can take up to an hour.

7. **Analyze your results. (See Figure 12-9.)**

In Figure 12-9, we have used PatScan to search the Human section of GenBank. Each occurrence of a DNA pattern that matches our request appears on a specific line, which indicates the GenBank/EMBL/DDBJ accession number and the coordinates of this match in the sequences.

The fact that a sequence corresponds to your pattern does *not* necessarily mean that it adopts the secondary structure you had in mind! PatScan checks that the hits are compatible with the specified secondary structure, but it doesn't make sure that this structure is the *most* stable one for these hits.

Figure 12-9:
Results of
the PatScan
server.

```
 PINE 4.10    MESSAGE TEXT          Folder: INBOX  Message 234 of 240 END

Subject: PatScan Results

The results for your PatScan search:

DATABASE:
human

PATTERN:
p1=8...9 3...8 ~p1

Maximum # of hits requested:    2000
# of hits found/reported:       4

O60641:[530,549] : AAAAAATT AATA AATTTTTT
O60641:[549,530] : AAAAAATT TATT AATTTTTT
Q9NTY7:[525,544] : AAAAAATT AATA AATTTTTT
Q9NTY7:[544,525] : AAAAAATT TATT AATTTTTT
COMPLETED REQUEST
```

Doing RNA Analysis for Free over the Internet

If you're interested in RNA, everything you need is on the Internet; many excellent resources are available online.

Studying evolution with ribosomal RNA

To study evolution, you need a gene that exists in all living organisms so that you can compare the rate of evolution of these organisms. For many years, biologists have used the ribosomal RNA (rRNA) for this purpose. This has become so commonplace that anytime you find a new bacterium, the first thing you want to do is sequence its ribosomal RNA so that you can see where it fits on the big tree of life.

Several laboratories have dedicated their entire work to helping the scientific community use rRNA sequences. Table 12-2 lists some of the most famous of these resources.

Table 12-2	Ribosomal RNA Resources on the Internet
Address	*Description*
rdp.cme.msu.edu	Provides data and services, including the possibility to make online phylogenetic analysis.
rrna.uia.ac.be/lsu	A European database on the larger of the two ribosomal subunits. It contains predicted structures. It is possible to query the database online. Features lots of online software.
rrna.uia.ac.be/ssu	Similar, but dedicated to the small ribosomal subunit.

Finding the small non-coding RNA you need

Small non-coding RNAs are poorly represented in major databases. Fortunately, excellent alternative databases make it possible to access the genes that you're interested in. Table 12-3 is a partial list of these resources.

Table 12-3	Some Non-Coding RNA Resources
Address	**Description**
mbcr.bcm.tmc.edu/smallRNA/smallrna.html	Dedicated to small non-coding RNAs
rna.wustl.edu/tRNAdb/	Dedicated to tRNAs
bighost.area.ba.cnr.it/BIG/UTRHome/	Dedicated to the untranslated regions of genes
www.indiana.edu/~tmrna/	Dedicated to the recently discovered tmRNA that are both transfer and messenger RNAs. (If you don't yet know what this is, you MUST take a look at this fascinating Web site!)

Generic RNA resources

Well maintained and usually up to date, these generic RNA resources (see Table 12-4) make it easy for you to locate the server that corresponds to your needs. On these lists of links you can find databases, software, and online servers.

Table 12-4	A List of Generic RNA Resources
Address	**Description**
www.imb-jena.de/RNA.html	RNA World, one of the most complete sites currently available
www.rnabase.org/links/	Another very complete site

Chapter 13

Building Phylogenetic Trees

Nothing in biology makes sense except in the light of evolution.

— Theodosius Dobzhansky (1973)

The purpose of this chapter is to show you how you can use a computer and a few online services to run *phylogenetic analyses*, the scientific procedure that lets you reconstruct the evolutionary history of a group of organisms or sequences. If you know what phylogenetic trees are all about but you don't have a clue how you can generate one with a computer, this chapter is for you.

Here we show you how you can select sequences, turn them into a multiple sequence alignment, and then turn this alignment into a tree. We show you two methods:

✔ **ClustalW:** This is an easy method that you must use as a black box without knowing really what's going on inside.

✔ **Phylip:** A more sophisticated method that enables you to control every parameter that plays a part in the making of your tree.

Whatever choice you make, bear in mind that a high-quality multiple sequence alignment is usually the real limiting factor. (If you're not sure what a multiple sequence alignment is, be sure to check out Chapter 9.)

Finding Out What Phylogenetic Trees Can Do for You

The purpose of *phylogeny* is to reconstruct the history of life and explain the present diversity of living creatures. This can be represented as a huge genealogic tree (the tree of life). The underlying principle of phylogeny is to try to group living creatures according to their level of similarity. In this context, we assume that the more similar two species are (such as human and ape), the closer they are to their common ancestor. Phylogeny is not a new subject and whether you trace back the birth of modern biology to Darwin and *The Origin of Species* or to Aristotle and his notion of categories, you can't escape this daunting fact: Biology is very much about classifying, and the best means of classification we have is phylogeny.

Phylogenetics is a special kind of phylogeny that relies on the comparison of equivalent genes coming from several species for reconstructing the genealogic tree of these species and finding out who is the closest relative of whom in the family. If necessary, you can also apply phylogenetics methods to the various genes of a gene family to reconstruct the history of the gene family by the same means. Take note that these trees make sense only if you believe in evolution!

The molecular stories you can uncover with phylogenetics are incredibly rich. In fact, we find it difficult *not* to see a parallel between the destiny of famous families (such as the Medicis, the Borgias, or the Kennedys) and the fates of gene families. When we unravel the chain of events that makes the story of a protein family, we find tales of mutations and deletions, duplications or speciation, loss and gain of function, inactivation, and all the other traumatic events that shaped the world as it is today. Nothing is ever taken for granted while life evolves!

Phylogenetics is here to let you discover all this. Phylogenetics is not a technique, nor even a discipline; phylogenetics is a science — and a major one. The mere task of giving you general ideas on phylogenetics is very much beyond the scope of this chapter, so, if this subject is new to you, we urge you to consult some of the excellent textbooks available on this subject.

In the context of bioinformatics analysis, there are three major reasons why you may want to use phylogenetics:

> ✔ **Determining the closest relatives of the organism that you're interested in:** For instance, if you're studying a new bacterium, you can sequence its ribosomal RNA and place it on a phylogenetic tree computed with all known ribosomal RNAs. This can give you a fairly good idea of who this bacterium really is.

✔ **Discovering the function of a gene:** If you're studying a gene, you can use phylogenetic trees to make sure that the gene you're interested in is orthologous to another well-characterized gene in another species.

✔ **Retracing the origin of a gene:** Most genes within a genome travel together through evolutionary time. However, from time to time, individual genes may jump from one species to another one — for instance, piggybacking a virus infection. Phylogenetic trees are great to reveal such events, which are called *horizontal* (or *lateral*) *transfers*.

Two genes are *orthologous* if they come from two different organisms, derive from a common ancestor gene, and have been separated only by speciation events (as opposed to gene duplications). In theory, two genes that are orthologous often have the same exact function (have similar roles) in the two different organisms they come from. On the other hand, when the two genes arose from gene duplication, we say that they are *paralogous*. Two genes that are paralogous belong to the same family but are more likely to have different functions than if they were orthologous.

To give you an image, if the joystick of a 747 Boeing were a gene, it would be orthologous with the steering wheel of your car. In contrast, the left front wheel of your car is paralogous with the right front wheel.

Preparing Your Phylogenetic Data

Biologists have been dealing with phylogeny for many decades now, hence the large number of different techniques available. Things have become a bit simpler since people started doing *phylogenetics* (phylogeny applied to gene data). Nonetheless, there are still many schools of thought around and ancient fights are still raging in the world of phylogeny. Be aware that when you choose a method you're also choosing sides! In this chapter, we try to focus on the methods that have attracted the smallest amount of controversy over the past few years.

The only one thing upon which everybody agrees is that data quality is critical. If you have very good data that's suitable for phylogenetic analysis, any reasonable method will lead to a reasonable tree. On the other hand, if your data isn't good, no method, however good that method, can rescue the story your data tells. A brutal summary of this observation exists: Garbage in, garbage out. This is true of most data analysis methods.

Phylogenetics uses genes to reconstruct evolutionary history. This type of data is much easier to interpret than the *morphology data* — derived from an analysis of an organism's form and structure — that biologists used to rely on. It's possible to compare gene sequences in a much more objective fashion than it is to compare morphological characters.

When you do phylogenetics, good data means *a highly accurate multiple sequence alignment that contains properly chosen sequences.* In the first part of this section, we show you how you can prepare this kind of data.

Choosing the right sequences for the right tree

When you build a phylogenetic tree, you make the assumption that the sequences you are comparing have a common ancestor. If your sequences are similar enough, this is a reasonable hypothesis. You want to use your tree to reconstitute the history of these sequences to find out who is the closest relative to whom.

Of course, this vision implies that evolution is always divergent. This means that if two sequences are similar, we rule out the possibility that this similarity may result from convergent evolution. In the vast majority of cases, this makes sense. At the molecular level, convergent evolution does exist, but it seems to be the exception rather than the rule.

Using DNA or protein sequences

To establish the relationship between two sequences, you want to measure the time that separates their divergence from their common ancestor. How long ago did they part? To answer this question, you need to compare your sequences and measure a distance or establish an evolutionary scenario. Should you do this on the protein or on the DNA sequence? We have a complex answer to this seemingly simple question:

✔ **If your DNA sequences are *more* than 70 percent identical:**

You can make a DNA multiple sequence alignment. If your sequences are coding for proteins, however, this is not recommended.

The problem with DNA sequence alignments is that the substitution matrices are not very good and the alignments take more time to compute because of the sequence lengths.

✔ **If your DNA sequences are *less* than 70 percent identical:**

If your sequences code for proteins, it is much safer to translate them into proteins and to build the multiple sequence alignment with the proteins.

If your sequences are too similar at the protein level, you can thread the DNA sequences back onto the protein alignment. No tool is available online to do this, but you can download one from

```
life.anu.edu.au/molecular/software/transalign/
```

After you've completed the threading, you have two possibilities:

- Most synonymous mutation sites are different. This is a typical case of saturation. In this case, the distance measurements you make at the protein level are more accurate.

- Synonymous sites are not saturated. Distance measurements made at the DNA level are more accurate in theory.

In practice, unless your sequences are almost identical, it is easier to keep working at the protein level. This may not be as accurate as working with DNA sequences, but, in most cases, you can expect the results to be reasonably good.

Choosing sequences to make either a gene tree or a species tree

Homologous genes are genes that derive from a common ancestor. They can have three types of relationships (Figure 13-1):

- **Orthologs:** They're only separated by speciation. A *speciation* is the phenomenon during which a common ancestor gives birth to two subgroups that slowly drift away to become distinct species. Assuming that the genomes are not rearranged in the two new species, two genes are orthologous when they correspond to the same ancestral gene in the ancestral genome. Biologists usually expect orthologs to have similar functions and structure. In Figure 13-1, A1 and A2 are orthologs, and so are B1 and B2.

- **Paralogs:** Paralogs are homologues separated by a duplication event, meaning that within a genome, a gene was duplicated. One of the duplicates may have kept the original function while the other duplicate could have acquired a new function. You can expect paralogs to have different but related functions.

- **Xenologs:** *Xeno* is a Greek word that means foreigner. Xenologs result from a lateral transfer between two organisms, where a lateral transfer is a direct DNA transfer between two species. This means that one of the species contains a gene that does not have the same history as the genome in which it is inserted.

When you select a group of homologous genes to make a phylogenetic tree, you always make what biologists name a *gene tree*. It is a tree that tells the story of the genes it contains.

If you select all the *paralogous* members of a large human gene family, your gene tree tells the story of this gene family only. You can use it to merely reconstruct the chain of duplications that led from one single ancestral gene to the current situation.

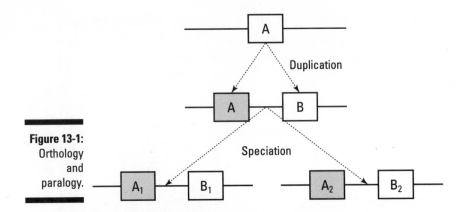

Figure 13-1:
Orthology
and
paralogy.

White T-shirts, spaghetti, evolution, and phylogeny

When we think of evolution, we tend to think of natural selection, survival of the fittest, struggle for life, and all these testosterone-loaded concepts. In molecular terms, gurus name this *neo-darwinism.* This popular theory states that the only reason mutations stay in a population is because they are helpful and increase the fitness of the individuals that carry them. This ancient theory doesn't account very well for synonymous mutations or other harmless DNA alterations.

In the sixties, the increased availability of molecular data led to the elaboration of neutralism by Kimura. *Neutralism,* also known as the neutral theory, challenges neo-darwinism by stating that most mutations are either lethal or neutral. Everybody agrees that lethal mutations are disposed of right away, but, for neutralists, the fact that a new non-lethal mutation spreads or not in the population depends mostly on chance. Even if these mutations are advantageous, chance is the driving force. Keep in mind

that neutral mutations aren't necessarily silent at the protein level and that they can also be mutations that slightly change a protein without altering its function. Of course, after these harmless but useless mutations are fixed, they can also find a niche and give some advantage to the individuals who carry them, later on. Neutralism is a very powerful theory because it accounts well for the way in which species build their adaptation potential before they need it.

If you're doing phylogeny, these neutral mutations are really the ones you're interested in. You do not want to see adaptation; you want to measure evolutionary distances and the best measure of evolutionary distance you have is counting the number of random and quasi-neutral events that occurred on your sequences since they separated from their ancestor. You can think of these mutations like each tick of a clock that has been working, one mutation at a time, for a million years.

To give you an image, imagine that this morning, like everyday, you took a white T-shirt and duplicated it for your two kids (Dave and Benny), and off they went. It was a long day; Dave and Benny both ate their spaghetti bolognese, both proving especially adept at emulating Michelangelo with the verve and passion with which they sketched out — in brilliant red sauce — a few Last Judgments on their white T-shirts. Each drop of colored-something landing on the immaculate shirts is a mutation, and the number of spots indicates how much the day has progressed. If you want to know how long Dave and Benny have been wearing these shirts, you can simply count the spots. This is true neutralism, and this is the ideal situation when you compare sequences.

Imagine now that there is a bit of selection going on: Dave and Benny may be very anxious about your reaction when they come back (okay, okay, probably unlikely, but bear with us for the sake of the argument), and they carefully clean every bit of color that lands on their shirts — except for those that you can't see or those that look "coooool". This would be the equivalent of selected mutations. When this goes on, you can no longer tell how long the shirt has been worn.

Non-selected mutations (neutral) are much better for evaluating distances because they accumulate smoothly. In coding DNA, the synonymous mutations are the more typical non-selected mutations. This is why DNA is better for measuring distances than protein sequences whose evolution is much more constrained. (These protein guys have work to do!)

Imagine now that the kids have had a second or third go at the spaghetti bolognese. Drops of sauce have landed on previous spots, and the shirts are now completely saturated with tomato sauce. What can you tell from this? How many spaghetti meals have Benny and Dave had? One? Two? Five? It's impossible to say! This is a typical saturation of our evolutionary time measure, and this is what happens when there have been one or more mutations on every neutral site of our DNA sequence. Determining how long this has been going on becomes impossible. Of course, the closer you are to saturation, the more imprecise your measurements are going to be.

One more thing before we rush to wash these filthy T-shirts. What if, on a certain day, Benny and Dave have not been doing the same thing? One may come home sparkling clean and the other muddy. Will you assume that they have spent their days in different time dimensions, or would you rather conclude that their shirts have evolved at different paces (or places)? Different paces sounds more likely, although if you ask the kids you may get a more confusing answer. Such differences also occur with genes and although we like to think that there is a universal molecular clock, we know it tends to tick at a different pace in every gene.

If you select a group of genes that are all *orthologous* from different species, the gene tree you get looks very much like a species tree, which lets you reconstruct the speciations that occurred while the species you're looking at (or their ancestors) were diverging. The best example of this type of gene tree is the ribosomal RNA phylogenetic tree that biologists use to reconstruct the big tree of life. Ribosomal RNA genes exist in every species and are clearly orthologous between species.

You can do things the other way round, of course, and use the tree to ask whether your sequences are putative orthologs or paralogs. This is a valid strategy if you know the true species tree.

Two genes that are homologous and come from two different species are not necessarily orthologous. They can also be paralogous like A1 and B2 are in Figure 13-1. Mixing orthologs and paralogs in a phylogenetic tree is a major source of trouble. Unfortunately, avoiding it is difficult.

In theory, there's no simple solution to the problem of establishing orthology — and demonstrating it is virtually impossible. However, in practice, authors have been using an empirical criterion that works reasonably well: the reciprocal BLAST best match. Here's how it's done:

1. **Choose a sequenceA from genomeA.**

2. **Carry out a BLAST search comparing your sequenceA and every sequence in another complete genomeB. (For more on BLAST searches, see Chapter 7.)**

 The BLAST search returns sequenceB as a top hit.

3. **Carry out a BLAST search comparing sequenceB and every sequence in genomeA.**

 If this search returns sequenceA as a top hit, you can treat sequenceA and sequenceB as orthologs from genomeA and genomeB.

 This result doesn't demonstrate that sequenceA and sequenceB *are* orthologous, but it does mean that you won't be doing something very wrong if you consider them so.

This strategy makes sense only when your two genomes are *complete*.

Eugene Koonin has assembled a high-quality collection of orthologous genes at NCBI. To do so, he used similar but slightly more sophisticated criteria. This collection is known as the COG (Clusters of Orthologous Groups), and you can find it online at www.ncbi.nlm.nih.gov.

Creating the perfect set

"Perfect" is meant to be ironic here, but we've listed in Table 13-1 a few points you should take into account when assembling your sequence set.

Table 13-1	Preparing a Set of Sequences for Making a Phylogenetic Tree
Problem	*Reason and Solution*
Avoid sequence fragments	Incomplete sequences make multiple sequence alignments and tree reconstruction methods very sick! You want to avoid these at all costs — or you at least want to use the same fragment for all the sequences.
Avoid Xenologs	Unless your purpose is to study them, avoid including genes that result from lateral transfer.
Avoid recombinant sequences	Some proteins result from the combination of several proteins. This is especially common in viruses. In terms of phylogeny, such proteins have two ancestors rather than one and standard tree methods are not equipped to represent this kind of relationship.
Avoid large complex families	Very large families that contain various domains and repeats can be very tricky to analyze. The ABC transporter is a good example of this kind of troubled family. If you can, stay away from these protein gangs, or try to work on smaller, more uniform subsets.
Try to make a small set	Phylogenetic reconstruction is one of the only cases where you're allowed to align very similar sequences and as many as you want. Unfortunately, it isn't because you need it that you can do it. If you have been through Chapter 9, you know this is not a good thing because most methods find it hard to deal with large datasets.
	Analyzing large sets means that you may not be able to use the most powerful methods, or that you will need to install these programs on your computer.

(continued)

Table 13-1 *(continued)*

Problem	Reason and Solution
Add an outgroup to your dataset	An outgroup is a sequence that you know has diverged long ago from the rest of the set. For instance, horse hemoglobin would be an outgroup if you're analyzing primate hemoglobin. This outgroup helps you by adding a root to your tree that symbolizes the first common ancestor of all your sequences.
	The choice of an outgroup relies mostly on biological criteria, yet you must also make sure that this sequence is similar enough to the rest of the set so that your multiple sequence alignment method can align it (see Chapter 9).

Living fossils do not exist! Unless you are using fossil DNA sequences, the sequences you are comparing all have the same age. Even if one of the species has retained more similarity with the common ancestor, its sequences have accumulated as many neutral mutations as the most sophisticated member of the family.

Preparing your multiple sequence alignment

Many phylogenetic tree reconstruction methods require a so-called *distance matrix*. The distance matrix is a table that contains the distances (or counts the number of evolutionary events) that separate each pair of sequences in your dataset.

In theory, you can build this matrix by comparing all your sequences two by two and fill up the distance matrix. In practice, this isn't correct because the pairwise comparison methods aren't accurate enough. (For more on pairwise comparisons, see Chapter 8.) The only way to accurately measure the distances between two sequences is to first make a multiple sequence alignment that includes these two sequences and then to measure the distances on their pairwise projection. The pairwise projection is a pairwise alignment that you *extract* from the multiple sequence alignment. It benefits from the information contained in ALL the other sequences.

The quality of your multiple sequence alignment is the real limiting factor when you make a tree, and there is no way you can make a good tree with a bad alignment. If you aren't sure what a multiple sequence alignment really is, make your way through Chapter 9.

Computing your multiple sequence alignment

To compute your multiple sequence alignment, you can use ANY of the methods we introduce in Chapter 9. These online servers include

- ✔ ClustalW: www.ebi.ac.uk/clustalw/index.html
- ✔ Dialign: bibiserv.techfak.uni-bielefeld.de/dialign/
- ✔ Tcoffee: igs-server.cnrs-mrs.fr/Tcoffee

Making sure you have the right multiple sequence alignment

Before using your multiple sequence alignment for building a tree, you want to make sure that it's as accurate as possible. In this section, we give you a few criteria and a list of things to do to improve the suitability of your multiple alignment. Figure 13-2 shows you the typical regions you may want to remove from your alignment.

Columns to remove

Figure 13-2:
Preparing a
multiple
sequence
alignment.

```
chite   - - - -ADKPKRPLSAYMLWLNSARESIKRENPDFK- VTEVAKKGGELWRGLKDAATAKQNYIRALQEYERNGG- -
wheat   - -DPNKPKRAPSAFFVFMGEFREEFKQKNPKNKSVAAVGKAAGERWKSLSEANKLKGEYNKAAAYNKGESA
trybr   KKDSNAPKRAMTSFMFFSSDFRS- - - -KHSDLS IVEMSKAAGAAWKELGPAEKDKERYKREM- - - - - - -
mouse   - - - - -KPKRPRSAYNIYVSESFQ- - -EAKDDS AQGKLKLVNEAWKNLSAPKDRIRYDNEMKSWEQMAE
                 ***. ::: .. .          :  . .        *  . *: *   :.*. :
```

To ensure that your multiple alignment is both accurate and suitable, follow these steps:

1. **Make sure there are as many gap-free columns as possible.**

 Gaps cause trouble for most phylogeny reconstruction methods. Some of these methods, such as ClustalW, can use the *complete-deletion* techniques and ignore every column that contains a gap.

2. **Remove the extremities of your multiple alignment.**

 The N-terminus and the C-terminus tend to be poorly conserved and therefore poorly aligned. You can safely remove them (refer to Figure 13-2).

3. **Remove the gap-rich regions of your alignment.**

 Internal gap-rich regions in a multiple sequence alignment often correspond to loops. Even if your program returns an alignment, it does not mean that this alignment is meaningful.

4. **Be sure to keep the most informative blocks.**

 The ideal multiple alignment for building a tree would be a high-quality alignment of sequences with a low level of identity, so that each position contains a trace of the family history.

 In Chapter 9, we show you what a good block looks like in a multiple sequence alignment. It is typically 20 to 30 amino acids long and contains a few conserved positions. Such blocks are ideal for producing high-quality trees.

 If you want, you can use the Tcoffee server to evaluate your multiple sequence alignment and remove columns that are unlikely to be correctly aligned. See Chapter 9 for more on how to do this.

 The best way to edit your multiple alignment is to use Jalview, the multiple alignment editor that we introduce in Chapter 10. However, there is a quick and dirty solution with Microsoft Word that lets you remove blocks very efficiently:

 a. While pressing the Alt key on your keyboard, use the mouse to select entire columns in your alignment.

 b. When you've selected everything that you want to remove, press the delete key to remove the selected block.

Building the Kind of Tree You Need

There are three major groups of methods you can use to build a tree: distance methods, parsimony methods, and likelihood methods. None of these methods is almighty. They all have their particular weaknesses and strengths.

This chapter focuses entirely on distance-based methods. They aren't necessarily the most accurate, but they are generally applicable and easy to set up. You can apply distance-based tree reconstruction methods on most situations, and they tend to generate very reasonable results.

When you decide to use a distance-based method to compute your phylogenetic tree, consider these four major ingredients:

- **Your multiple alignment:** Which method? Which sequences?

- **The distance measure:** Which substitution matrix?

 ✔ **The tree reconstruction algorithm:** UPGMA, NJ, Fitch, or Kitsch?

 ✔ **The type of tree you want to display:** A cladogram without distances or a phenogram that shows distances? A rooted or an unrooted tree?

In this section, we show you how you can use the Internet to compute a tree and to display it in any format that you need.

Computing your tree

Combining the right ingredients for making your tree is much like cooking. There is no such thing as *the best recipe ever*, and the best tree you can produce is mostly a matter of taste and the mood of the day.

Just like when it comes to food, there are two very different ways to produce trees. The first one uses ClustalW; it's quick, hassle free, and somehow very similar to (good) fast food. We show you in a steps list how you can simply cut and paste your sequences to obtain a tree while avoiding embarrassing questions.

The second steps list is for those of you out there who love to buy fresh vegetables on the market and make your own salad dressing — sticklers for detail, if you catch our drift. With this steps list, you can control every ingredient that goes into the making of your tree. Phylip is the name of this package that you can use to do state-of-the art phylogeny. This is serious stuff and world-renowned phylogeny specialists use it to do their work. Despite this intimidating introduction, Phylip is surprisingly easy to use.

Producing an (almost) instant phylogenetic tree

If you need a tree and you have firmly decided that you do not want to know *anything* about this barbaric kind of gardening, ClustalW is definitely for you. ClustalW is above all a multiple sequence alignment package, but some of the ClustalW servers let you use this program to produce phylogenetic trees.

On the EBI ClustalW server, and on most ClustalW Web servers, if you feed ClustalW with unaligned sequences, the program returns a tree. This tree is **NOT** a phylogenetic tree; it is a guide tree that ClustalW uses to assemble the multiple sequence alignment. In general, any tree with a .dnd extension returned by ClustalW is NOT a phylogenetic tree.

You CANNOT use this tree in place of a phylogenetic tree. Doing so would be a mistake.

Here is how you can use the EBI ClustalW server to produce a phylogenetic tree. First and foremost, you need to gather your sequences:

1. **Point your browser to** `www.expasy.ch/sprot/sprot-retrieve-list.html.`

2. **Select the FASTA radio button.**

3. **Enter the Accession Number of your sequences.**

 Be sure to enter just one accession number per line. If you don't have any accession numbers of your own, you can enter the following:

 O12957, O02672, O02683, O02690, O02681, O02687, O02673, O02688, O12959, O02677, O02689, O02682, O02676, O02680, O12954

 These are homologous fragments of the gamma fibrinogen. They have been sequenced in several species for comparative studies. We can use fragments rather than entire sequences because these fragments are all homologous.

4. **Click the Create FTP File button.**

5. **Copy the sequences into a word processor.**

6. **Rename your sequences.**

 Unless you have a formidable memory, trees with species names are much easier to read. Here is how we renamed the sequences whose accession numbers we entered in Step 3:

 O12957 (Sheep), O02672 (Moose), O02683 (Giraffe)

 _O02690 (Chevrotain), O02681 (Beluga) O02687 (Sperm_whale)_

 O02673 (Rorqual), O02688 (Pig), O12959 (Peccary)

 O02677 (Dromedary), O02689 (Tapir), O02682 (Horse)

 O02676 (Hyena), O02680 (Coyote), O12954 (Hippopotamus).

With sequences in hand, we can now move on to the real work:

1. **Point your browser to** `www.ebi.ac.uk/clustalw.`

2. **Paste your sequences into the main Sequences window (near the bottom of the page).**

3. **Click the Run Button.**

4. **Wait until your results return.**

 ClustalW returns a long output that includes

 - _The program log_ (top of the output)

 - _The multiple alignment_ (your multiple alignment)

- *The guide tree* at the bottom of the page, as shown in Figure 13-3. This guide tree tells you the order that ClustalW used to align your sequences. Note that this guide is, phylogenetically speaking, roughly correct although it isn't really accurate for species like hippopotamus. We can do much better with a real phylogenetic tree.

Figure 13-3:
Guide tree produced by ClustalW.

As an alternative to the rest of this steps list, you can also click the Jalview button (near the top of the page) and use this multiple alignment editor to produce your phylogenetic tree. With Jalview open, simply request a tree by choosing Calculate⇨Neighbor Joining Tree Using PID from Jalview's main menu.

5. **Copy the multiple sequence alignment onto the Clipboard.**

 Do not forget to include the top line that starts with *CLUSTAL W.*

6. **Click the Back button of your browser.**

7. **Click the Reset button on the ClustalW form.**

8. **Paste your multiple alignment into the Sequence window, as shown in Figure 13-4.**

 You can enter your alignment in any format you like, including ALN (the ClustalW format), MSF, or FASTA.

9. **Choose NJ from the Phylogenetic Tree: Tree Type drop-down menu, as shown in Figure 13-5.**

 This menu enables you to choose between different methods for computing your tree. Specialists generally agree that NJ, the Neighbor Joining method, is the best one for the task at hand.

If you set the Phylogenetic Tree: Tree Type drop-down menu to anything other than None, you must provide ClustalW with aligned sequences. "Why?" you might ask? Well, when you choose anything but None from the drop-down menu, ClustalW *stops computing alignments* and assumes that you have provided it with aligned sequences. If you don't do this and give unaligned FASTA sequences to the server, it simply assumes these sequences are aligned and computes a tree. Of course (unless your sequences can be aligned without gaps), this tree will be wrong!

Figure 13-4:
Computing a
phylogenetic
tree with
ClustalW.

Figure 13-5:
The
Phylogenetic
Tree: Tree
Type drop-
down menu
in ClustalW.

10. **Choose On from the Correct Dist. drop-down menu (located directly beneath the Phylogenetic Tree: Tree Type drop-down menu).**

 This option makes it possible to correct for multiple substitutions. It tends to elongate the branches of your tree and is mostly useful for distantly related sequences.

11. **Choose On from the Ignore Gaps drop-down menu.**

 This causes ClustalW to ignore every column of your multiple alignment that contains gaps. Turning this option on ensures that all the sequences are compared on the same number of residues.

12. **Choose Phylogram from the Tree Type drop-down menu.**

 Don't confuse this drop-down menu with the Tree Type menu you see under the Phylogenetic Tree heading. The Tree Type drop-down menu for Step 12 is directly underneath the Matrix drop-down menu.

In a phylogram, the branch lengths represent real distances, while in a cladogram, branches indicate only the topology (branching order) of your tree.

The EBI server sometimes has problems displaying cladograms and it is therefore better to request a phylogram. You can always change this later by using the phylodendron server. (We show you how at the end of this chapter.)

13. Click the Run button.

Wait until ClustalW returns your NJ tree, as shown in Figure 13-6.

Figure 13-6:
An NJ tree produced by ClustalW.

This tree is much more accurate than the guide tree shown in Figure 13-3. For starters, we can clearly see now the relation that exists between the hippopotamus and the whales as postulated by Higgins and Grauer a few years ago.

14. Try to save your results.

Easier said than done. ClustalW displays your tree in a nice graphic, but you cannot save this tree. If you want to keep a record of the graphic, you must either make a screen shot of the image or print it into a PDF file.

15. If screen shots or PDFs are not what you want, click the hyperlink above the tree.

This displays your tree in a text format known as Newick format — shown in Figure 13-7 — famous for its many parentheses.

The Newick format is the most standard way to store phylogenetic trees, and most phylogenetic packages can use this format. If you want to keep only one file related to your tree, keep this one!

```
(
 (
  (
   (
    (
     (
      Tapir:0.08900,
      Horse:0.14784)
     :0.02087,
     (
      Hyena:0.07996,
      Coyotte:0.11741)
     :0.06795)
    :0.03214,
    (
     (
      Pig:0.06880,
      Pecari:0.03646)
     :0.04934,
     Dromedary:0.08224)
    :0.01391)
   :0.02323,
   (
    (
     Beluga:0.02632,
     Rorqual:0.02632)
    :0.01579,
```

Figure 13-7:
The Newick
Tree format.

Making a tree with Phylip

When it comes to phylogenetic trees, Phylip is a household item. Along with PAUP, it is one of the most widely used resources for computing highly accurate phylogenetic trees.

One of the nice things with Phylip is that it makes it easy for you to *bootstrap* (check its reliability; see the sidebar, "What is bootstrapping?") your tree, thus making sure that the topology you're looking at is well supported by your multiple sequence alignment.

If you decide to become a specialist, Phylip has all you need and you can use it to build almost any kind of tree with any method you like. In the following steps list, we show you how to produce a phylogenetic tree with bootstrap information.

Before going onto this server, you must have a high quality multiple sequence alignment ready. For the purpose of this steps list, we use the alignment we produced for the previous steps list (the homologous fragments of the gamma fibrinogen). You must also be ready to provide a valid e-mail address.

1. **Point your browser to**
 `bioweb.pasteur.fr/seqanal/phylogeny/phylip-uk.html`.

 The Phylogeny: Phylip Programs page appears, as shown in Figure 13-8. This page shows all the Phylip programs installed on this server. You can produce a phylogenetic tree by using these programs one after the other. The only trick is to use them in the right order.

Phylogeny : Phylip programs

Documentation.
FAQ (Frequently Asked Questions).

- **Programs for molecular sequence data [sequence.doc]**

 o DNA
 dnadist [advanced form] [dnadist.doc]
 Distances from DNA sequences.
 dnapars [advanced form] [dnapars.doc]
 Parsimony method for DNA.
 dnaml
 (Maximum likelihood method) has been removed ; please use rather fastDNAml, which is much faster
 and equivalent.
 o Proteins
 protdist [advanced form] [protdist.doc]
 Distances from protein sequences.
 protpars [advanced form] [protpars.doc]
 Parsimony method for protein sequences.

- **Programs for distance matrix data [distance.doc]**

Figure 13-8:
Phylip
online at the
Pasteur
Institute.

2. **Under the Proteins heading, right next to the protdist link, click the advanced form link.**

 The Phylip: Protdist page appears.

 To make a distance-based tree, you first need to generate a distance matrix that contains the pairwise distances between all your sequences, as measured on the multiple sequence alignment.

 Some methods enable you to do this on either DNA or on proteins. There are also different programs for closely or distantly related sequences.

 Protdist is a method that uses a substitution matrix to measure the distance between your aligned sequences.

 Protpars does the same thing, but it tries to evaluate the number of mutations at the DNA level required for the observed amino-acid differences. Protpars works better with closely related sequences

3. **Enter your e-mail address in the Your E-Mail field of the Phylip: Protdist page.**

 If the server is overloaded or if you are analyzing a large alignment, it sends your results by e-mail rather than online.

 The e-mail you receive contains a URL. When you open this URL, your browser takes you to a page similar to what's shown in Figure 13-11.

 Do not give bogus e-mail addresses to the server! As soon as it times out, the server decides to return your results by e-mail. You cannot control this; if you give a phony e-mail address, however, the server keeps running on a job whose result it cannot send you. This sounds very much like wasting everybody's time. In our experience, it never took more than a few minutes before receiving the e-mail.

4. **Paste your multiple alignment into the Sequence window, as shown in Figure 13-9.**

 Do not forget to copy the header line that starts with CLUSTALW along with your alignment.

5. **Click the Bootstrap options link.**

 You are brought to the Bootstrap Options section of the Web page.

6. **Select the Perform a Bootstrap Before Analysis check box, as shown in Figure 13-10.**

 Bootstrapping your data is a way of assessing the quality of your tree; see the sidebar "What is bootstrapping?"

Reset | Run protdist | ● bioinformatics4dummies@wiley.com | your e-mail
(● = required, ● = conditionally required)

● Alignement File : please enter either :

1. the name of a **file**: [] Browse...

 CLUSTAL W (1.82) multiple sequence alignment

 Tapir
2. *or* the **actual data** here: RFGSYCPTTCGIADFLSTYQTXVDKDLQVLEDILNQAENKTSEAKELIKA

(sequence format)

Jones-Taylor-Thornton matrix ▾ Distance model (P)
Gamma distribution of rates among positions (G) ? ○ [default] ○ No ○ Yes ○ Gamma+Invariant

Bootstrap options

Weight options

Figure 13-9: Protdist server.

Bootstrap options

☑ Perform a bootstrap before analysis

| Bootstrap | ▾ | Resampling methods |

⦿ 3 ⬜ Random number seed (must be odd)

2 ⬜ How many replicates

[Return to the main part with your favorite browser's Back function]

Figure 13-10:
The Bootstrap options of protdist.

7. **Set the seed to some odd number such as *1* or *3*.**

 The *seed* controls the generation of random numbers during the bootstrap. Different seeds lead to different results.

8. **Set the number of replicates to *2*.**

 The number of replicates is the number of bootstrap cycles that you do. Normally, this number should be at least 100.

 The value we use here is only meant to give you an idea of what this function does for you. For real analysis, you need at least 100 replicates. With a value of 2, don't expect a meaningful bootstrap value!

 If you require a large number of replicates, the server usually prefers returning the results by e-mail.

9. **Scroll back to the top of the page and click the Run Protdist button.**

 Eventually, you'll see results similar to what's shown in Figure 13-11. If your browser times out, the server will send you an e-mail that contains a URL. If you open this URL, you're taken to a page of results similar to what's shown in Figure 13-11.

Phylip : protdist - Program to compute distance matrix from protein sequences (Felsenstein)

Results:

outfile

| neighbor ▾ | Run the selected program on outfile |

params

seqboot.params

protdist.out

standard error file

From now, this files will remain accessible for 10 days at:
http://bioweb.pasteur.fr/seqanal/tmp/protdist/A31954610272762/

You can save them individually by the **Save file** function if needed.

Figure 13-11:
Protdist results.

10. **Choose Neighbor from the first drop-down menu on the result page.**

 Neighbor is a program that turns your distance matrix into a Neighbor Joining tree.

 If you want, you can use another tree reconstruction method, such as Fitch, which uses the mean least square method. These days, people usually agree that NJ provides a very good trade-off between all the available methods.

11. **Click the Run the Selected Program on Outfile button.**

12. **Select the type of tree you want to produce by selecting the radio button next to Neighbor-Joining.**

 Neighbor Joining and UPGMA are related methods, but Neighbor Joining produces trees that are much more accurate.

 UPGMA produces rooted trees.

 Neighbor Joining produces unrooted trees that you can root later.

13. **Select the Bootstrap options link.**

14. **Select the Analyze Multiple Data Sets option.**

15. **Indicate the number of data sets you're providing, as shown in Figure 13-12.**

Figure 13-12:
The Bootstrap option of Neighbor.

Bootstrap options

☑ Analyze multiple data sets (M)
◉ 2 _____ How many data sets
☑ Compute a consensus tree

[Return to the main part with your favorite browser's Back function]

16. **Set the Random Seed to an odd number like *1* or *3*, and check the Randomize box.**

 These are the replicates you requested in Step 8. This way you are telling Neighbor how many trees it must generate.

17 **Select the Compute a Consensus Tree option.**

 This option tells Neighbor to return the tree that has the best agreement with the replicate data sets you are providing it with.

 This option makes sense only if you're providing multiple datasets.

18. **Scroll down to the Other Parameters section and choose an outgroup.**

 Choosing an outgroup only matters if you selected Neighbor Joining in Step 10.

You can specify which outgroup you want to use according to its position in the multiple alignment, and Neighbor uses this outgroup as a root.

In your tree, the *root* represents the common ancestor of all the sequences. Remember, however, that living fossils do not exist! This outgroup is not the real root, but the real root is necessarily between the outgroup and its closest node.

Depending on the server and the default options, ClustalW sometimes changes the order of your sequences when it aligns them. It is the order in the multiple alignment that prevails when you specify your outgroup. Take a look at your alignment to ensure that you're choosing the right outgroup.

19. **Scroll to the top of the page and click the Run Neighbor button.**

 Your results appear in a new page as a series of hyperlinks.

20. **Open the consensus tree files.**

 The two first files on the output list are the consensus tree files. They're hyperlinked, so you can simply click them to display their contents.

 `outfile.consense` (see Figure 13-13) contains a text version of your consensus phylogenetic tree.

 The consensus tree is always unrooted. The number above each branch indicates its stability.

 `outtree.consense` is the same tree but in the Newick format.

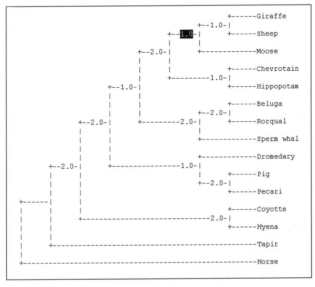

Figure 13-13: A consensus tree output by Neighbor.

21. Open the normal trees.

Neighbor also outputs the trees it used to build the consensus in `outfile` and `outtree`.

`outfile` contains a text representation of your trees. It contains as many trees as you requested replicates for in Steps 8 and 16.

If you checked Neighbor Joining in Step 12, the trees are rooted with the outgroup.

If you requested UPGMA, the trees have been automatically rooted.

`outtree` contains the trees in Newick format.

22. Save and re-display your output.

It is difficult to display a tree with bootstrap values using online tools. If you want to keep these values, the best thing to do is to save the `outfile.consense` file into a tree. This is a standard text file that you can save with the File➪Save As option of your browser.

If you want to redisplay your tree by using another graphic format, the most convenient thing to do is to save the file `outtree.consense` and to turn it into a GIF or a PDF file with the Phylodendron server. (See the last part of this section for more on how to do this.)

Knowing what's what in your tree

When you work with phylogenetic trees, there's always a bit of special vocabulary involved. Figure 13-14 recapitulates the names of the various components in your tree. Some of these names may sound a bit technical, but it's worth knowing their meaning because phylogenetic jargon is becoming more and more common in mainstream publications.

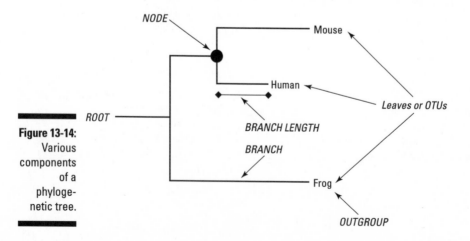

Figure 13-14:
Various components of a phylogenetic tree.

Leaves are known as *OTUs* (Operational Taxonomic Units). Each *node* represents an ancestral OTU. A *clade* is a group of OTUs that includes several sequences and their common ancestor nodes. The *root* is the common ancestor of all the OTUs.

TECHNICAL STUFF

What is bootstrapping?

Because your computer is able to produce a tree doesn't necessarily mean that the tree is correct. Many things can go wrong in the tree's construction, including faulty data or an incorrect alignment. In most cases, while globally correct, your tree will contain a few inaccurate branches. This is where *bootstrapping* comes in handy. Tree bootstrapping is a way to check whether your tree is biologically meaningful by assessing its *robustness* — that is, checking whether every portion of the alignment equally supports your tree. The logic behind bootstrapping is that you measure the solidity of each branch in your tree so as to determine whether you can trust it when drawing your conclusions.

Bootstrapping is a multiple step process. The program starts by sampling columns in your initial alignment. The purpose is to generate a *bootstrap alignment* in which some columns are missing and others are duplicated. Here's how it's done:

1. **Choose a column at random from the initial multiple alignment obtained with your sequences.**

 This column is the first column of your *bootstrap alignment.*

2. **Select a new column in the initial alignment and copy it to extend the bootstrap alignment.**

 By *extend* here, we mean making the alignment one column longer than the column you just copied from the initial alignment.

3. **Repeat Step 2 until your bootstrap alignment contains as many columns as your initial alignment.**

 Note that some of the columns are either never selected or appear several times in the final bootstrap alignment. This is *random sampling.*

For instance, here is how you could use an initial alignment to generate two bootstrap alignments:

```
Initial Alignment
Column 1 2 3 4 5 6 7 8 9
seq1   A B C D E F G H I
seq2   A A B B C B A C A
seq3   C C A C B A C A B

Bootstrap Alignment 1
1 1 8 1 2 5 1 8 2
A A H A B E A H B
A A C A B C A C A
C C A C C B C A C

Bootstrap Alignment 2
1 4 5 6 6 3 4 1 7
A D E F F C D A G
A B B C C B B A A
C C B A A A C C C
```

Note that the first bootstrap alignment contains Column #1 of the initial alignment four times.

This procedure generates many multiple alignments that look more or less like the original. The purpose of all this is to check whether all the columns in your initial alignment tell a similar evolutionary story (that should be the case).

(continued)

(continued)

During the next step, each random alignment is turned into a distance matrix, and each matrix is turned into a tree. To build the consensus tree, Phylip takes the average of all the trees it has generated from the bootstrap alignments. To assess the quality of each branch in the consensus tree, Phylip counts the number of bootstrap trees that contain this branch. Good branches are those that appear in every bootstrap tree. They are the ones on which you can strap your swing!

Take a look at the consensus tree in Figure 13-13. On this consensus tree, every branch comes along with a number between 1 and 2. This value tells you how solid your branch is. This value falls between 1 and 2 because, in Steps 8 and 15, we chose to have 2 bootstrap cycles. Neighbor generated 2 bootstrap trees with these 2 bootstrap alignments, and it generated a consensus tree that's the average of these 2 bootstrap trees.

In a tree, a branch always separates your data into two groups: sequences on the left and sequences on the right side of the branch. The numbers above the branches in the consensus tree indicate how many branches exist in your two trees that split the data in exactly the same way as the branch you're looking at.

For instance, if you look at the consensus tree in Figure 13-13, you find that the branch containing giraffe and sheep has a value of only 1. The reason for this is that in one of the bootstrap trees, giraffe and moose are in the same group while in the other tree, giraffe is in the same group as sheep (and moose is alone). In theory, you could conclude that the tree indicates some uncertainty on whether sheep and giraffe are more closely related to one another than each is related to moose. In practice, however, with only 2 bootstraps, you really can't say anything of the kind! You need at least 100 bootstrap cycles before you start convincing farmers to breed giraffes for their wool.

A *branch* defines the relation between a clade or an OTU and the rest of the tree. A tree can be *scaled,* in which case the branch length means something in terms of evolutionary time, or it can be *unscaled* in which case it is only the topology of the tree that is informative.

The tree may be rooted or unrooted. *Unrooted* trees do not identify evolutionary paths. If your tree is unrooted, you accept the possibility that any of the nodes or OTUs could be closer to the common ancestor than to any of the other OTUs. If you have an *outgroup,* you know that the root must be inserted in the branch between this outgroup and the other clades.

Displaying your phylogenetic tree

Phylodendron is a powerful Web-based server that lets you turn a text file in Newick format into a graphic display of your tree. You can control every

minute aspect of the display and generate images in any format you fancy. In the following steps, we show you how to use Phylodendron to turn a phylogenetic tree into a GIF picture that you can include in any document:

1. **Point your browser to** www.es.embnet.org/Doc/phylodendron/ treeprint-form.html.

 The Phylodendron Phylogenetic Tree Printer page, run by EMBnet, appears.

2. **Paste your tree in Newick format in the Input window.**

3. **On the top row under Tree Styles, select the Phenogram radio button.**

 If you select Tree Diag, you must make sure that you're providing the program with an unrooted tree; otherwise, the program will hang.

4. **In the Extra Options section, choose GIF image from the Format drop-down menu.**

5. **If you want horizontal distances to have a meaning, select the Use Node Lengths check box in the Tree Growth section.**

6. **Click the Submit button.**

7. **Choose File⇨Save As from your browser's main menu to save your results as a GIF file that you can use for inclusion in other documents.**

Doing Phylogeny for Free over the Internet

Conducting phylogeny over the Internet isn't yet as developed as database research or other kinds of sequence analysis methods. Fortunately, over the last few years, the community has made some high-quality resources available. These resources might not run on very powerful servers, but they're perfectly designed if all you need to do is build a tree once in a while.

Finding online resources

A vast amount of phylogeny resources are available over the Internet. Yet most of these programs aren't available online; you need to install them on your own computer. Table 13-2 lists a few resources that you can use online.

Table 13-2 Online Sites for Making Phylogenetic Trees

Address	Description
`www.ebi.ac.uk/clustalw/`	You can use ClustalW to build multiple alignments and compute NJ trees.
`www.genebee.msu.ru/ clustal/basic.html`	The Genebee server can produce genuine phylogenetic trees in one step.
`www.es.embnet.org/Doc/ jalview/ help.html`	You can use Jalview to produce NJ trees. It is a very powerful tool that can combine alignment edition with tree computation.
`sdmc.krdl.org.sg:8080/ ~lxzhang/phylip/`	A Web interface for Phylip. Many other phylogeny-related programs are also available online at the Pasteur site.
`bioweb.pasteur.fr/seqanal/ phylogeny/phylip-uk.html`	A Web interface for Phylip.
`www.genebee.msu.ru/services/ phtree_reduced.html`	Very powerful interface for a new tree reconstruction method.

Finding generic resources

Table 13-3 lists some of the most important resources in phylogeny. In most situations, you should be able to find what you need a few clicks away from these pages. We also include two resources that should cover your immediate need if you want to find out for yourself the basics of phylogeny.

Table 13-3 Generic Phylogenetic Resources on the Internet

Address	Description
`evolution.genetics. washington.edu/phylip/ software.html`	Joe Felsenstein's pages, where Phylip lives; it's also one of the most extensive collections of resources available. Truly a legendary site!
`www.ucmp.berkeley.edu/ subway/phylo/phylosoft.html`	A very complete list of phylogeny resources.

Address	Description
paup.csit.fsu.edu/ index.html	The home of PAUP, the other legendary phylogeny package. Although PAUP is a commercial package, it's reasonably priced and worth every penny, according to specialists.
www.ncbi.nlm.nih.gov/ About/primer/phylo.html	The NCBI primer on phylogeny.
www.techfak.uni-bielefeld. de/bcd/Curric/MathAn/ mathan.html	A high-quality course on tree reconstruction methods.

Collections of orthologous genes

Collections of orthologous or homologous genes are very useful if you want to ask functional or phylogenetic questions. Table 13-4 lists some databases that provide this type of information.

Table 13-4	Collections of Orthologous Sequences
Address	**Description**
www.ncbi.nlm.nih.gov/COG/	Cluster of orthologous sequences maintained by the NCBI. Each cluster contains proteins from bacterial genomes.
pbil.univ-lyon1.fr/ databases/hovergen.html	A collection of orthologous vertebrate genes.
pbil.univ-lyon1.fr/ databases/hobacgen.html	A collection of orthologous bacterial genes.
systers.molgen.mpg.de	Another collection of homologous sequences.
www.cme.msu.edu/ RDP, rrna.uia.ac.be	Two extensive collections of ribosomal RNA sequences, which are very useful for classifying new organisms and come with appropriate phylogenetic tools.

Part V
The Part of Tens

The 5th Wave — By Rich Tennant

"Well, that proves it. He's definitely from your side of the family."

In this part . . .

This part is just what it says on the box: It contains tables of links, resources, and various things useful but difficult to classify. If you haven't found what you were looking for anywhere else in this book yet, you'll probably find it here, or a couple of mouse clicks away from one of the addresses we give you. In this part, we also give you some useful guidelines if you're planning to let your research depend on Internet resources. Better read this before it's too late!

Chapter 14

The Ten Commandments When Using Servers

- -

In This Chapter

▶ Making sure you don't divulge confidential data

▶ Ensuring that you're able to reproduce your work

▶ Using appropriate data

▶ Making sure you save the right files

- -

I never did give anybody hell. I just told the truth, and they thought it was hell.

— Harry S. Truman

The nicest thing about using Web-based servers is that nothing is ever a problem on the Internet. You can run any program you fancy. You don't need to update or maintain any database, and, if the server is broken, you can simply go and use another server. On the Web, you've got it so easy that you might even forget what you're doing and keep pushing buttons until something nice pops onto your screen. A bit like what you — and we — do in front of the television!

That's all fine and dandy, but if you need to use these bioinformatics tools for doing some research, you need to be a bit more careful or this virtual paradise will rapidly turn into real hell. Nothing is more frustrating than a good result you can't reproduce, especially in science.

In this chapter, we give you a few pieces of advice designed to make life easy for you if you want to produce results that your work can really depend on.

Your Data is Never Secure on the Web

When you submit a sequence to an Internet server, you send it out into the world. With the exception of a few Java applets that you can use offline, never treat an Internet submission as a safe process.

While the people who run these servers are renowned for their honesty, they rarely have the means to ensure tight security for the data you send them. This means that your data may be intercepted during transit.

We personally think that nobody really gives a damn about the sequences you send over the Internet. However, your employers probably think that the risk is real, and if you want to keep your job, you'd better share this fear!

If you need secrecy, you'll just have to install the programs you need on your own computer. No ifs, ands, or buts possible here.

Remember the Server, the Database, and the Program Version You Used

If you run the same program (like ClustalW) on two different servers, it does not at all mean that you're doing the same thing twice. The setup of the servers or even the program version can be different. This is the reason why *"I ran ClustalW over the Internet"* doesn't give enough information for someone to reproduce your results — aside from the fact that it doesn't sound very professional either.

To avoid any confusion, write down the name of the server as well as the version of the program that it runs. For instance, you should not expect a server that runs ClustalW 1.77 to return the same results as a server that runs ClustalW 1.81.

If your program (such as BLAST) uses a database (such as SWISS-PROT), you should write down the name and the version of the database you're searching.

Servers also change with time. They get upgraded, improved, and so on. On average, this happens every six months.

Write Down the Sequence Identification Numbers

When you analyze a sequence, make sure that you have enough information on this sequence. Try to keep its ID (identification) and its AC (Accession) numbers, as well as any other information that seems to be relevant. Remember that the ID of an entry can change from one database release to the next, but the AC number never changes.

Write Down the Program Parameters

When you run a server, you often need to slightly alter the default parameters so that the program does exactly what you need it to do. Many programs supplement their output with a file that contains all the parameters you used. This file often has a name that contains the word *log*. For BLAST, these parameters are at the bottom of the result page. If your program outputs such a file, be sure to save it.

As an alternative, you can use the PrntScrn facility of Windows to make a screen shot. (See the next commandment, "Save Your Internet Results the Right Way," for tips on how to do this.) PrntScrn can make exactly reproducing your request much easier for you.

Save Your Internet Results the Right Way

Saving Internet results can be trickier than it seems at first glance. This is especially true when you try to save documents that contain images like BLAST results. We cannot give you a foolproof procedure because this changes all the time, depending on your Web browser, the way the browser is installed on your computer, and the kind of data you're downloading. When it comes to saving files, Java is a major source of trouble, and often the only solution is to make a screen shot.

Never assume that an important result has been saved correctly; always double-check. The latest version of Internet Explorer and Netscape makes it possible to save complete pages (including pictures).

If you can't save your page, try to print it as a PDF file.

If printing into a PDF file fails as well, make a screen shot by using the Windows PrntScrn utility:

1. **Press the PrntScrn key on your keyboard.**

 This puts a copy of your screen onto the Clipboard.

2. **Open Microsoft Paint by choosing Start⇨Programs⇨Accessories⇨Paint.**

3. **Press Ctrl+V and say Yes to the message that proposes that you resize the bitmap.**

4. **Save as usual or print.**

 Save using the `.jpg` format to generate a compressed picture.

Whenever possible, also save a text version of your results! Many programs output a staid text version of the results along with the flashy graphic output. This text version is much easier for storing and exploiting your results. In many cases, it is the only format that other programs can use.

Use E-values

E-values (expectation values) tell you how many times a result as good as the one you're looking at could have been reached by chance alone. On its own, this has no special biological meaning. However, the rationale behind the use of E-values in biology is that a good result is something that chance could not have produced without lots of help from nature.

A good E-value is a low E-value. For instance, an E-value of 10^{-32} is better than an E-value of 10^{-4}

Make Sure You Can Trust Your Alignments

Just because you have an alignment doesn't mean that you can trust it implicitly! Before you build an elaborate theory on your alignment, make sure that it's biologically correct. Measure the percent identity and look for conserved blocks or important conserved residues. Also, be critical and use the guidelines we give you in Chapter 9!

Use Different Programs to Check Borderline Results

If you're not sure of your results, try to reproduce them by using different methods. Robustness is the best way to check whether you're looking at a true artifact or not. Genuine artifacts are *not* robust. For instance, if you generate a phylogenetic tree with two methods and you get two very different trees, there's a very good chance that none of these trees is in fact the correct one.

When you use different programs, make sure that you double-check with a method that relies on a different principle. For instance, there's no point in comparing an NJ tree from ClustalW with another NJ tree from Phylip. What you want to do is confront your NJ tree with a Fitsch or an ML tree.

As an alternative, you can also change some of the key parameters of your program and check how they affect the results you're interested in. In a multiple alignment, for instance, you could change the gap penalties.

Stay Away from Unpublished Methods!

This isn't to say that you should never try something new, but using well-established methods is the best way to produce results that people trust. Never use an unknown program that's not published and whose principle you don't understand. In brief, stay away from any student's midterm project pages!

Databases Are Not Like Good Wine

Databases are not like good wine: They do not age well! Avoid using data that a colleague downloaded last year and left to rot away on a PC.

Anytime you need a database or some sequence data, always make sure that you're using fresh data.

Just Because It Looks Free Doesn't Mean It Is Free . . .

Public data isn't always available for you to use anyway you see fit. Before you start using some data on a server, make sure that you know whom this data belongs to. If you work in a company, you may have to purchase a license before using it.

Many companies advertise themselves by providing free data or services. This lasts only as long as the company's management types see fit. If your work depends on such resources, an abrupt discontinuation can mean disaster! We aren't saying that you should never use private resources, but you should always know how long the resources are going to last.

Biting the Bullet at the Right Time

The reason why we've dedicated this entire book to online resources is because online resources are easy to use and don't require any kind of complicated setup. They let you concentrate on biology, not on your computer! This is exactly what you need . . . if you use bioinformatics only occasionally. Don't expect to be able to assemble and annotate your own viral or bacterial genome, however, without some serious investment in bioinformatics and knowledgeable people!

If you find that you start relying heavily on some servers, you'll have to bite the bullet and install these programs on your own computer. Installing and maintaining software is easier than it seems. Such software, however, works best on Linux or Unix systems. If you start from scratch, learning all you need to install and use Linux would probably take you a couple of weeks. Even if your time is tight, consider this a safe investment.

How will you know when that time has come? Finding yourself doing 20 BLASTs in a row using an online server is a sure sign that you need the independence and flexibility that comes with having the software installed on your own computer!

Chapter 15

Some Useful Bioinformatics Resources

In This Chapter

▶ Ten important databases

▶ Some very important programs

▶ Our favorite link collections and generic resources

How did you know that I knew you knew?

— Typical example of knowledge-based programming

*O*ne of the very nice and unique things about bioinformatics is that (almost) everything you need is available online. In this chapter, we show you around the main resources. If you have a fairly good idea of what's behind every link we list here, consider yourself a fairly well-trained fellow in the field of bioinformatics.

Remember that this field is moving rapidly. Being operational in bioinformatics is not so much about knowing things but about knowing where to find the knowledge you need. This book is not exhaustive at all, but it offers you many gateways. Do not hesitate to go further and explore. While on this fascinating journey, keep new things in the corner of your mind until you need them.

Ten Major Databases

Bioinformatics is about exploring biological information. This information is safely kept in databases. These databases are our modern museums, cyber-reference collections where all the knowledge of an era is carefully stored and classified. Table 15-1 lists ten of these databases. It isn't at all an exhaustive list, but a biased vision that this chapter title limits to just ten items!

Table 15-1	Ten Important Bioinformatics Databases	
Name	*Address*	*Description*
GenBank/DDBJ/EMBL	www.ncbi.nlm.nih.gov	Nucleotide sequences
Ensembl	www.ensembl.org	Human/mouse genome
PubMed	www.ncbi.nlm.nih.gov	Literature references
NR	www.ncbi.nlm.nih.gov	Protein sequences
SWISS-PROT	www.expasy.ch	Protein sequences
InterPro	www.ebi.ac.uk	Protein domains
OMIM	www.ncbi.nlm.nih.gov	Genetic diseases
Enzymes	www.chem.qmul.ac.uk	Enzymes
PDB	www.rcsb.org/pdb/	Protein structures
KEGG	www.genome.ad.jp	Metabolic pathways

Ten Major Bioinformatics Software Programs

Data is nothing if you don't know how to analyze it. This is what bioinformatics software programs are made for. You can use these programs to search for your data, to analyze it, and to display it. Table 15-2 lists the five main categories of programs that you may come across in bioinformatics, along with what we believe to be the most characteristic pieces of software in each of these categories. Each of these programs is introduced somewhere else in this book, as indicated in the table.

Table 15-2 also contains the address of an online server for each program. In most cases, you can find alternative addresses in the corresponding chapter.

Table 15-2 Ten Important Software Programs in Bioinformatics

Category	Name	Address	Description
Database Search	SRS	`srs.ebi.ac.uk`	Database search (Chapter 4)
	Entrez	`www.ncbi.nlm.nih.gov`	Database search (Chapter 3)
	BLAST	`www.ncbi.nlm.nih.gov`	Homology search (Chapter 7)
	DALI	`www.ebi.ac.uk`	Structure database search (Chapter 11)
Multiple alignment	ClustalW	`www.ebi.ac.uk`	Multiple sequence alignment (Chapter 9)
	Tcoffee	`igs-server.cnrs-mrs.fr/Tcoffee`	Multiple sequence alignment (Chapter 9)
Prediction	GenScan	`genes.mit.edu`	Gene prediction (Chapter 5)
	PredictProtein	`cubic.bioc.columbia.edu`	Protein structure prediction (Chapter 11)
	Mfold	`www.bioinfo.rpi.edu`	RNA structure prediction (Chapter 12)
Phylogenetics	Phylip	`bioweb.pasteur.fr/intro-uk.html`	Tree reconstruction (Chapter 13)
Editing/Visualization	Jalview	`www.es.embnet.org`	Alignment editor (Chapter 10)
	Rasmol	`www.bernstein-plus-sons.com/software/rasmol/`	Structure visualization (Chapter 11)

Ten Major Resource Locators

Bioinformatics changes fast. State-of-the-art programs become obsolete and established programs disappear, institutes are renamed, and new programs come out. That's the way of the world. If you find this ongoing change a bit unsettling, here's some reassuring news: Anything new you might be looking for is probably only one or two clicks away from the generic resources we list in Table 15-3. In this list, the last four links take you to some very good online resources where you can find almost anything you want on bioinformatics related topics, including what's going on.

Table 15-3	Ten Bioinformatics Resource Locators	
Name	*Address*	*Description*
ExPASy	www.expasy.ch	Dedicated to proteins
ArrayExpress	www.ebi.ac.uk/ microarray/	DNA chips
Swbic	www.swbic.org/	Miscellaneous links
Pasteur	bioweb.pasteur.fr/ intro-uk.html	Miscellaneous links; many online tools
RNA World	www.imb-jena.de/ RNA.html	RNA-related links
Phylip	evolution.genetics. washington/edu/	Everything on phylogeny
NCBI primers	www.ncbi.nlm.nih.gov/ education	Very good primers on many subjects
Bielefeld	bibiserv.techfak. uni-bielefeld.de/ intro/dist.html	Awesome online course
Bio-informer	www.ebi.ac.uk/ Information/News/	The EBI online news
Coffee Corner	www.ncbi.nlm.nih.gov	Online protein news

Glossary

· ·

algorithm: Mathematical formulation of (the logic behind) a computer program. Often used synonymously for *program*, *software*, or *package*.

alignment: Representation of two or more protein or nucleotide sequences where homologous amino acids or nucleotides are in the same columns while missing amino acids or nucleotides are replaced with gaps.

BankIt: A computer program developed by the NCBI for submitting your own sequences to GenBank.

bioinformatics: A collection of methods used for analyzing molecular biology data with a computer.

bit: Contraction of *binary digit*. The basic unit of information in a binary numbering system — the kind of system used by computers.

BLAST: Popular program for comparing a sequence with all the sequences contained in a database.

BLOSUM: Block Scoring Matrix, a popular substitution matrix for aligning protein sequences.

bootstrap: Method for checking the robustness of the branches in a phylogenetic tree.

CDS: CoDing Segment. The portion of a gene that precisely encodes a protein.

clade: Group of related species and their common ancestors.

cladogram: Phylogenetic tree showing the relationship between the species.

ClustalW: The most popular program for making multiple sequence alignments.

clustering: Action of grouping sequences (or other objects) according to their similarity.

codon: A word of three nucleotides, in the context of an ORF; the basic element of the genetic code relating nucleotide sequences with protein sequences in a one-to-one correspondence.

consensus: A pseudo sequence where each residue summarizes (using the majority rule) the content of a column in a multiple sequence alignment.

comparative genomics: Analysis based on the comparisons of entire genomes.

Covariance analysis: The study of a sequence that follows the evolution of two different positions and looks for some correlation. This kind of analysis often relies on multiple sequence alignments, looking for columns that contain correlated mutations. Ribosomal RNA is the best known example.

DDBJ: DNA Data Bank of Japan.

dendrogram: Another name for *phylogenetic tree*. The ClustalW guide tree is often referred to as a dendrogram. It is a file with a .dnd extension.

Dialign: A popular multiple sequence alignment package.

distance matrix: Matrix that contains all the pairwise distances between sequences in a data set. Distance matrices are often used to compute phylogenetic trees. Not to be confused with a substitution matrix.

distribution: A file distributed over the Internet that contains everything needed for the installation of a computer program on your machine.

domain: The structural definition of a domain is a protein unit of at least 50 amino acids that can fold on its own. Sequence analysts also use the word *domain* for protein sequences that occur in various contexts. A domain can also be the portion of a transmembrane protein that's inside or outside the cell.

dot matrix: *See* dot plot.

dot plot: Method for representing the similarity between two sequences without using an alignment.

Dotlet: Popular tool for comparing two sequences with a dot matrices.

Dotter: Name of a dot-plot program.

Dottup: Name of a dot-plot program.

EBI: European Bioinformatics Institute. The European homologue of the NCBI in the U.S.

EMBL: European Molecular Biology Laboratory. This acronym often refers to the nucleotide database the laboratory maintains.

EMBnet: European Molecular Biology Network. An international network of laboratories that provides online services and organizes bioinformatics courses.

Emboss: A popular collection of bioinformatics programs.

Ensembl project: A European project devoted to the annotation of the human genome.

Entrez: The NCBI database querying system that's similar to SRS at the EBI. Best known in the context of the PubMed/Medline bibliographical databases.

Entry: An elementary component of a database, accessible at once by using its unique identifier. Entries are usually subdivided into fields.

E-value: Expectation value. Given a database and the score of a hit, the E-value tells you how many times you could have expected such a result just by chance. In sequence analysis, good E-values must be very low (around 10-5 or even lower).

ExPASy: A server maintained by the Swiss Institute of Bioinformatics. ExPASy is the home of SWISS-PROT, the annotated protein database.

FASTA: One of the first popular programs for searching databases. By extension, FASTA has become the name of the sequence format used by the FASTA program.

field: In the context of databases, a well-delimited part of an entry, containing information of a precise nature (author, date, and so on). You can normally search databases by explicitly targeting specific fields.

gaps: Representation of insertions or deletions in sequence alignments.

GenBank: The main nucleotide database, which is co-maintained by the NCBI, the EMBL, and DDBJ.

Genetic code: The (quasi-) universal correspondence table between the 64 (4x4x4) different codons and the 20 amino acids. Only three codons (*Stop codons*) do not correspond to an amino acid: TAA, TAG, and TGA. These are used to signify the end of protein-coding regions.

Gibbs sampler: Local multiple sequence alignment method that uses a stochastic algorithm.

global alignment: An alignment of two sequences where no amino acid or nucleotide is discarded. They're all either aligned with other amino acids/ nucleotides or aligned with gaps.

guide tree: Tree used to guide the order in which sequences are aligned when making a progressive alignment. The guide tree is not a phylogenetic tree.

hit: Refers to a sequence similar to your query that you find while conducting a database search.

HMM: Hidden Markov Model. A mathematical formulation of a succession of hidden, mutually exclusive properties (such as intron or exon) associated with one sequence or a multiple sequence alignment. In the context of proteins, HMM is used interchangeably with the words *motifs* or *profiles*. In the context of DNA, HMMs are used to predict the location and structure of genes.

Hobacgen: Extensive collection of homologous bacterial genes.

homologues: Two genes are homologous if they evolved from the same common ancestor.

hydrophilic sketches: A series of hydrophilic amino acids in a protein sequence. Such sketches often indicate a surface loop.

identity: Percentage of identical amino acids or nucleotides in the alignment of two sequences. Usually, the identity is measured on the aligned residues (gaps are ignored).

InterPro: A federative database of profiles, patterns, and HMMs for detecting protein domains.

iterative: Said of a method that runs several times on the same dataset before delivering its result.

Jalview: A popular tool for editing and analyzing multiple sequence alignments.

Java applet: A small piece of software that your browser installs automatically and that runs on your computer.

KEGG: Kyoto Encyclopedia of Genes and Genomes. A world-famous Japanese database on genomes and biochemical pathways.

Lalign: A popular tool for finding the ten (or more) best local alignments between two sequences.

lateral transfer: Said of a gene that an organism acquired from another organism, instead of inheriting it from its ancestor.

local alignment: An alignment of the most similar segments between sequences. Dissimilar segments are not taken into account and are removed in the final output. Smith-Waterman and BLAST do local alignments.

low-complexity filtering: Removal of low-complexity sequences when doing a database search.

low-complexity sequence: A sequence that contains a few elements repeated many times. PIFGPPPIPPIP is an example of a low-complexity sequence.

match: *see* hit.

Medline: A collection of bibliographic references maintained by the NCBI. *See* PubMed.

multiple alignment: An alignment of more than two sequences.

NCBI: National Center for Biotechnology Information. A component of the U.S. National Institute of Health dedicated to bioinformatic research, software development, and the service and maintenance of leading public resources such as the GenBank (sequences) and PubMed (bibliography) databases. The United States' homologue of the EBI in Europe.

Newick format: A popular format for phylogenetic trees.

NJ: Neighbor Joining. It is the most popular method for reconstructing phylogenetic trees. UPGMA is the other main alternative.

NR: The Non Redundant Protein database, which contains all the putative protein sequences contained in the nucleotide databases. Its European equivalent is TrEMBL.

ORF: Open reading frame. A part of a DNA sequence without Stop codons, thus allowing the (putative or real) translation of a protein sequence.

pairwise alignment: Alignment of two (a pair of) sequences.

PAM: Point Accepted Mutation. A popular substitution matrix for aligning proteins.

parsimony: A technique for reconstructing phylogenetic trees.

patterns: Conserved residues that one can use as a functional signature.

PDB: Protein Data Base. A database that contains every available 3-D structure. It contains mostly proteins but also a few DNA and RNA structures.

Pfam: A collection of profiles for detecting domains and protein families.

Phylip: A powerful package for building phylogenetic trees.

phylogenetic tree: Tree representation of the family history of a set of sequences that share a common ancestor.

phylogenetics: Reconstruction of the evolution history of a gene family based on gene comparison.

PIR: Protein Information Resource. An annotated protein database similar to SWISS-PROT. PIR is also the name of a sequence format similar to FASTA.

polymorphism: Alternative forms of genes and other sequences. They are often constituted of punctual mutations, similar to SNP.

Prodom: A collection of automatically generated profiles for detecting domains and protein families.

profile: In bioinformatics, this word can mean several things. Often, a profile is a special formulation of a multiple alignment that lets you scan databases for new members of the family. PROSITE and Pfam are collections of profiles. HMM is often used interchangeably with profile. Do not confuse this meaning of profile with that of a *protein profile*.

PROSITE: A popular collection of protein domains and patterns.

protein profile: A tool for visualizing a particular property (hydropathy, charge, local composition, and so on) along a sequence by using a sliding window technique.

PubMed: NCBI's efficient implementation of the Medline bibliographical database produced by the U.S. National Library of Medicine. PubMed, Medline, and Entrez are sometimes used interchangeably.

query: Question asked when searching a database. If you make a similarity search, your query is a sequence. If you use SRS or Entrez, your query is a keyword. More complicated queries may involve several keywords, field restricted searches, and limits (such as dates).

RasMol: Popular software package for visualizing 3-D structures.

similarity: Percent of similar amino acids in the alignment of two sequences. Two amino acids are similar if they have similar physicochemical properties.

Smith-Waterman: A popular program for searching databases with local alignments.

SNP: Single Nucleotide Polymorphism. SNPs are point mutations observed when comparing different genomes of the same specie. *See* polymorphism.

SRS: Sequence retrieval system. The system used at the EBI to search databases with keywords. It is similar to Entrez at NCBI.

stochastic: Describes a program that uses chance to compute its results. Genetic algorithms, the Gibbs sampler, or simulated annealing are all stochastic algorithms.

submission: The data you send to a server.

substitution matrix: Matrix that contains a numerical score associated with the cost (or reward) of each possible mutation (or conservation). Unlikely mutations are penalized with a negative score.

SWISS-PROT: One of the most extensive annotated protein databases available.

Tcoffee: A package for computing, evaluating, and combining multiple sequence alignments. The successor of ClustalW.

TIGR: The Institute for Genome Research. A non-profit private organization world famous for its work in bacterial genomics.

TrEMBL: Short for Translated EMBL, which contains all the putative protein sequences contained in the nucleotide databases. Its U.S. equivalent is NR.

twilight zone: Describes the situation where the similarity between two sequences is so low (less than 25 percent) that it's impossible to tell whether they're homologous or not.

UPGMA: Unweighted Pair Group Method with Arithmetic Mean. A method for reconstructing phylogenetic trees. Not as accurate as NJ.

Index

• E •

Notes

Notes

FOR DUMMIES®

A world of resources to help you grow

HOME, GARDEN & HOBBIES

0-7645-5295-3

0-7645-5130-2

0-7645-5106-X

Also available:

Auto Repair For Dummies
(0-7645-5089-6)

Chess For Dummies
(0-7645-5003-9)

Home Maintenance For Dummies
(0-7645-5215-5)

Organizing For Dummies
(0-7645-5300-3)

Piano For Dummies
(0-7645-5105-1)

Poker For Dummies
(0-7645-5232-5)

Quilting For Dummies
(0-7645-5118-3)

Rock Guitar For Dummies
(0-7645-5356-9)

Roses For Dummies
(0-7645-5202-3)

Sewing For Dummies
(0-7645-5137-X)

FOOD & WINE

0-7645-5250-3

0-7645-5390-9

0-7645-5114-0

Also available:

Bartending For Dummies
(0-7645-5051-9)

Chinese Cooking For Dummies
(0-7645-5247-3)

Christmas Cooking For Dummies
(0-7645-5407-7)

Diabetes Cookbook For Dummies
(0-7645-5230-9)

Grilling For Dummies
(0-7645-5076-4)

Low-Fat Cooking For Dummies
(0-7645-5035-7)

Slow Cookers For Dummies
(0-7645-5240-6)

TRAVEL

0-7645-5453-0

0-7645-5438-7

0-7645-5448-4

Also available:

America's National Parks For Dummies
(0-7645-6204-5)

Caribbean For Dummies
(0-7645-5445-X)

Cruise Vacations For Dummies 2003
(0-7645-5459-X)

Europe For Dummies
(0-7645-5456-5)

Ireland For Dummies
(0-7645-6199-5)

France For Dummies
(0-7645-6292-4)

London For Dummies
(0-7645-5416-6)

Mexico's Beach Resorts For Dummies
(0-7645-6262-2)

Paris For Dummies
(0-7645-5494-8)

RV Vacations For Dummies
(0-7645-5443-3)

Walt Disney World & Orlando For Dummies
(0-7645-5444-1)

Available wherever books are sold. Go to www.dummies.com or call 1-877-762-2974 to order direct.

FOR DUMMIES®

Plain-English solutions for everyday challenges

COMPUTER BASICS

0-7645-0838-5

0-7645-1663-9

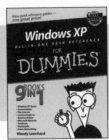

0-7645-1548-9

Also available:

PCs All-in-One Desk Reference For Dummies (0-7645-0791-5)

Pocket PC For Dummies (0-7645-1640-X)

Treo and Visor For Dummies (0-7645-1673-6)

Troubleshooting Your PC For Dummies (0-7645-1669-8)

Upgrading & Fixing PCs For Dummies (0-7645-1665-5)

Windows XP For Dummies (0-7645-0893-8)

Windows XP For Dummies Quick Reference (0-7645-0897-0)

BUSINESS SOFTWARE

0-7645-0822-9

0-7645-0839-3

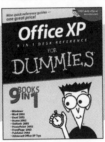

0-7645-0819-9

Also available:

Excel Data Analysis For Dummies (0-7645-1661-2)

Excel 2002 All-in-One Desk Reference For Dummies (0-7645-1794-5)

Excel 2002 For Dummies Quick Reference (0-7645-0829-6)

GoldMine "X" For Dummies (0-7645-0845-8)

Microsoft CRM For Dummies (0-7645-1698-1)

Microsoft Project 2002 For Dummies (0-7645-1628-0)

Office XP For Dummies (0-7645-0830-X)

Outlook 2002 For Dummies (0-7645-0828-8)

Get smart! Visit www.dummies.com

- **Find listings of even more *For Dummies* titles**

- **Browse online articles**

- **Sign up for Dummies eTips™**

- **Check out *For Dummies* fitness videos and other products**

- **Order from our online bookstore**

Available wherever books are sold. Go to www.dummies.com or call 1-877-762-2974 to order direct.